Carbohydrates: Integrated Research on Glycobiology and Glycotechnology

Volume I

Carbohydrates: Integrated Research on Glycobiology and Glycotechnology Volume I

Edited by **Sydney Marsh**

New York

Published by Callisto Reference,
106 Park Avenue, Suite 200,
New York, NY 10016, USA
www.callistoreference.com

Carbohydrates: Integrated Research on Glycobiology and Glycotechnology
Volume I
Edited by Sydney Marsh

International Standard Book Number: 978-1-63239-107-0 (Hardback)

Printed in the United States of America.

Contents

Preface

This book has been compiled for those interested in the study of carbohydrates. It has many topics which have been addressed by experts from diverse disciplines of microbiology, chemistry, botany, zoology and biotechnology. It encompasses the fundamentals of carbohydrates along with the tools, technologies and experiences for those who are involved in glycobiology and related fields. The book covers organic reactions of carbohydrates, analysis of carbohydrate derivatives, studies of DC-SIGN antagonists and the biosynthesis of carbohydrates in a microorganism. This is a comprehensive book which would cater to the needs of different kinds of readers.

The information shared in this book is based on empirical researches made by veterans in this field of study. The elaborative information provided in this book will help the readers further their scope of knowledge leading to advancements in this field.

Finally, I would like to thank my fellow researchers who gave constructive feedback and my family members who supported me at every step of my research.

Editor

Section 1

Chemistry and Biochemistry

Carbohydrate Microarray

Chuan-Fa Chang

Additional information is available at the end of the chapter

1. Introduction

Glycosylation adorns more than one half of the proteins in eukaryotic cells [1,2]. This post-translational modification plays an indispensible role in many important biological events, especially on cell surface [1,3]. Alterations in carbohydrate structures are known to correlate with the changes in protein stability and clearance, as well as various physiological functions including cell-cell adhesion, inflammation, tumor metastasis, and infection of bacteria and viruses [4-8]. Although glycosylation is essential for the formation and progression of various diseases, study of this subject is hampered by lack of effective tools available to date, in addition to structural heterogeneity and complexity of carbohydrates. A number of techniques have been developed to analyze the binding interactions between carbohydrates and proteins [2,9]. For instances, lectin blotting/binding assay has become a routine method to determine the glycan-protein interactions [10], but the relatively low sensitivity and the necessity of multiple wash steps/time-consuming have restricted the sensitivity and application. Surface plasmon resonance is another highly sensitive method which monitors the interactions in real time and in a quantitative manner [11-16]. However, sometimes the sensitivity is relatively low toward the use of low molecular-weight carbohydrates, though the problem can be overcome by labeling sugars with heavy metal ions [17]. In addition, fluorescence polarization and two-photon fluorescence correlation have been applied to study lectin-glycan interactions [18-21]. The most applicable technique is carbohydrate microarrays which immobilize oligosaccharides to a solid supports are developed and widely used to measure the carbohydrate binding properties of proteins, cells, or viruses [22-25]. For example, a high-content glycan microarray is developed by a robotic microarray printing technology in which amine-functionalized glycans are coupled to the succinimide esters on glass slides [26,27]. These microarrays have also been subjected for profiling the carbohydrate binding specificities of lectins, antibodies, and intact viruses.

2. Carbohydrate microarray

In our recent work, we have developed two novel carbohydrate microarrays: solution microarray [28] and membrane microarray [29]. Carbohydrate solution assay is a high-throughput, homogenous and sensitive method to characterize protein-carbohydrate interactions and glycostructures by in-solution proximity binding with photosensitizers (**Figure 1**). The technology, also called AlphaScreen™, is first described by Ullman et al., and has been used to study interactions between biomolecules [30-34]. In these assays, a light signal is generated when a donor bead and an acceptor bead are brought into proximity. This method usually provides good sensitivity with *femto*-mole detection under optimized conditions, relying on the binding affinity between analytes. All the procedures are carried out in 384-welled microtiter plates, thus qualifying the protocol as high-throughput. Two particles of 200 nm are involved in this technology including streptavidin-coated particles (donor beads) and protein A-conjugated particles (acceptor beads). Biotinylated polyacrylamide (biotin-PAA)-based glycans that are immobilized on donor beads can be recognized by lectins or antibodies, and connected with acceptor beads through specific antibodies (**Figure 1**). A number of carbohydrate binding proteins, including eleven lectins and seven antibodies, are profiled for their carbohydrate binding specificity to validate the efficacy of this developed technology. This assay is performed in homogeneous solutions and does not require extra wash steps, preventing the loss of weak bindings that often occur in the repeating washes of glycan microarray. However, antigen/ligand excess effect may happen in the homogeneous solution assay if the concentrations of carbohydrate epitopes, proteins, or antibodies are too high. One mg of biotin-PAA-sugar can be applied for fifty thousand assays because minimal amount of materials are needed in this microarray system (a range of nano-gram is required per well). Although the detection limit of biotin-PAA-sugar is good (2 ng per well), the linear range is too narrow for quantitative application.

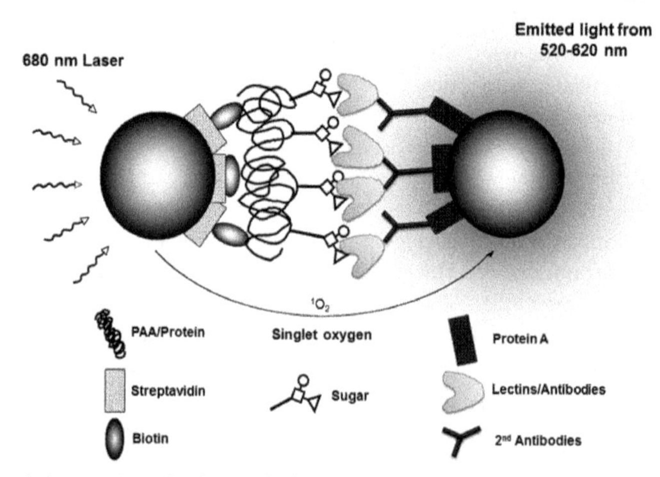

Figure 1. In-solution proximity binding with photosensitizers which was developed to characterize the protein-carbohydrate interactions [28].

Carbohydrate membrane microarray is fabricated by immobilization of the biotin-conjugated PAA-based glycans on aldehyde-functionalized UltraBind via streptavidin. Streptavidin interacted strongly with biotin and formed covalent linkage with membranes after reductive amination, which prevented the loss of glycans from membrane during repeated wash steps. The use of PAA also avoided the nonspecific interactions that take place in other studies between some lectins and non-glycosylated proteins (e.g. HAS or BSA) [35]. The operation of this carbohydrate membrane microarray is similar to that of Western blotting and can be performed easily by anyone without prior intensive training.

3. Applications

3.1. Carbohydrate binding profiles of lectins and antibodies (solution microarray)

Fifty-four biotinylated polyacrylamide backboned glycans (biotin-PAA-glycans) (**Table 1**) are collected in total to examine fifteen carbohydrate-binding proteins, including eight lectins (Con A, DBA, GS-I, PNA, SBA, UEA-1, WFA and WGA), and six antibodies (anti-Le^a, Le^b, Le^x, Le^y, sialyl Le^a and sialyl Le^x). The resulting signals are indicated with bars as relative intensities (**Figures 2 and 3**). The natural carbohydrate ligands for these lectins are listed in **Table 2**. All of the lectins showed nearly the same carbohydrate binding preferences as those in literatures. For example, concanavalin A (Con A) bound preferentially to mannose (No. 3) and biantennary *N*-glycan (No. 53), and very weakly to 3- and 6-sulfated galactosides (No. 19, 23 and 25). DBA, a GalNAc-binding lectin, recognized GalNAcα1-3Gal-containing epitopes (No. 11 and 39). ECA interacted with LacNAc disaccharide, Galβ1-4(6-sulfo)GlcNAc, and Galβ1-4(α1-2Fuc)GlcNAc (No. 17, 24, 31 and 47), and weakly bound to Le^c (Galβ1-3GlcNAc, No. 20). GS-I preferred interacting with Gal/GalNAc that contains α1-3 or 1-4 linkage (No. 11, 13, 14, 16, 40 and 42). MAA, in this study, recognized mainly to 3′-sulfated Galβ1-3GlcNAc, 3′-sulfated Galβ1-4GlcNAc and LacNAc and weakly to 3-sialylated galactosides (No. 26, 37 and 53). PNA interacted with Galβ1-3GalNAc (No. 15) and bound to some galactosides weakly (No. 12, 16, 20, 45 and 46). SBA preferentially interacted with α-linked galactosides (No. 16 and 42) and *N*-acetylgalactosaminoside (No. 11). SNA, a well-known α2-6 sialoside-binding lectin, interacted strongly to 6′-sialyl lactose and sialylated diantennary *N*-glycan (No. 36 and 53). UEA-1 specifically bound with Fucα1-2Gal-containing glycans (No. 18, 31 and 49). Due to weak interaction with PAA, WFA is the only one lectin showing higher background signals than the others. It recognized nearly half of the glycans on the glycan library, such as GlcNAc- and NeuAcα2-3-Gal/NeuAcα2-6-Gal containing saccharides. WGA also bound to terminal Gal or GalNAc epitopes (GalNAcα1-3Gal, No. 11 and Galβ1-4($6HSO_3$)GlcNAc, No. 24) according to some minor signals. Interestingly, WGA showed better interactions with chitotriose than with chitobiose and GlcNAc. In addition, the binding specificities of monoclonal anti-carbohydrate antibodies also revealed some interesting features. As shown in **Figure 3**, anti-Le^a antibody bound tightly with Le^x, but less with Le^b and sialyl Le^a. Anti-Le^b antibody represented specificities for both Le^b and

Le^{a}, but less for Le^{x} and sialyl Le^{x}. Anti-Le^{y} antibody not only binding to Le^{y}, but also recognized lactose, Le^{x}, sialyl Le^{x} and H type 2 structures. We also compared the binding patterns of lectins with the results reported by Blixt and coworkers at CFG in which 264 different glycans are studied by using the printed microarray (Ver. 2) (http://www.functionalglycomics.org/glycomics/publicdata/ primaryscreen.jsp). There are forty-seven glyco-epitopes are found to be identical in both analyses. Even the different principles and procedures of the two systems, the binding patterns of eight lectins are nearly the same, except for a few minor differences. For example, our characterized patterns of WFA and WGA show 90% similarity to the CFG data. Nevertheless, the interactions of SBA, WFA and WGA to β-GalNAc (No. 2) in the CFG's printed microarray are not observed in our system. Both of our method and the printed microarray indicate that MAA preferentially binds to sulfated glycans [36]. Because of the observed consistency shown by the two very different methods, we conclude the protein-glycan binding interactions are not affected by the PAA linker, the assay procedure (washing vs. non-washing) and the interacting microenvironment (2D for printed microarray vs. 3D for our solution microarray).

No.	Glycan Name	No.	Glycan Name
1	PAA-biotin	28	GlcNAcβ1-4GlcNAcβ1-4GlcNAcβ, sp= NHCOCH2NH-
2	β-GlcNAc	29	GlcNAcβ1-3Galβ1-4GlcNAcβ
3	α-Mannose	30	Fucα1-2Galβ1-3GlcNAcβ, Le^{d} (H type1)
4	β-GlcNAc	31	Fucα1-2Galβ1-4GlcNAcβ (H type2)
5	β-GalNAc	32	Galβ1-3(Fucα1-4)GlcNAcβ (Le^{a})
6	α-Fuc	33	Galβ1-4(Fucα1-3)GlcNAcβ (Le^{x})
7	α-NeuAc	34	3-HSO$_3$-Galβ1-4(Fucα1-3)GlcNAcβ (3'sulfate Le^{x})
8	α-NeuGc	35	NeuAcα2-3Galβ1-3GlcNAcβ (3'Sialyl Le^{c})
9	Glcα1-4Glcβ	36	NeuAcα2-6Galβ1-4Glcβ (6'Sialyl Lactose)
10	GlcNAcβ1-4GlcNAcβ	37	NeuAcα2-3Galβ1-4Glcβ (3'Sialyl Lactose)
11	GalNAcα1-3Galβ	38	NeuAcα2-3(NeuAcα2-6)GalNAcα
12	Galβ1-4Glcβ (Lactose)	39	GalNAcα1-3(Fucα1-2)Galβ (Blood Group A)
13	Galα1-3Galβ	40	Galα1-3(Fucα1-2)Galβ (Blood Group B)
14	Galα1-3GalNAcβ	41	3-HSO$_3$-Galβ1-3(Fucα1-4)GlcNAcβ (3'sulfate Le^{a})
15	Galβ1-3GalNAcβ	42	Galα1-4Galβ1-4Glcβ
16	Galα1-4GlcNAcβ (αLacNAc)	43	NeuAcα2-3Galβ1-4GlcNAcβ
17	Galβ1-4GlcNAcβ (LacNAc)	44	NeuAcα2-3Galβ1-3GalNAcα
18	Fucα1-2Galβ	45	Galβ1-3(NeuAcα2-6)GalNAcα
19	3-HSO$_3$-Galβ1-4GlcNAcβ	46	Galβ1-3GlcNAcβ1-3Galβ1-4Glcβ
20	Galβ1-3GlcNAc (Le^{c})	47	Galβ1-4GlcNAcβ1-3Galβ1-4Glcβ
21	NeuAcα2-6GalNAcα	48	Fucα1-2Galβ1-3(Fucα1-4)GlcNAcβ (Le^{b})
22	NeuGcα2-6GalNAcα	49	Fucα1-2Galβ1-4(Fucα1-3)GlcNAcβ (Le^{y})
23	3-HSO$_3$-Galβ1-3GlcNAcβ	50	NeuAcα2-3Galβ1-3(Fucα1-4)GlcNAcβ (sialyl Le^{a})
24	Galβ1-4(6-HSO$_3$)GlcNAcβ	51	NeuAcα2-3Galβ1-4(Fucα1-3)GlcNAcβ (sialyl Le^{x})
25	6-HSO$_3$-Galβ1-4GlcNAcβ	52	(NeuAcα2-8)$_{5-6}$
26	NeuAcα2-3Gal	53	(NeuAcα2-6Galβ1-4GlcNAcβ1-2Man)$_2$α1-3,6Manβ1-4GlcNAcβ1-4GlcNAcβ
27	NeuAcα2-3GalNAcα	54	H_2O

Table 1. List of biotin-PAA-glycans (fifty-two) used in glycan solution microarray [28].

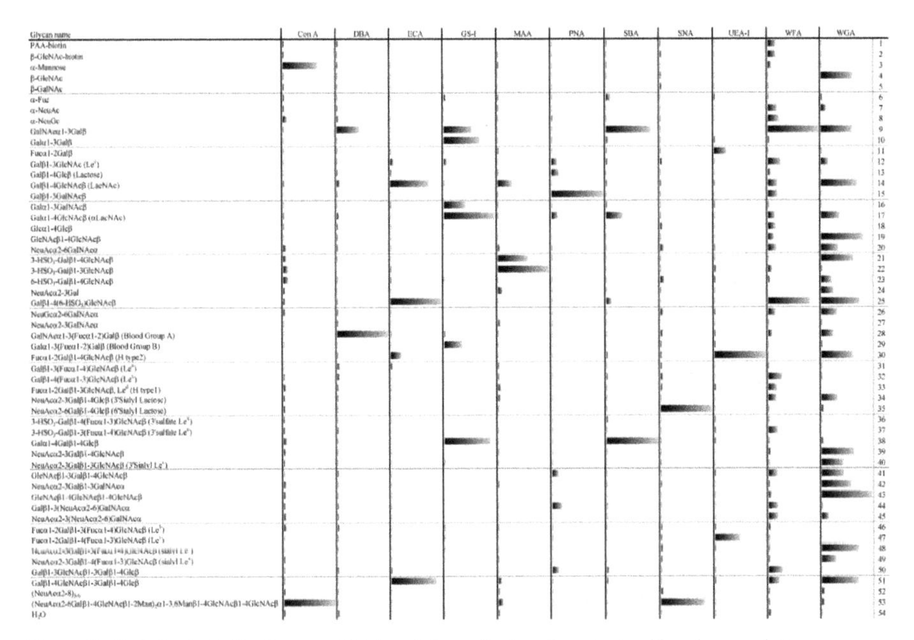

Figure 2. Carbohydrate binding specificities of eleven lectins characterized by glycan solution microarray [28].

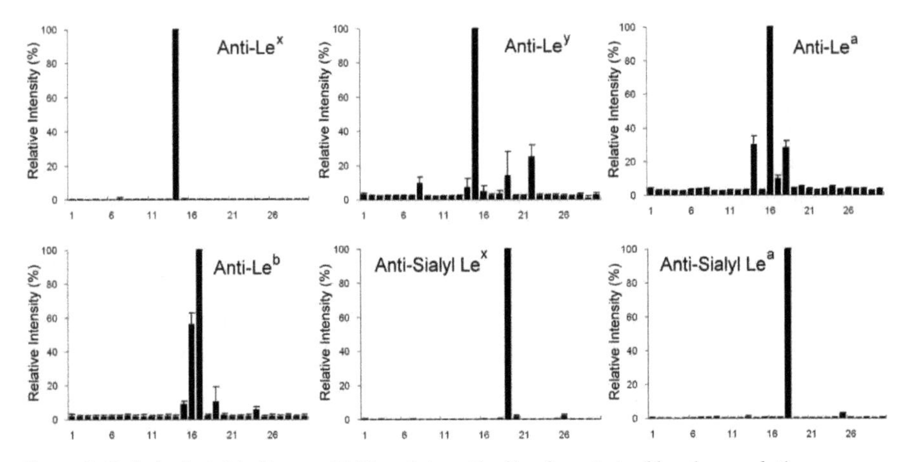

Figure 3. Carbohydrate binding specificities of six antibodies characterized by glycan solution microarray [28].

Lectins	Binding specificities	Lectins	Binding specificities
Con A	Man, Glc, GlcNAc	**WGA**	GlcNAcβ1-4GlcNAc, Neu5Ac
DBA	GalNAc	**ECA**	Galβ1-4GlcNAc
PNA	Galβ1-3GalNAc	**MMA**	Gal
SBA	GalNAc/Gal	**GS-I**	Gal
UEA-1	Fuc	**SNA**	Neu5Acα2-6
WFA	GalNAc		

Table 2. Carbohydrate ligands of commercial available lectins

3.2. Binding patterns of seventeen lectins and four antibodies (membrane microarray)

The principle and procedures of carbohydrate membrane microarray are showed in **Figure 4**. The western blotting like procedures not only reduces the time and interference, but also increases the application of this platform. In order to look deep inside the carbohydrate binding preferences of proteins and microorganisms, the collections of biotin-PAA-glycans were to increased eighty-eight different structures (**Table 3**). The glycan binding specificities of sixteen lectins (six alkaline phosphatase (AP)-conjugated lectins, four FITC-conjugated lectins, six unconjugated lectins) and four Lewis blood-group antibodies are evaluated and showed in **Figures 5 and 6**. All the lectins recognized the glycans that are consistent with the literature. For instance, ECA preferentially interacted with LacNAc, lactose, GalNAc, and Gal terminal sugars; PNA specifically bound to the Galβ1-3GalNAc structure; SBA dominantly recognized α-linked GalNAc epitopes; 3-sulfate LacNAc is ligand for MAA [36]. Compare the patterns of unconjugated lectins with conjugated lectins (AP- or FITC-attached) indicated that the glycan preferences of ECA and PNA are not interfered by conjugation. More binding signals are observed in the binding profiles of AP-conjugated MAA, SBA and WGA compared with unconjugated or FITC-attached ones. Additionally, the binding patterns of DBA, ECA, GS-I, MAA, SBA and VVA are highly consistent with those reported by Consortium for Functional Glycomics (CFG, printed microarray Ver. 2,). However, few inconsistencies are also observed in the study of MPA, PNA, UEA and WGA. Furthermore, the binding patterns of four Lewis blood group antibodies represented very high specificities (**Figure 6**).

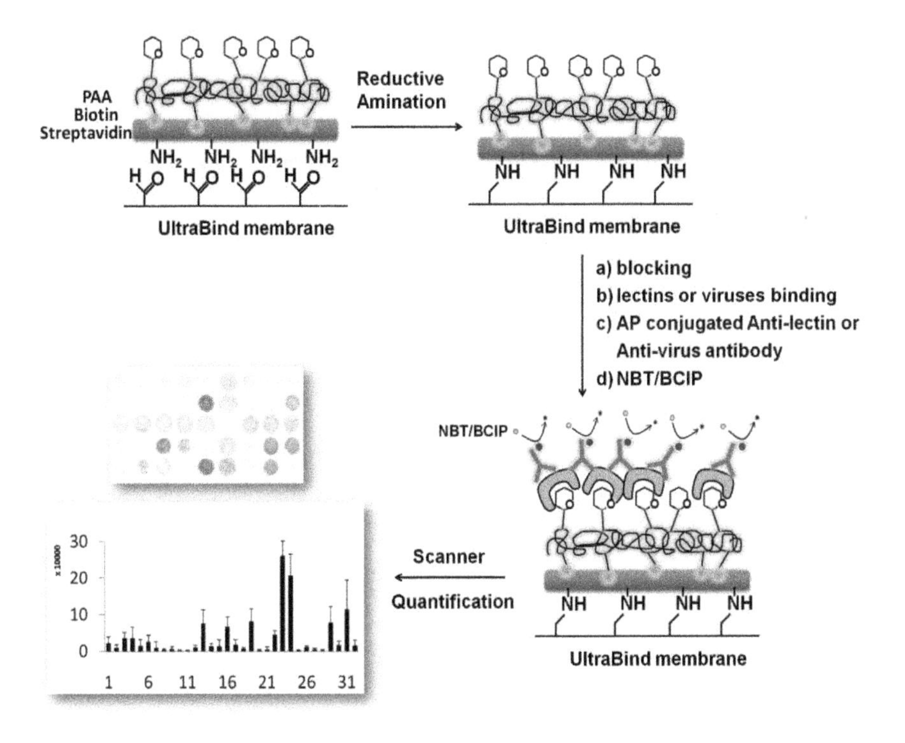

Figure 4. Fabrication, principle, and procedures of carbohydrate membrane microarray [29].

S-1	Blank-PAA-biotin	S-46	Neu5Acα2-6Galβ-PAA-biotin
S-2	β-GlcNAc-sp-biotin	S-47	Neu5Gcα2-6GalNAc-PAA-biotin
S-3	α-Mannose-PAA-biotin	S-48	Neu5Acα2-3GalNAcα-PAA-biotin
S-4	β-GlcNAc-PAA-biotin	S-49	Blood Group A-tri-PAA-biotin
S-5	β-GalNAc-PAA-biotin	S-50	Blood Group B-tri-PAA-biotin
S-6	α-L-Fuc-PAA-biotin	S-51	H(type2)-PAA-biotin
S-7	α-Neu5Ac-PAA-biotin	S-52	Le^a-PAA-biotin
S-8	α-Neu5Ac-OCH$_2$C$_6$H$_4$-p-NHCOOCH$_2$-PAA-biotin	S-53	Le^x-PAA-biotin
S-9	MDP(muramyl dipeptide)-PAA-biotin	S-54	Le^d(H type1)-PAA-biotin
S-10	α-Neu5Gc-PAA-biotin	S-55	3'Sialyl-Lactose-PAA-biotin
S-11	β-D-Gal-3-sulfate-PAA-biotin	S-56	6'Sialyl-Lactose-PAA-biotin
S-12	β-D-GlcNAc-6-sulfate-PAA-biotin	S-57	3-HSO$_3$-Le^x-PAA-biotin
S-13	GalNAcα1-3Galβ-PAA-biotin	S-58	3-HSO$_3$-Le^a-PAA-biotin
S-14	Galα1-3Galβ-PAA-biotin	S-59	Galα1-4Galβ1-4Glcβ-PAA-biotin
S-15	Fucα1-2Galβ-PAA-biotion	S-60	Galα1-3Galβ1-4Glcβ-PAA-biotin
S-16	Le^c(Galβ1-3GlcNAc)-PAA-biotin	S-61	GlcNAcβ1-2Galβ1-3GalNAcα-PAA-biotin
S-17	Galβ1-4Glcβ-PAA-biotin (Lactose)	S-62	Neu5Acα2-3Galβ1-4GlcNAcβ-PAA-Biotin
S-18	LacNAc-PAA-biotin	S-63	3'Sialyl-Le^c-PAA-biotin
S-19	Fucα1-3GlcNAcβ-PAA-biotin	S-64	Galα1-3Galβ1-4GlcNAcβ-PAA-biotin, sp=-NHCOCH$_2$NH-
S-20	Fucα1-4GlcNAcβ-PAA-biotin	S-65	GlcNAcα1-3Galβ1-3GalNAcα-PAA-biotin
S-21	GalNAcα1-3GalNAcα-PAA-biotin	S-66	GlcNAcβ1-3Galβ1-3GalNAcα-PAA-biotin
S-22	Galα1-3GalNAcα-PAA-biotin	S-67	Galβ1-3(GlcNAcβ1-6)GalNAcα-PAA-biotin
S-23	Galβ1-3GalNAcβ-PAA-biotin	S-68	Blood type A (tri)-PAA-biotin, sp=(CH$_2$)$_3$NHCO(CH$_2$)$_5$NH-
S-24	Galα1-3GalNAcβ-PAA-biotin	S-69	Blood type B (tri)-PAA-biotin, sp=(CH$_2$)$_3$NHCO(CH$_2$)$_5$NH-
S-25	Galβ1-3Galβ-PAA-biotin	S-70	GlcNAcb1-3Galβ1-4GlcNAc-PAA-biotin
S-26	GlcNAcβ1-3Galβ-PAA-biotin	S-71	Neu5Acα2-3Galβ1-3GalNAcα-PAA-Biotin
S-27	αLacNAc-PAA-biotin	S-72	GlcNAcβ1-3(GlcNAcβ1-6)GalNAcα-PAA-biotin
S-28	Glcα1-4Glcβ-PAA-biotin	S-73	Galα1-4Galβ1-4GlcNAcβ-PAA-biotin
S-29	Galβ1-3GalNAcα-PAA-biotin, sp=-p-OC$_6$H$_4$-	S-74	GlcNAcβ1-4GlcNAcβ1-4GlcNAc-PAA-biotin
S-30	Galα1-2Galβ-PAA-biotin	S-75	Galβ1-3(Neu5Acα2-6)GalNAcα-PAA-biotin
S-31	GlcNAcβ1-4GlcNAc-PAA-biotin	S-76	Neu5Acα2-3(Neu5Acα2-6)GalNAc-PAA-biotin
S-32	GlcNAcβ1-4GlcNAcβ-PAA-biotin, sp=-NHCOCH$_2$NH-	S-77	Galβ1-4GlcNAcβ1-3GalNAcα-PAA-biotin
S-33	Neu5Acα2-6GalNAc-PAA-biotin	S-78	Le^b-PAA-biotin
S-34	H(type 3)-PAA-biotin	S-79	Le^y-PAA-biotin
S-35	3-HSO$_3$-Galβ1-4GlcNAc-PAA-biotin	S-80	Sialyl Le^a-PAA-biotin
S-36	3-HSO$_3$-Galβ1-3GlcNAcβ-PAA-biotin	S-81	Sialyl Le^x-PAA-biotin
S-37	Galα1-6Glcβ-PAA-biotin (melibiose)	S-82	GlcNAcβ1-3(GlcNAcβ1-6)Galβ1-4Glcβ-PAA-biotin
S-38	Neu5Acα2-8Neu5Acα-sp**-PAA-biotin, (Neu5Ac)$_2$	S-83	Galα1-3(Fucα1-2)Galβ1-4GlcNAc-PAA-biotin
S-39	Galβ1-2Galβ-PAA-biotin	S-84	Galβ1-3GlcNAcβ1-3Galβ1-4Glcβ-PAA-biotin
S-40	6-HSO$_3$-Galβ1-4GlcNAc-PAA-biotin	S-85	Galβ1-4GlcNAcβ1-3Galβ1-4Glcβ-PAA-biotin
S-41	Neu5Acα2-3Gal-PAA-biotin	S-86	(NeuAcα2-8)5-6-PAA-biotin
S-42	Galβ1-4(6-HSO$_3$)GlcNAcb-PAA-biotin	S-87	Galβ1-4GlcNAcβ1-3(Galβ1-4GlcNAcβ1-6)GalNAcα-PAA-biotin
S-43	3-HSO$_3$-Galβ1-3GalNAcβ-PAA-biotin (sulfate-TF)	S-88	α2-6 sialylated diantennary N-glycans-PAA-biotin
S-44	GlcNAcβ1-3GalNAcα-PAA-biotin	S-89	GalNAc-α-Ser-PAA-biotin
S-45	GlcNAcβ1-6GalNAcα-PAA-biotin	S-90	H_2O

α2-6 sialylated diantennary N-glycans :(NeuAcα2-6Galβ1-4GlcNAcβ1-2Man)$_2$α1-3,6Manβ1-4GlcNAcβ1-4GlcNAc

Table 3. List of biotin-PAA-glycans (eighty-eight) used in carbohydrate membrane microarray [29].

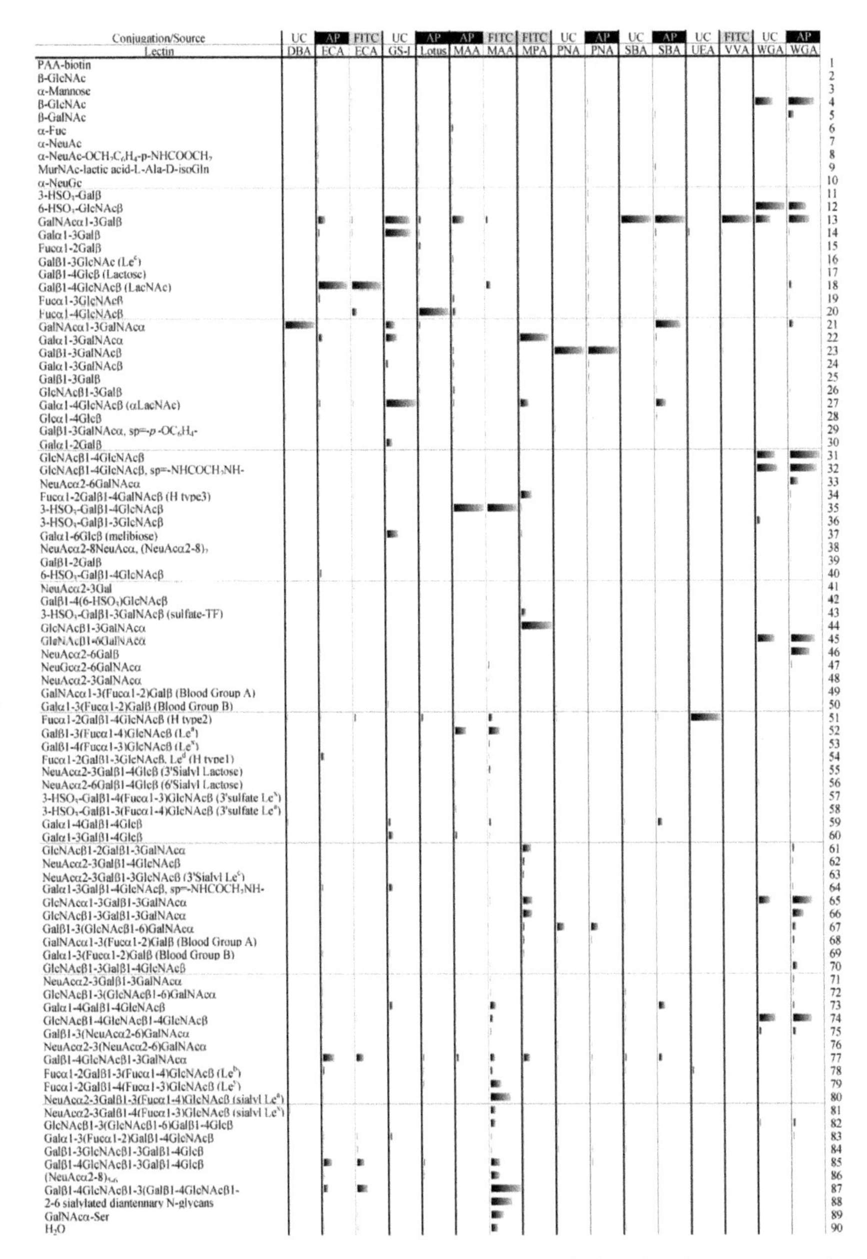

Figure 5. Carbohydrate binding specificities of sixteen lectins (six alkaline phosphatase-conjugated lectins, four FITC-conjugated lectins, and six unconjugated lectins) characterized by carbohydrate membrane microarray [29].

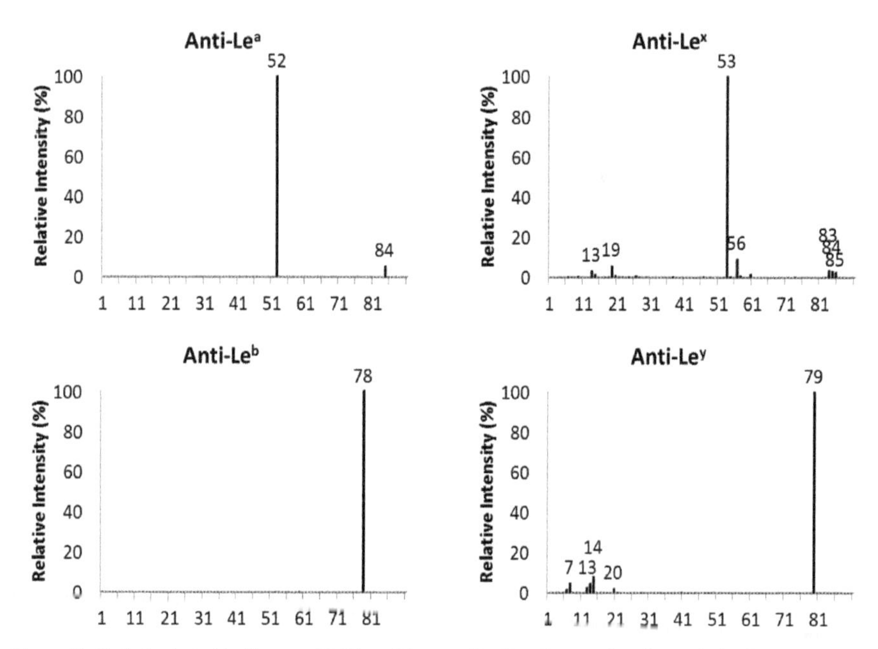

Figure 6. Carbohydrate binding specificities of four antibodies characterized by carbohydrate membrane microarray [29].

3.3. Surveillance of seasonal and pandemic H1N1 influenza A viruses (membrane microarray)

Several methods are applied for exploring the carbohydrate-binding preference of influenza viruses or recombinant HA, including cell-based assays [37-39], immobilization of virus [40-46] and glycan microarray technology [47-49]. In our study, 30 biotin-PAA-glycans (**Table 4**, most of them were sulfated and sialylated) were selected and fabricated on membrane and applied for the carbohydrate ligand surveillance of influenza clinical isolates. We subjected this influenza membrane microarray to characterize the glycan-binding features of five seasonal influenza A (H1N1) and seven A(H1N1)pdm09 clinical isolates. The binding patterns of all studied viruses are successfully profiled, with each virus exhibiting different and clear patterns, thereby enabling characterization (**Figure 7, Lane 1 to 12**). The majority of the seasonal H1N1 viruses bound strongly to 6'-sialyl lactose and sialyl biantennary *N*-glycan (**Figure 7, Lane 1 to 5**). A/Taiwan/1156/2006, A/Taiwan/510/2008, and A/Taiwan/289/2009 isolates interacted with the two glycans, and with numerous α2-3 sialylated structures, including 3'-sialyl lactose, NeuAcα2-3(NeuAcα2-6)GalNAc, sialyl Le^a, and sialyl Le^x. In addition, A/Taiwan/1156/2006 isolate also recognized NeuAc (linked to *p*-aniline), NeuGc, some of the sulfated glycans (3'-sulfated Le^a, 3'-sulfated Le^x, 6-sulfate LacNAc and 6'-sulfate LacNAc), NeuAcα2-6GalNAc, and a NeuAcα2-8 dimer. Contrary to seasonal H1N1 isolated in 2009, pandemic (H1N1) isolated in 2009 accepted more substrates

(**Figure 7, Lane 6 to 9**). California/07/2009 and A/Taiwan/2024/2009 bound with 6-sulfate GlcNAc, 3'-sulfated Le^a, 3'-sulfated Le^x, all of the 2-3 sialylated glycans with the exception of NeuAcα2-3GalNAc, 6'-sialyl lactose, sialyl biantennary *N*-glycan, and α2-8 oligomers. A/Taiwan/942/2009 favored four sulfated glycans (3-sulafe Gal, 6-sulfate GlcNAc, 3'-sulfated Le^a and 3'-sulfated Le^x), most of the 2-3 sialylated glycans (except NeuAcα2-3Gal and NeuAcα2-3GalNAc) and three 2-6 sialylated glycans (NeuAcα2-6Gal, 6'-sialyl lactose and sialyl biantennary *N*-glycan). A/Taiwan/987/2009 recognized most of the PAA-sugar substrates tested, with the exception of the PAA backbone, muramyl dipeptide, αNeuAc, 3-sulafe Galβ1-4GlcNAc, 3-sulafe Galβ1-3GlcNAc and Galβ1-3(NeuAcα2-6)GalNAc. In addition, we found that pandemic (H1N1) viruses isolated in 2009 represented broader substrate specificities than the pandemic viruses isolated in 2010, especially to sulfated sugars. A/Taiwan/395/2010 and A/Taiwan/1477/2010 recognized NeuAcα2-3Gal, 3'sialyl lactose, NeuAcα2-3Galβ1-3GlcNAc, sialyl-Le^a, NeuAcα2-6GalNAc, NeuAcα2-6Gal, 6'-sialyl lactose, and sialyl biantennary *N*-glycan (**Figure 7, Lane 11 to 12**). A/Taiwan/257/2010 accepted only six glycans including NeuAc (linked to *p*-aniline), 3'sialyl lactose, NeuAcα2-6GalNAc, NeuAcα2-6Gal, 6'-sialyl lactose, and sialyl biantennary *N*-glycan (**Figure 7, Lane 9**).

I-1	Blank-PAA-biotin	I-17	Neu5Gcα2-6GalNAc-PAA-biotin
I-2	α-Neu5Ac-PAA-biotin	I-18	Neu5Acα2-3GalNAcα-PAA-biotin
I-3	α-Neu5Ac-$OCH_2C_6H_4$-p-$NHCOOCH_2$-PAA-biotin	I-19	3'Sialyl-Lactose-PAA-biotin
I-4	MDP(muramyl dipeptide)-PAA-biotin	I-20	6'Sialyl-Lactose-PAA-biotin
I-5	α-Neu5Gc-PAA-biotin	I-21	3-HSO_3-Le^x-PAA-biotin
I-6	β-D-Gal-3-sulfate-PAA-biotin	I-22	3-HSO_3-Le^a-PAA-biotin
I-7	β-D-GlcNAc-6-sulfate-PAA-biotin	I-23	Neu5Acα2-3Galβ1-4GlcNAcβ-PAA-Biotin
I-8	Neu5Acα2-6GalNAc-PAA-biotin	I-24	3'Sialyl-Le^c-PAA-biotin
I-9	3-HSO_3-Galβ1-4GlcNAc-PAA-biotin	I-25	Neu5Acα2-3Galβ1-3GalNAcα-PAA-Biotin
I-10	3-HSO_3-Galβ1-3GlcNAcβ-PAA-biotin	I-26	Galβ1-3(Neu5Acα2-6)GalNAcα-PAA-biotin
I-11	Neu5Acα2-8Neu5Acα-sp**-PAA-biotin, $(Neu5Ac)_2$	I-27	Neu5Acα2-3(Neu5Acα2-6)GalNAc-PAA-biotin
I-12	6-HSO_3-Galβ1-4GlcNAc-PAA-biotin	I-28	Sialyl Le^a-PAA-biotin
I-13	Neu5Acα2-3Gal-PAA-biotin	I-29	Sialyl Le^x-PAA-biotin
I-14	Galβ1-4(6-HSO_3)GlcNAcb-PAA-biotin	I-30	(NeuAcα2-8)5-6-PAA-biotin
I-15	3-HSO_3-Galβ1-3GalNAcβ-PAA-biotin (sulfate-TF)	I-31	α2-6 sialylated diantennary N-glycans -PAA-biotin
I-16	Neu5Acα2-6Galβ-PAA-biotin	I-32	H_2O

α2-6 sialylated diantennary N-glycans :$(NeuAc\alpha2\text{-}6Gal\beta1\text{-}4GlcNAc\beta1\text{-}2Man)_2\alpha1\text{-}3,6Man\beta1\text{-}4GlcNAc\beta1\text{-}4GlcNAc$

Table 4. List of biotin-PAA-glycans (thirty) fabricated on carbohydrate membrane microarray for carbohydrate binding surveillance of influenza clinical isolates [29] [50].

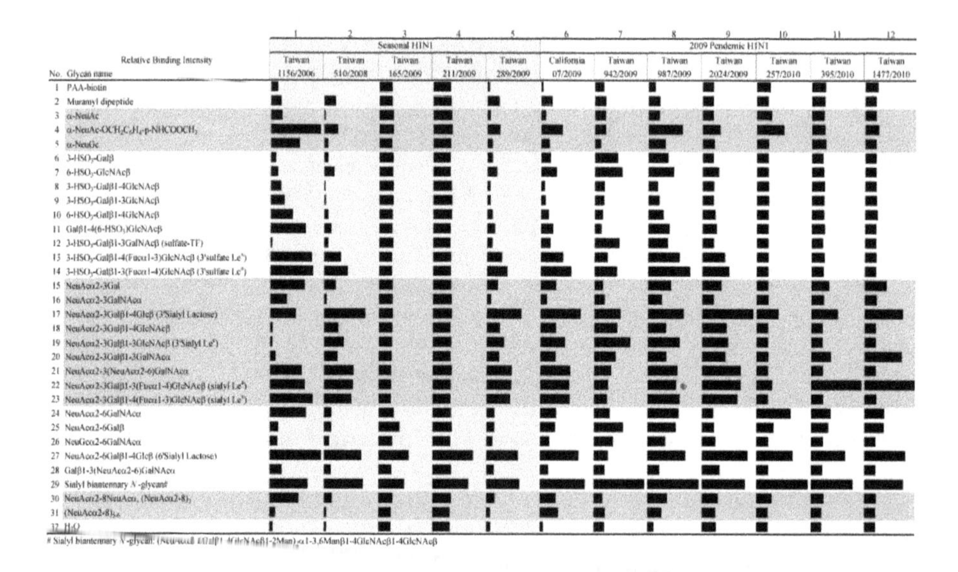

Figure 7. Carbohydrate binding profiles of twelve influenza A clinical isolates characterized by carbohydrate membrane microarray [50].

3.4. Surveillance of influenza B clinical isolates (membrane microarray)

The influenza carbohydrate membrane microarray is subsequently subjected for the surveillance of twelve influenza B (IB) clinical isolates [50]. All of the viruses are collected in the Clinical Virology Laboratory of National Cheng Kung University Hospital between 2001 and 2004. The clinical isolates are amplified in MDCK cells and purified by a sucrose gradient. The binding patterns of the influenza B viruses are investigated and are all found to exhibit clear binding profiles (**Figure 8**). Three common glycans, including α-Neu5Ac (*p*-anilinyl linked), 6'sialyl-lactose, and α2-6 sialylated biantennary *N*-glycans are bound by all of the influenza B viruses. Furthermore, some additional bindings are also observed. For instance, B/Taiwan/314/2001 and B/Taiwan/1729/2002 also recognized sulfated sugars (3-sulafe Gal, 6-sulfate GlcNAc, Galβ1-4(6-HSO_3)GlcNAc, and 3'-sulfated Le^a) and two α2-6 sialosides (NeuAcα2-6GalNAc and NeuAcα2-6Gal). B/Taiwan/288/2002 and B/Taiwan/1902/2004 portrayed minor interactions with α2-3 (3'-sialyl lactose, NeuAcα2-3Galβ1-4GlcNAc, 3'-sialyl Le^c and sialyl Le^x) and α2-6 sialylated sugars (NeuAcα2-6GalNAc and NeuAcα2-6Gal). Although the signal is relatively low, B/Taiwan/872/2004 and B/Taiwan/913/2004 also interacted with sulfated sugars. Surprisingly, seven viruses showed weak bindings to α2-8 linked sialic acids, including B/Taiwan/314/2001, B/Taiwan/174/2002, B/Taiwan/288/2002, B/Taiwan/1729/2002, B/Taiwan/872/2004, and B/Taiwan/913/2004.

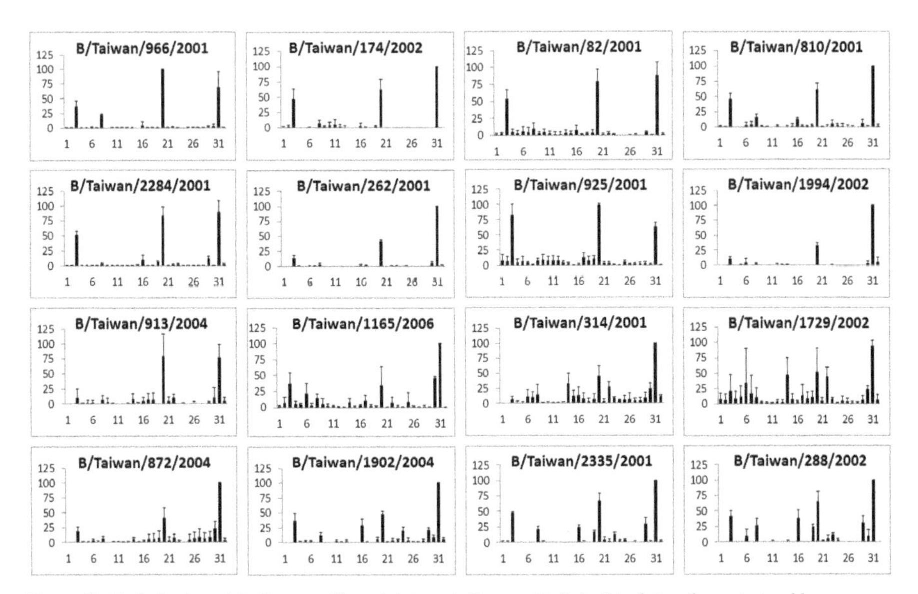

Figure 8. Carbohydrate binding profiles of sixteen influenza B clinical isolates characterized by carbohydrate membrane microarray [50].

The DNA sequences of hemagglutinin of the twelve influenza B clinical isolates were analyzed, translated into sequences of amino acids, and aligned with B/HongKong/8/73, Victoria (B/Victoria/02/87), and Yamagata (B/Yamagata/16/88) lineages using online software EBI ClustalW2. Surprisingly, the twelve influenza B clinical isolates could all be categorized into two groups depending on protein sequence alignment. Group 1 contained eight clinical isolates (B/Taiwan/314/2001, B/Taiwan/174/2002, B/Taiwan/288/2002, B/Taiwan/1729/2002, B/Taiwan/1994/2002, B/Taiwan/872/2004, B/Taiwan/913/2004 and B/Taiwan/1902/2004, **Figure 8, Table 5**) and Group 2 contained four viruses (B/Taiwan/262/2001, B/Taiwan/925/2001, B/Taiwan/966/2001, B/Taiwan/2284/2001). We found that all of the viruses in group 2 were isolated in 2001. In addition, there were twelve amino acid differences between the two groups. Most of the amino acids were located at the chain region and subunit interfaces, while four amino acids were located within the antigenic site (loop 160, amino acid number of B/HK/73) [51,52]. Additionally, four amino acids changed the charge property from group 1 to group 2 viruses (**Table 5**, No. 56, 116, 181 and 182, amino acid number of B/Yamagata/16/88). Different carbohydrate binding properties were also observed between some viruses of the two groups. Specifically, group 2 viruses interacted with three major sialylated glycans, while some of group 1 viruses bound not only strongly to the three glycans, but also weakly to sulfated and α2-3 sialylated glycans. Additionally, the number of charged amino acids was higher in group 2 than in group 1 viruses and B/HongKong/8/73 (**Table 5**).

	Amino acid number of B/HK/8/73	29	48	56	75	116	162	*	*	176	180	181	217	Number of charged amino
	B/HK/8/73	V	Q	N	T	N	K			Y	K	G	V	2
	B/Victoria/02/87	V	K	K	T	N	K	D	N	N	E	G	V	5
	B/Yamagata/16/88	V	K	N	T	N	R	D		Y	K	G	V	4
Group1	B/Taiwan/314/2001	V	R	D	A	N	K	D	N	H	E	G	V	5
	B/Taiwan/174/2002	V	R	D	T	N	K	D	N	Y	E	G	V	
	B/Taiwan/288/2002	V	R	D	T	N	K	D	N	Y	E	G	V	
	B/Taiwan/1729/2002	V	R	D	T	N	K	D	N	Y	E	G	V	
	B/Taiwan/1994/2002	V	R	D	T	N	K	D	N	Y	E	G	V	
	B/Taiwan/872/2004	V	R	D	T	N	K	D	N	Y	E	G	V	
	B/Taiwan/913/2004	V	R	D	T	N	K	D	N	Y	E	G	V	
	B/Taiwan/1902/2004	V	R	D	T	N	K	D	N	Y	E	G	V	
Group2	B/Taiwan/262/2001	A	K	T	I	K	R	E	N	H	K	E	I	7
	B/Taiwan/925/2001	A	K	T	I	K	R	D	N	H	K	E	I	
	B/Taiwan/966/2001	A	K	T	I	K	R	E	N	H	K	E	I	
	B/Taiwan/2284/2001	A	K	T	I	K	R	E	N	H	K	E	I	
	Amino acid number of B/Yamagata/16/88	29	48	56	75	116	162	163	*	177	181	182	218	

Table 5. Critical amino acid differences between the two groups of influenza B clinical isolated viruses [50].

According to the binding preference for α2-6 and α2-3 sialylated oligosaccharides displayed by these clinical isolated influenza viruses, these clinical isolates might infect not only the surface of nasal mucosa, pharynx, larynx, trachea and bronchi (that are resided in the upper respiratory tract to express α2-6 sialosides predominantly), but also alveoli, bronchi (that are located in the lower respiratory tract), eyes, and tissues in the gastrointestinal tract (that contain both α2-3 and α2-6 sialosides) [53,54]. After we surveyed more clinical isolates, factors which affect the binding preference of IB could be dissected. We believe that application of the carbohydrate microarray for detailed analysis of sugar-binding structures will eventually build the connection between clinical symptoms and the genetic specification of influenza viruses.

3.5. Characterization the glycan structure of glycoproteins (solution microarray)

The aforementioned lectins are further applied to characterize the glycan structures of six biotinylated proteins including ovalbumin, porcine mucin, human serum albumin, human transferrin, fetuin and asialofetuin. Distinctive glycopatterns are generated in accordance with the analysis of lectins (**Figure 9**). Porcine mucin is known to have *O*-linked fucosylated glycans with terminal GalNAc, GlcNAc and NeuAc residues. Our results showed positive signals in the tests with the ECA, DBA, UEA-1 and WGA lectins, indicating that the mucin contains the determinants of Galβ1-4GlcNAc, GalNAc, Fucα1-2Gal and GlcNAc/NeuAc, respectively. Manα1-3(Fucα1-6)GlcNAc and NeuAcα2-6Galβ1-4GlcNAc containing biantennary or triantennary *N*-glycans have been reported as the major glyco-structures of human transferrin [55-58]. Strong Con A and SNA signals are shown in our binding assay, but the relative intensity of GlcNAc-binding lectins is not as good as those of Con A and SNA, which is the contribution of the interference by the terminal sialic acids [59]. All of the results are similar to the analyses from lectin microarray or dot blot analysis [55-58].

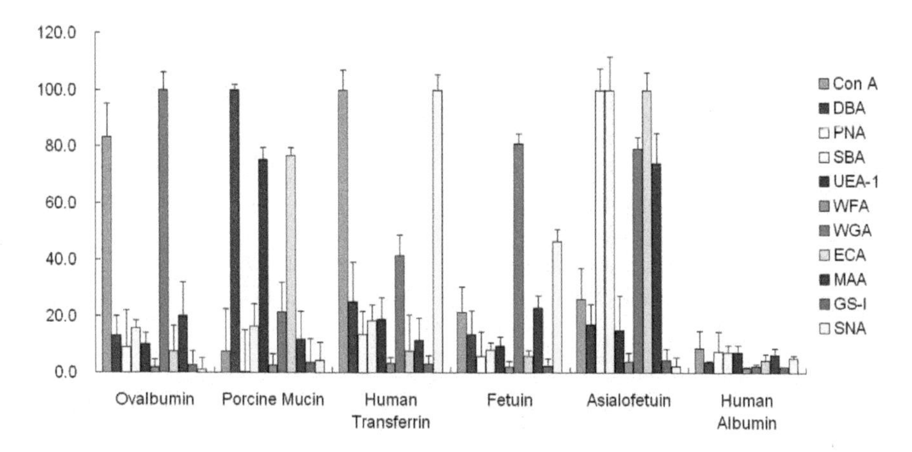

Figure 9. Glycopattern profiling of six biotinylated proteins with ten lectins by glycan solution microarray. The six proteins include ovalbumin, porcine mucin, human serum albumin, human transferrin, fetuin and asialofetuin [28].

4. Perspective

In summary, carbohydrate-protein interactions are the key steps for many physiological and pathological evens. Hence, development of new carbohydrate microarrays is important for detecting these activities. Our studies have demonstrated a rapid, highly sensitive, and reliable method to characterize carbohydrate-protein interactions with minimized materials (in the range of ng per well). Lectins and antibodies are evaluated and most of the results are coherent to previous reports, including those of CFG. Solution microarray is the first homogeneous and washless carbohydrate microarray which offers a reliable alternative for characterizing sugar-binding features of lectins and proteins, as well as antibodies. In addition, membrane microarray is also easy handled with low cost (USD$0.25 for one membrane microarray). The easy-handling feature avoids the necessity of repetitive training and accelerates the screening efficiency. Both of the microarrays represent a convenient and reliable way to examine the carbohydrate-binding features of various proteins, high-throughput drug screening, and the glycan binding surveillance of influenza viruses.

Author details

Chuan-Fa Chang
Department of Medical Laboratory Science and Biotechnology,
College of Medicine, National Cheng Kung University, Taiwan

5. References

[1] Varki, A. (1993) *Glycobiology* 3, 97-130

[2] Ratner, D. M., Adams, E. W., Disney, M. D., and Seeberger, P. H. (2004) *Chembiochem* 5, 1375-1383

[3] Fukuda, M. (2000) *Molecular and Cellular Glycobiology*, Oxford University Press

[4] Magnani, J. L., Brockhaus, M., Smith, D. F., Ginsburg, V., Blaszczyk, M., Mitchell, K. F., Steplewski, Z., and Koprowski, H. (1981) *Science* 212, 55-56

[5] Sacchettini, J. C., Baum, L. G., and Brewer, C. F. (2001) *Biochemistry* 40, 3009-3015

[6] Kansas, G. S. (1996) *Blood* 88, 3259-3287

[7] Geijtenbeek, T. B., Torensma, R., van Vliet, S. J., van Duijnhoven, G. C., Adema, G. J., van Kooyk, Y., and Figdor, C. G. (2000) *Cell* 100, 575-585

[8] Skehel, J. J., and Wiley, D. C. (2000) *Annu Rev Biochem* 69, 531-569

[9] Raman, R., Raguram, S., Venkataraman, G., Paulson, J. C., and Sasisekharan, R. (2005) *Nat Methods* 2, 817-824

[10] Wu, A. M. (2001) *The Molecular Immunology of Complex Carbohydreats-2*, Kluwer Academic/Plenum Publishers

[11] Ratner, D. M., Adams, E. W., Su, J., O'Keefe, B. R., Mrksich, M., and Seeberger, P. H. (2004) *Chembiochem* 5, 379-382

[12] Smith, E. A., Thomas, W. D., Kiessling, L. L., and Corn, R. M. (2003) *J Am Chem Soc* 125, 6140-6148

[13] Vila-Perello, M., Gutierrez Gallego, R., and Andreu, D. (2005) *Chembiochem* 6, 1831-1838

[14] Ofokansi, K. C., Okorie, O., and Adikwu, M. U. (2009) *Biol Pharm Bull* 32, 1754-1759

[15] Sletmoen, M., Dam, T. K., Gerken, T. A., Stokke, B. T., and Brewer, C. F. (2009) *Biopolymers* 91, 719-728

[16] Tian, P., Engelbrektson, A., and Mandrell, R. (2008) *Appl Environ Microbiol* 74, 4271-4276

[17] Beccati, D., Halkes, K. M., Batema, G. D., Guillena, G., Carvalho de Souza, A., van Koten, G., and Kamerling, J. P. (2005) *Chembiochem* 6, 1196-1203

[18] Pohl, W. H., Hellmuth, H., Hilbert, M., Seibel, J., and Walla, P. J. (2006) *Chembiochem* 7, 268-274

[19] Khan, M. I., Surolia, N., Mathew, M. K., Balaram, P., and Surolia, A. (1981) *Eur J Biochem* 115, 149-152

[20] Lee, Y. C. (2001) *Anal Biochem* 297, 123-127

[21] Sorme, P., Kahl-Knutsson, B., Huflejt, M., Nilsson, U. J., and Leffler, H. (2004) *Anal Biochem* 334, 36-47

[22] Fukui, S., Feizi, T., Galustian, C., Lawson, A. M., and Chai, W. (2002) *Nat Biotechnol* 20, 1011-1017

[23] Huang, C. Y., Thayer, D. A., Chang, A. Y., Best, M. D., Hoffmann, J., Head, S., and Wong, C. H. (2006) *Proc Natl Acad Sci U S A* 103, 15-20

[24] Feizi, T., Fazio, F., Chai, W., and Wong, C. H. (2003) *Curr Opin Struct Biol* 13, 637-645

[25] Shin, I., Park, S., and Lee, M. R. (2005) *Chemistry* 11, 2894-2901

[26] Blixt, O., Head, S., Mondala, T., Scanlan, C., Huflejt, M. E., Alvarez, R., Bryan, M. C., Fazio, F., Calarese, D., Stevens, J., Razi, N., Stevens, D. J., Skehel, J. J., van Die, I., Burton,

D. R., Wilson, I. A., Cummings, R., Bovin, N., Wong, C. H., and Paulson, J. C. (2004) *Proc Natl Acad Sci U S A* 101, 17033-17038
[27] Wong, C. H. (2005) *J Org Chem* 70, 4219-4225
[28] Chang, C. F., Pan, J. F., Lin, C. N., Wu, I. L., Wong, C. H., and Lin, C. H. (2011) *Glycobiology* 21, 895-902
[29] Lao, W. I., Wang, Y. F., Kuo, Y. D., Lin, C. H., Chang, T. C., Su, I. J., Wang, J. R., and Chang, C. F. (2011) *Future Med Chem* 3, 283-296
[30] Ullman, E. F., Kirakossian, H., Singh, S., Wu, Z. P., Irvin, B. R., Pease, J. S., Switchenko, A. C., Irvine, J. D., Dafforn, A., Skold, C. N., and et al. (1994) *Proc Natl Acad Sci U S A* 91, 5426-5430
[31] Ullman, E. F., Kirakossian, H., Switchenko, A. C., Ishkanian, J., Ericson, M., Wartchow, C. A., Pirio, M., Pease, J., Irvin, B. R., Singh, S., Singh, R., Patel, R., Dafforn, A., Davalian, D., Skold, C., Kurn, N., and Wagner, D. B. (1996) *Clin Chem* 42, 1518-1526
[32] Warner, G., Illy, C., Pedro, L., Roby, P., and Bosse, R. (2004) *Curr Med Chem* 11, 721-730
[33] Bouchecareilh, M., Caruso, M. E., Roby, P., Parent, S., Rouleau, N., Taouji, S., Pluquet, O., Bosse, R., Moenner, M., and Chevet, E. (2010) *J Biomol Screen* 15, 406-417
[34] Yi, F., Zhu, P., Southall, N., Inglese, J., Austin, C. P., Zheng, W., and Regan, L. (2009) *J Biomol Screen* 14, 273-281
[35] Manimala, J. C., Roach, T. A., Li, Z., and Gildersleeve, J. C. (2006) *Angew Chem Int Ed Engl* 45, 3607-3610
[36] Bai, X., Brown, J. R., Varki, A., and Esko, J. D. (2001) *Glycobiology* 11, 621-632
[37] Rogers, G. N., and D'Souza, B. L. (1989) *Virology* 173, 317-322
[38] Suzuki, Y., Nagao, Y., Kato, H., Suzuki, T., Matsumoto, M., and Murayama, J. (1987) *Biochim Biophys Acta* 903, 417-424
[39] Suptawiwat, O., Kongchanagul, A., Chan-It, W., Thitithanyanont, A., Wiriyarat, W., Chaichuen, K., Songserm, T., Suzuki, Y., Puthavathana, P., and Auewarakul, P. (2008) *J Clin Virol* 42, 186-189
[40] Gambaryan, A. S., Tuzikov, A. B., Piskarev, V. E., Yamnikova, S. S., Lvov, D. K., Robertson, J. S., Bovin, N. V., and Matrosovich, M. N. (1997) *Virology* 232, 345-350
[41] Gambaryan, A. S., and Matrosovich, M. N. (1992) *J Virol Methods* 39, 111-123
[42] Sauter, N. K., Hanson, J. E., Glick, G. D., Brown, J. H., Crowther, R. L., Park, S. J., Skehel, J. J., and Wiley, D. C. (1992) *Biochemistry* 31, 9609-9621
[43] Auewarakul, P., Suptawiwat, O., Kongchanagul, A., Sangma, C., Suzuki, Y., Ungchusak, K., Louisirirotchanakul, S., Lerdsamran, H., Pooruk, P., Thitithanyanont, A., Pittayawonganon, C., Guo, C. T., Hiramatsu, H., Jampangern, W., Chunsutthiwat, S., and Puthavathana, P. (2007) *J Virol* 81, 9950-9955
[44] Yamada, S., Suzuki, Y., Suzuki, T., Le, M. Q., Nidom, C. A., Sakai-Tagawa, Y., Muramoto, Y., Ito, M., Kiso, M., Horimoto, T., Shinya, K., Sawada, T., Usui, T., Murata, T., Lin, Y., Hay, A., Haire, L. F., Stevens, D. J., Russell, R. J., Gamblin, S. J., Skehel, J. J., and Kawaoka, Y. (2006) *Nature* 444, 378-382
[45] Gambaryan, A., Yamnikova, S., Lvov, D., Tuzikov, A., Chinarev, A., Pazynina, G., Webster, R., Matrosovich, M., and Bovin, N. (2005) *Virology* 334, 276-283

[46] Gambaryan, A. S., Tuzikov, A. B., Pazynina, G. V., Webster, R. G., Matrosovich, M. N., and Bovin, N. V. (2004) *Virology* 326, 310-316
[47] Gall, A., Hoffmann, B., Harder, T., Grund, C., Hoper, D., and Beer, M. (2009) *J Clin Microbiol* 47, 327-334
[48] Stevens, J., Blixt, O., Paulson, J. C., and Wilson, I. A. (2006) *Nat Rev Microbiol* 4, 857-864
[49] Stevens, J., Blixt, O., Glaser, L., Taubenberger, J. K., Palese, P., Paulson, J. C., and Wilson, I. A. (2006) *J Mol Biol* 355, 1143-1155
[50] Wang, Y. F., Lao, W. I., Kuo, Y. D., Guu, S. Y., Wang, H. C., Lin, C. H., Wang, J. R., Su, I. J., and Chang, C. F. (2012) *Fut Virol* 7, 13
[51] Wang, Q., Cheng, F., Lu, M., Tian, X., and Ma, J. (2008) *J Virol* 82, 3011-3020
[52] Wang, Q., Tian, X., Chen, X., and Ma, J. (2007) *Proc Natl Acad Sci U S A* 104, 16874-16879
[53] Kumlin, U., Olofsson, S., Dimock, K., Arnberg, N. (2008) *Influenza and Other Respiratory Viruses* 2, 8
[54] Kogure, T., Suzuki, T., Takahashi, T., Miyamoto, D., Hidari, K. I., Guo, C. T., Ito, T., Kawaoka, Y., and Suzuki, Y. (2006) *Glycoconj J* 23, 101-106
[55] Kolarich, D., and Altmann, F. (2000) *Anal Biochem* 285, 64-75
[56] Pilobello, K. T., Krishnamoorthy, L., Slawek, D., and Mahal, L. K. (2005) *Chembiochem* 6, 985-989
[57] Yang, S. N., Liu, C. A., Chung, M. Y., Huang, H. C., Yeh, G. C., Wong, C. S., Lin, W. W., Yang, C. H., and Tao, P. L. (2006) *Hippocampus* 16, 521-530
[58] Karlsson, N. G., and Packer, N. H. (2002) *Anal Biochem* 305, 173-185
[59] Angeloni, S., Ridet, J. L., Kusy, N., Gao, H., Crevoisier, F., Guinchard, S., Kochhar, S., Sigrist, H., and Sprenger, N. (2005) *Glycobiology* 15, 31-41

Chapter 2

Boron-Carbohydrate Interactions

Brighid Pappin, Milton J. Kiefel and Todd A. Houston

Additional information is available at the end of the chapter

1. Introduction

Boron-polyol interactions are of fundamental importance to human health [1], plant growth [2] and quorum sensing among certain bacteria [3]. Such diversity is perhaps not surprising when one considers boron is one of the ten most abundant elements in sea water and carbohydrates make up the planet's most abundant class of biomass. Several boronic acids matrices are commercially available for the purification of glycoproteins by affinity chromatography [4], and boronic acids are also useful carbohydrate protecting groups.[5,6] Recently, complexes between boron and sugars have become a lynchpin for the development of synthetic carbohydrate receptors.[7] These complexes involve covalent interactions that are reversible in aqueous solution. This chapter reviews current understanding of these processes, provides a historical perspective on their discovery, identifies methods for studying these complexes and classifies these interactions by carbohydrate type. Such information is key to the design and synthesis of synthetic lectins, also termed "boronolectins" when containing boron [7].

The very nature of the reversible binding between boron acids and alcohols has been exploited in many different ways. The use of boronic acid carbohydrate recognition molecules could provide an avenue for the selective detection of specific sugars for future use in early diagnostics. By targeting cell-surface sugars, a boron-based probe could recognize particular characteristic epitopes for the identification of diseases leading to earlier treatments. In this chapter we not only review some of the fundamental aspects of boron-carbohydrate interactions but also discuss how this translates into the design of synthetic carbohydrate receptors.

2. Boron-carbohydrate interactions

2.1. Discovery of boron-sugar interactions

The first hint of the marriage between boron and polyols was detected by Biot in his seminal studies on optical rotation. In 1832 he noted that the rotation of tartaric acid changed in the presence of boric acid.[8] It would be a century later before interaction of boron acids (boric, boronic and borinic) and monosaccharides was studied in detail. In 1913, Böeseken first noted that glucose increased the acidity of boric acid solutions.[9] It was nearly another half century before Lorand and Edwards published work quantifying the affinity of boric and phenylboronic acids for simple diols (e.g.-ethylene glycol, catechol) and common monosaccharides (i.e.-glucose, fructose, mannose, galactose).[10] The covalent product between a boronic acid and a diol is termed a boronate ester, analogous to a carboxylate ester. These interactions are favoured at basic pH ranges where the tetrahedral boronate ester is formed (Figure 1). The interchange between boron acids and divalent ligands in aqueous solution can be complex and varied depending on pH.

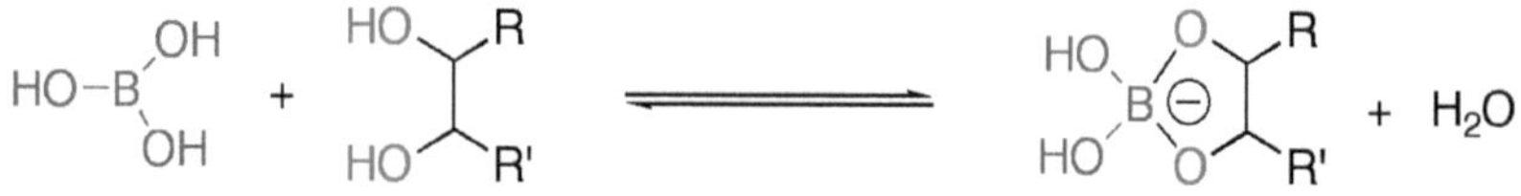

Figure 1. Boric acid interactions with vicinal diol of sugar.

2.2. Fundamentals of boron-diol exchange

There are two general organoboron families of boric acid descent that can form esters with diols through loss of water. These are boronic acids--where one hydroxy group of the parent boric acid is substituted by carbon--and borinic acids, where two hydroxy groups are substituted by carbon-based substituents.

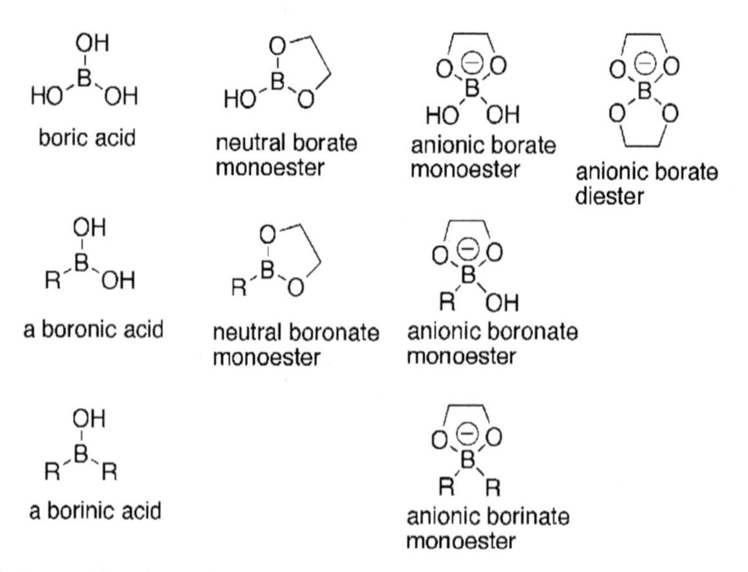

Figure 2. Boron acids and possible esters with ethylene glycol.

Boric and boronic acids can form either neutral or anionic esters depending on the pH. Diol binding by boron acids is favoured at basic pH, while esterification of boron by hydroxycarboxylic acids is favoured in acidic pH ranges. Borinic acids can only form anionic borinate esters upon dehydrative condensation with a diol or divalent ligand. Boric acid can also form an anionic, tetrahedral diester with diols and related divalent ligands (Figure 2). While boronates can form neutral esters in non-polar solvents, they tend to form anionic boronate esters in water (Figure 3). Boronate ester formation is not favoured near physiologic pH and is completely cleaved under strongly acidic conditions.

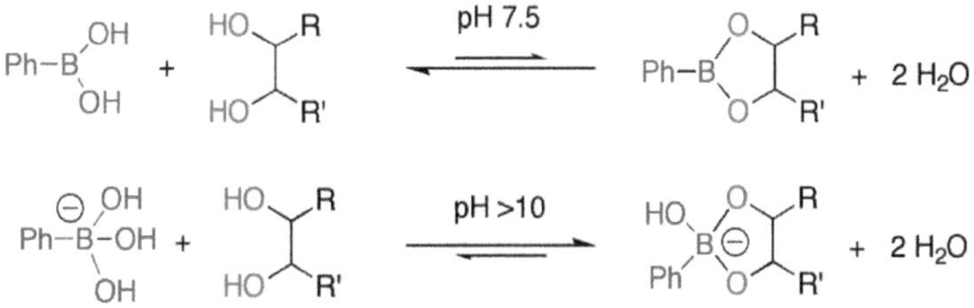

Figure 3. Diol exchange with phenylboronic acid at varied pH.

This is because the neutral boronate ester is generally more Lewis acidic than the parent boronic acid—i.e. pK_a (acid) > pK_a (ester), (Scheme 1).[11] Thus, boronate ester formation is favoured at higher pH where elevated hydroxide concentrations ensure the boronate ester is "trapped" in its more stable tetrahedral form. However, depending on the specific monosaccharide, its boronate esters are not always more Lewis acidic than the free boronic acid.[12] Rate constants for esterification of simple boronates by diols fall in the range of 10^2-10^3 $M^{-1}s^{-1}$.[13] Ishihara uncovered evidence it is the trigonal boronic acid that exchanges most rapidly with diols irregardless of pH.[14] The relative affinity of boronates for diols in most carbohydrates is of the order: *cis*-1,2-diol > 1,3-diol >> *trans*-1,2-diol. Thus, certain monosaccharides have an intrinsically higher affinity for boron acids.

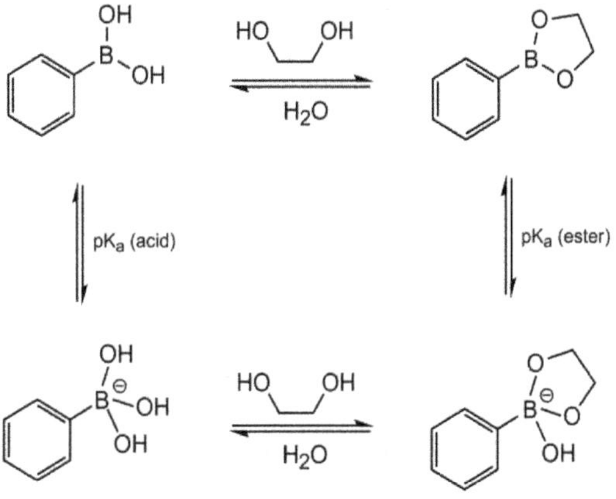

Figure 4. Multiple equilibria involved in diol exchange with phenylboronic acid.

2.3. Detection and elucidation of boron-sugar complexes

Methods for identifying boron-polyol interactions include the following, listed in roughly chronological order of the introduction of their use in this area:

a. Optical rotation/ORD: 1832 [8,15,16]
b. pH change/titration: 1913 [9]
c. Conductivity: 1928
d. Temperature jump: 1969 [17]
e. X-ray crystallography: 1973 [18,19]
f. ^{11}B-NMR: 1973 [19-22]
g. Fluorescence/CD: 1990's [23-25]
h. ESI/MALDI Mass Spectrometry: 1990's [26-28]

The second half of the list defines techniques that are most frequently used today in the study of boron-carbohydrate interactions. Obviously, X-ray crystallography provides the least ambiguous information about the structure of the boronate ester of interest. However, these boronate-sugar adducts are often amphiphilic in nature and do not lend themselves to the production of suitable crystals. The relatively slow exchange between boron acids and diols on the NMR time scale often makes it difficult to study by proton NMR. However, ^{11}B-NMR can be quite useful due to the dramatic shift of the boron resonance when it is converted from its neutral, trigonal form as a boronic acid to its anionic, tetrahedral form as a boronate ester. [20-22]

Optical methods such as fluorescence and circular dichroism (CD) are powerful tools for detecting boron-carbohydrate binding interactions. Yoon and Czarnik reported the first fluorescent boronate designed to detect binding to monosaccharides.[23] James, Shinkai and co-workers reported the first fluorescent boronates to function by photoinduced electron transfer (PET) to generate an increased fluorescence output upon carbohydrate binding.[24] This type of "turn-on" system tends to be most useful in a biological setting where background fluorescence quenching can be a problem for fluorophores that respond by fluorescence quenching ("turn-off") to ligand binding. A more complete understanding of the aminoboronate PET fluorescence mechanism has been developed by the groups of Wang [29] and Anslyn [30]. They have demonstrated that solvent insertion disrupting any dative boron-nitrogen interaction is responsible for the increased fluorescence output upon ligand binding. The Shinkai group has also designed a number of CD-active boronate receptors for oligosaccharides and have used this method to detect binding of target substrates.[25]

Advances in mass spectrometry (MS) over the past few decades, particularly electrospray ionization (ESI) and matrix-assisted laser desorption/ionization (MALDI), have revolutionized the application of this instrumental method to the study of host-guest and protein-ligand interactions. Certainly, the field of boron-based carbohydrate receptors has also benefited from the substantial improvement and refinement of these and other MS techniques. However, the tendency of boronates to dehydrate and/or oligomerize to varying degrees depending on their solvation can complicate MS analysis of boronate-carbohydrate esters. We have found that use of a glycerol matrix for fast atom bombardment (FAB) ionization is particularly useful for mass spectrometric characterization of diboronate species.[31] While other techniques are used in the study of boron-polyol complexes, those mentioned here are among the most common routinely used in the field today.

3. Boron-based carbohydrate receptors

3.1. Boron-based monosaccharide receptors

The 1992 work of Yoon and Czarnik first demonstrated the potential of boronic acids as fluorescent carbohydrate receptors for sensing applications.[23] In the past 20 years, a great deal of research has focused on the development of boron-based glucose receptors for incorporation as sensors in blood sugar monitors for diabetics.[7, 32] This has lead to the commercial development of contact lenses that can signal when circulating glucose levels drop by changing the colour of the lense to alert the wearer.[33] The affinity of mono-boronates for glucose is low at physiologic pH, but bis-boronates offer a substantial improvement in binding affinity. A landmark study from the Shinkai group involved development of a chiral glucose sensor capable of discriminating between enantiomers of glucose.[34] This utilized aminoboronates as PET sensors around a chiral binaphthol core (Figure 5). The Singaram group has developed bis-boronate bipyridinium salts (viologens) that can be tuned for selective binding of glucose (Figure 6).[35] These compounds coupled with anionic dyes are also in commercial development as blood glucose sensors.

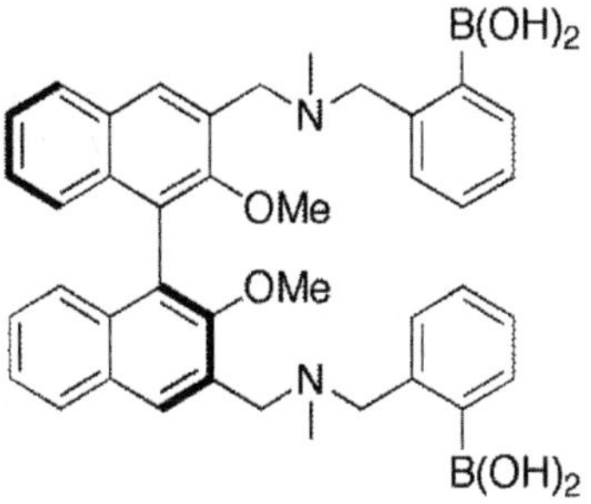

B(OH)$_2$
N
OMe
OMe
N
B(OH)$_2$

Figure 5. Shinkai's chiral binaphthol glucose sensor.

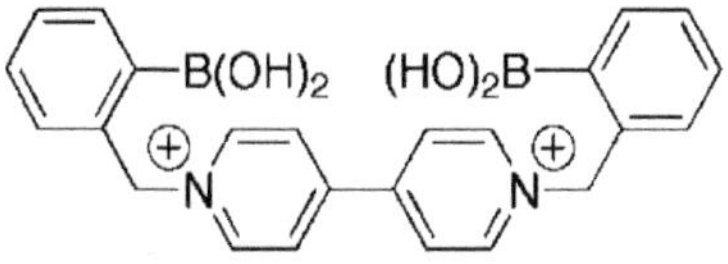

B(OH)$_2$ (HO)$_2$B
N N

Figure 6. One isomer of Singaram's family of glucose sensors.

It was initially presumed that two boronotes could bind to the C-1/C-2 diol and the C-4/C-6 diol of glucose in its hexopyranoside form.[5] However, it has been shown that boronic acids have a much higher affinity for the furanoside form of free hexoses.[36] In fact, boronates have virtually no affinity for methyl glyocsides locked in their pyranoside form at physiologic pH. This means that boronates would not be useful components in synthetic carbohydrate receptors for many cell surface carbohydrates. Mammalian cell-surface glycoconjugates, in particular, are dominated by hexopyranoside structures. The Hall group has provided an important solution to this problem when they showed that benzoboroxoles can bind methyl hexopyranosides in water at pH 7.5.[37] For glucopyranosides, the only significant binding site is the C-4/C-6 diol as all vicinal diols in this system are of a *trans* relationship. In galactopyranosides, there is an additional possible binding site: the C-3/C-4 *cis*-diol (Figure 7):

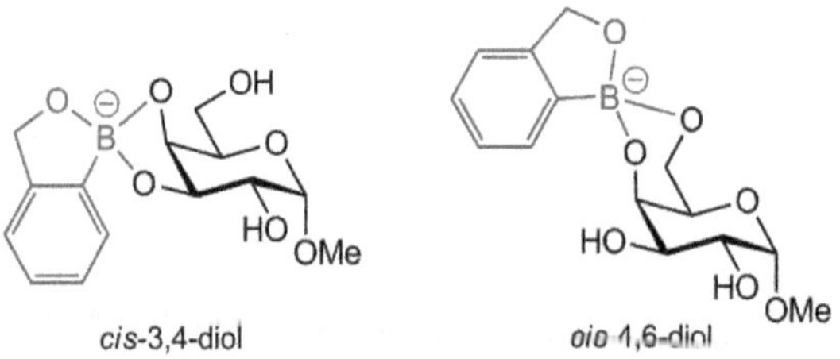

Figure 7. Potential binding modes between benzoboroxole (blue) and methyl-galactoside.

While a significant amount of research has been dedicated to the study of boronate-monosaccharide interactions, very little has been invested in borinate-monosaccharide exchange. Taylor has recently reported that borinic acids have substantial affinity for catechols and α-hydroxycarboxylates,[38] greater than that of 2-fluoro-5-nitrophenyboronic acid,[39] a boronate that is able to bind sugars at neutral pH. The affinity of this boronate for monosaccharides is greater than the affinity of a borinic acid for the same sugars, but this affinity in the latter case is still significant. Whether borinates can effectively bind to hexopyranosides under the same conditions still needs to be defined.

3.2. Boron-based sugar acid receptors

Boron-tartaric acid interactions were studied throughout the 20th century beginning with a report in 1911 on the ability of tartrate to increase the solubility of boric acid.[40] The design of sophisticated boron based receptors for tartrate did not arise until near the end of the century when Anslyn reported the first in 1999.[41] This receptor (Figure 8) also binds citrate with what is perhaps the highest association constant reported for a small molecule with a boron-based receptor ($K_a = 2x10^5$).[42] In 2002, we showed that Shinkai's binaphthol glucose receptor (Figure 5) has a high affinity for tartrate as well.[43] Bis-boronates such as this can bind simultaneously to both α-hydroxycarboxylates. James further showed that chiral discrimination between tartrate enantiomers can also be obtained with this receptor as was the case with monosaccharide enantiomers.[44] While the history of study surrounding boron-tartaric acid interactions is long and varied, study of the interaction of boron with sugar acid monosaccharides such as sialic acid and glucuronic acid has arisen much more recently. Fundamental to the understanding of these complexes is the fact that, unlike esterification with diols, boronate esterification by α-hydroxycarboxylic acids is favoured below pH 7.[44] We have recently provided a short review on the subject of boron:α-hydroxycarboxylate interactions used in sensing and catalysis.[45]

Shinkai first reported a boron-based sugar acid receptor containing a metal chelate that has significant affinity for glucuronic acid (log K_a = 3.4) and galacturonic acid (log K_a = 3.1) while the affinity for sialic acid was an order of magnitude lower (log K_a = 2.3).[46] Presumably the carboxylate of the sugar acid can coordinate to the chelated zinc while the boron binds to a vicinal diol on the monosaccharide. Smith and Taylor used a combination of electrostatic interaction and a boronate anchor within a polymeric system to bind to sialic acid selectively.[47] In 2004, Strongin identified a boronate that offered a colorimetric response to the presence of sialic acid.[48]

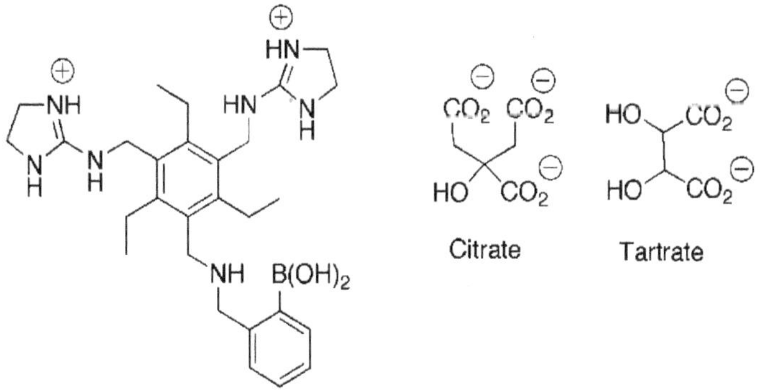

Figure 8. Anslyn's guanidino-boronate receptor and high affinity ligands.

We have recently communicated the development of a bis-boronate that can bind to sialic acid at both its α-hydroxycarboxylate-type group at the anomeric centre and its glycerol tail (Figure 9).[49] Elevated levels of free sialic acid in the blood can be indicative of the presence of certain cancers. This system uses a unique combination of boronates whose esterification has an opposing affect on the overall fluorescence output of the receptor. This diminishes signals from competing ligands such as glucose that are present at much higher concentration in the blood but cannot span both binding site to strongly quench fluorescence.

As discussed in the following sections, several groups have taken advantage of the affinity of boronates for the glycerol tail of sialic acid to target glycoconjugates on cell surfaces. Several of these synthetic compounds display lectin-like biological characteristics that offer promise of the future development of bioactive boron-based molecules.

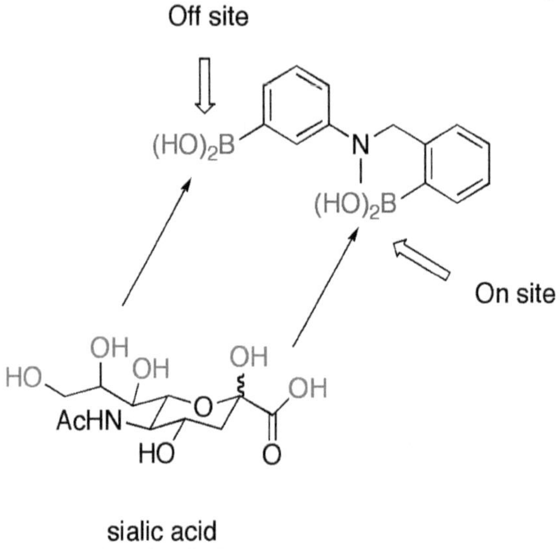

Figure 9. Our divergent response fluorescent receptor for sialic acid.

3.3. Boron-based oligosaccharide receptors

Creating receptors for oligosaccharides offers additional levels of complexity relative to monosaccharides. An obvious difference is the increased degrees of freedom available to oligosaccharides, particularly those that contain 1,6-linkages. In 2000, Shinkai reported development of a meso-meso-linked porphyrin scaffold where distance between two boronates was tuned to selectively bind to a tetrasaccharide of maltose (maltotetrose) over other oligomers containing from two to seven glucose units (Figure 10).[50] Binding of the two boronates must take place at both the reducing and non-reducing termini of the oligosaccharide. This is due to the fact that the C-1/C-2 diol at the reducing end and the C-4/C-6 at the non-reducing end are the only potential binding sites with an appreciable affinity for boronates. The ability to bind the tetramer selectively stems from the rigidly defined distance between the two boronates, i.e.-the tetrasaccharide offers the optimal fit to bridge these two boronates. The Shinkai group has also reported a similar strategy to bind to the important cell-surface trisaccharide, Lewis X.[51] In this case interaction is not with a reducing sugar but presumably with diols on both the galactose and fucose residues.

Heparin is a natural polysaccharide used clinically for its anti-coagulant properties. In 2002, Anslyn communicated a colorimetric sensing ensemble for detection of heparin.[52] As heparin has a high anionic charge density, the receptor was designed with a number of complementary cationic amino groups alongside boronic acids. This group has further reported success in sensing heparin within serum using a second generation receptor.[53] Schrader has recently developed a fluorescent polymeric heparin sensor that can quantify this polysaccharide with unprecedented sensitivity (30 nM).[54] Coupling of boron-carbohydrate interactions with electrostatic attraction in a multivalent manner is responsible for the high affinity of this receptor for its substrate. In spite of this avidity, the interaction can be controlled in a biologically relevant manner. Binding of the polymer to heparin can be reversed by the addition of protamine, similar to reversal of the complex between heparin and its natural target, anti-thrombin III. Other examples of biological mimicry by boron-based systems are delineated below and in the next section.

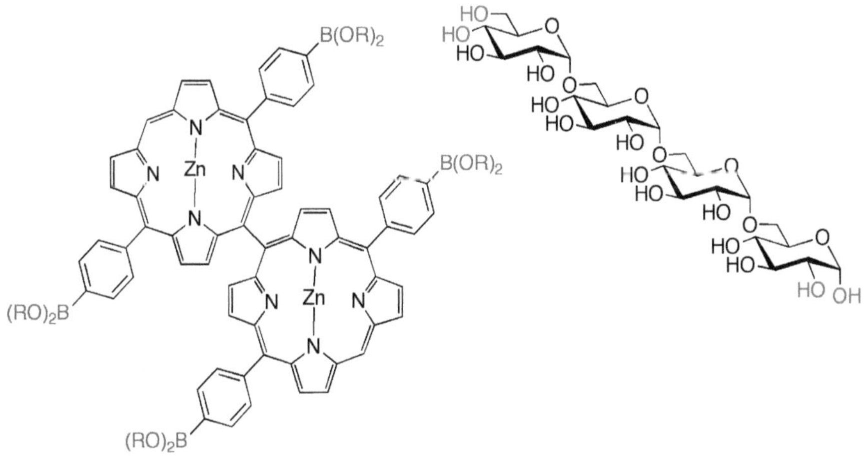

Figure 10. Shinkai's oligosaccharide receptor and maltotetrose.

The demonstrated affinity of benzoboroxoles for hexopyranosides makes these boron derivatives attractive components of receptors designed to target mammalian oligosaccharides. In 2010, Hall reported development of a bis-benzoboroxole receptor for the Thomsen-Friedenreich (TF) antigen, a tumor marker composed of consecutive galactose-based residues Figure 11.[55] This receptor was optimized within a combinatorial library constructed to add additional H-bonding and hydrophobic interactions between host and its oligosaccharide guest. A natural lectin receptor for this disaccharide, peanut agglutinin lectin (PNA), binds the TF antigen quite strongly relative to other protein-carbohydrate interactions (K_d = 10^7). However, the synthetic bis-benzoboroxole inhibits binding of PNA to TF-antigen labelled protein at low micromolar concentrations. The bis-benzoboroxole has a higher affinity for the disaccharide than the corresponding bis-boronate although the latter still has significant affinity highlighting the importance of the additional H-bonding and hydrophobic interactions in this system. This work indicates it should be possible to target multiple hexopyranoside structures with other oligomeric benzoboroxole systems.

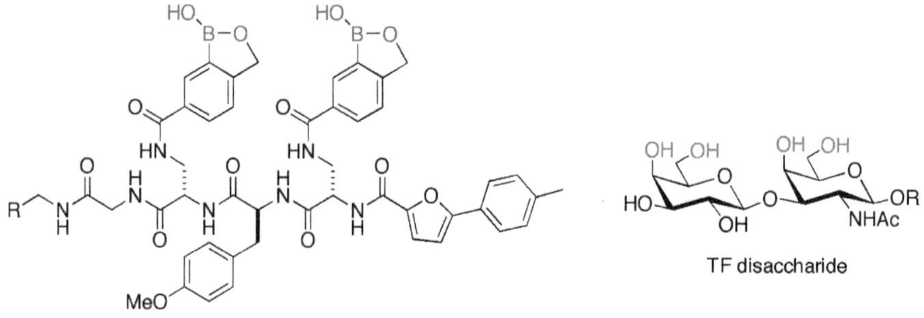

Figure 11. Hall's bis-benzoboroxole receptor for the TF antigen.

4. Boron-cell surface interactions

One ultimate goal of research into synthetic carbohydrate receptors is the development of compounds that can bind directly to cell surface glycocojugates. Such synthetic lectins may serve as diagnostics to monitor changes in cell surface structure associated with disease progression such as cancer. Additionally, they may be used as drug-targeting agents to deliver chemotherapeutic agents to specific cell types. An early and initially underappreciated demonstration of the targeting of cell-surface structures was the work of Hageman with fluorescent dansyl boronates shown to associate with *Bacillus subtilis*.[56] They were also able to show a diboronate could display other lectin like properties such as promoting the agglutination of erythrocytes. Not long after that, Gallop developed a method he defined as "boradaption" using boronates to transfer lipophilic dyes and probes into cells.[57] Although the precise mechanism was not delineated, some boronates were shown to alter the latter stages of *N*-linked glycoprotein processing.[58] A great deal of work has also gone into the development of lipophilic boronic acids as membrane transport agents for hydrophilic molecules such as sialic acid and its derivatives.[59] There is commercial interest in such artificial transporters for the extraction of monosaccharides, such as glucose and fructose, and disaccharides like lactose from natural sources.[28]

In 2002, Weston and Wang reported the ability to target a specific oligosaccharide epitope of a cell surface glycoconjugate.[60, 61] Use of a fluorescent bis-boronate to label the cancer-related antigen sialyl Lewis X on hepatocellular carcinoma cells was an important achievement. They used a combinatorial approach to optimize a bis-boronate in targeting the sialic acid and fucose residues on the tetrasaccharide. The receptor did not label cells that contained Lewis Y antigens lacking sialic acids, or were treated with fucosidase, to remove fucose from the cell surface. This indicates that both components are necessary for interaction with the synthetic receptor. The study marked the first time, as far as we are aware, that two different monosaccharide types had been targeted by design with a boronolectin on a cell surface (Figure 12).

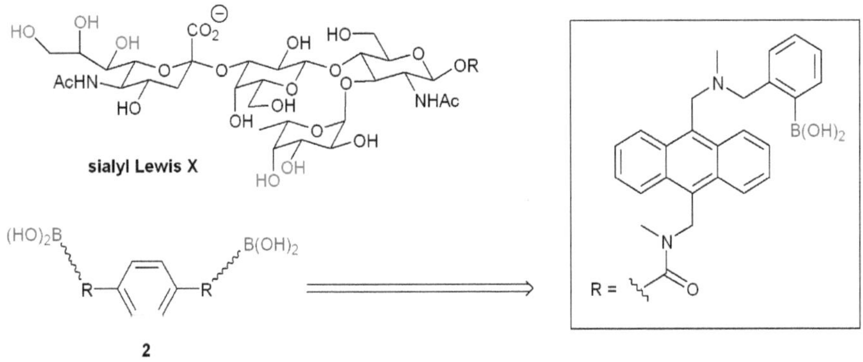

Figure 12. Wang's sialyl Lewis X receptor. Reprinted with permission. John Wiley & Sons, ©2010.

The Kataoka group has used boronic acids on a number of platforms to target cell surface sialic acids to engender a biological or analytical response. They have shown that polymeric boronates can cause the induction of lymphocytes in the same way as natural lectins do.[62] In addition these polyboronates can out-compete natural sialic acid-specific lectins for a cell surface. In collaboration with Miyahara, a powerful method for the direct determination of cell-surface sialic acid levels has been developed.[63, 64] Use of a self-assembled monolayer on a gold electrode allows a coating of boronates to be applied. Potentiometric measurements in the presence of cell suspensions containing either 0, 15, 30, or 100% metastatic cells are readily distinguishable (Figure 13).[64]

The study, application and manipulation of boron-carbohydrate interactions continues to expand into its third century. The properties of oligomeric and polymeric boronic acids in a cellular setting demonstrate that the terms "boronolectin" and/or synthetic lectin are appropriate. What remains for the field to advance are more examples targeting cell-surface carbohydrate structures beyond those containing sialic acid. The Hall group's receptor for the TF antigen marks a seminal step in this direction.[54] For a more comprehensive review of boron-based carbohydrate receptors in the context of other synthetic and biologic sugar binding systems, readers are directed to the recent publication of Wang.[7]

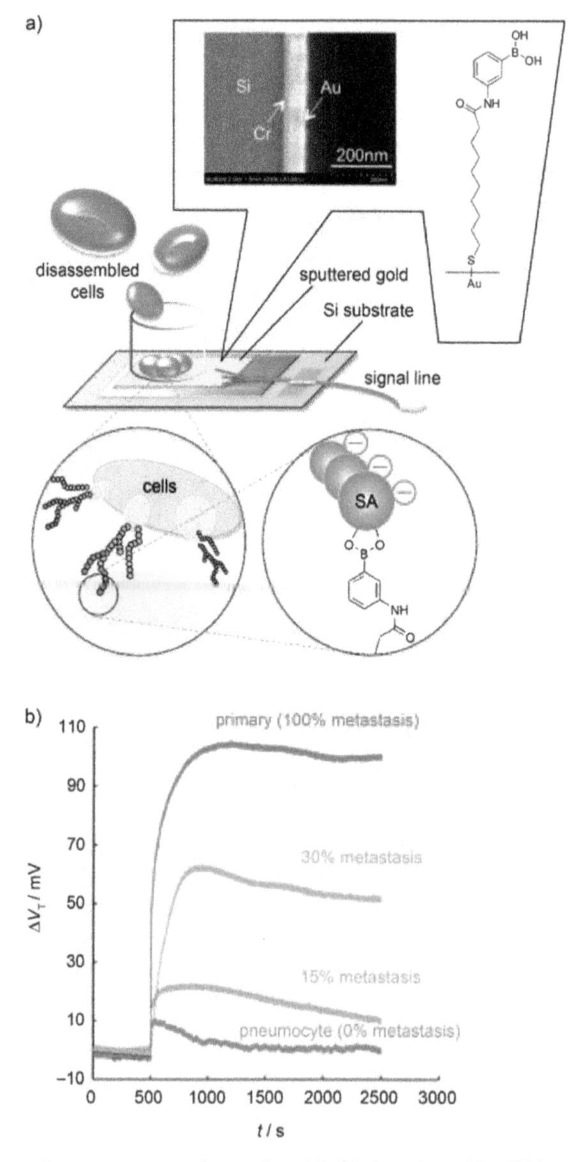

Figure 13. a) Schematic representation of potentiometric SA detection with a PBA-modified gold electrode. An SEM image of a cross-section of the electrode is shown at the top next to the chemical structure of the PBA-modified self-assembled monolayer introduced onto the electrode surface. b) Change in the threshold voltage (VT) of the PBA-modified FET as a function of time upon the addition of cell suspensions (10^6 cells/mL) with various degrees of metastasis. [64] Reprinted with permission. John Wiley & Sons, ©2010.

5. Conclusions

In spite of its long and rich history, understanding of boron acid interactions with carbohydrates continues to increase into the 21st century. In the past 20 years, much fundamental knowledge has been gained, principally from the development of boronate-based glucose receptors for application toward blood sugar monitoring in diabetics. Currently, however, significant effort is being dedicated to the development of boron-based receptors for more complex oligosaccharides. This challenge is being undertaken by an increasing number of research groups throughout the world. These designer receptors may find application in diagnostics for cancer or infectious diseases, in drug targeting, or in providing a more fundamental understanding of the biochemical roles of cell-surface carbohydrates. The ability of boron acids to distinguish between closely related polyols either stereoselectively or chemoselectively makes them an obvious choice for anchoring synthetic carbohydrate receptors.[65] Engineering these interactions to target specific oligosaccharides is currently a difficult challenge as witnessed by the limited number of boron-based oligosaccharide receptors that have been developed at this stage. However, coupling boron-carbohydrate interactions with several additional non-covalent interactions—electrostatic, H-bonding, hydrophobic—offers the best chance of success. Future endeavours will determine the scope and limitations of boron-based carbohydrate receptors and sensors.

Author details

Brighid Pappin
Institute for Glycomics, Griffith University, Gold Coast, Australia

Milton J. Kiefel
Institute for Glycomics, Griffith University, Gold Coast, Australia

Todd A. Houston
Institute for Glycomics, Griffith University, Gold Coast, Australia

School of Biomolecular and Physical Sciences, Griffith University, Nathan, Australia

6. References

[1] Scorei IR (2011) Boron Compounds in the Breast Cancer Cells Chemoprevention and Chemotherapy. In Gunduz E, Gunduz M, editors. Breast Cancer – Current and Alternative Therapeutic Modalities. Rijeka: InTech. pp. 91-114.

[2] O'Neill MA, Eberhard S, Albersheim P, Darvill AG (2001) Requirement of Borate Cross-Linking of Cell Wall Rhamnogalacturonan II for Arabidopsis Growth. *Science* 294: 846-849.

[3] Chen X, Schauder S, Potier N, Dorsselaer AV, Pelczer I, Bassler BL, Hughson FM (2002) Structural identification of a bacterial quorom-sensing signal containing boron. *Nature* 415: 545-549.

[4] Hageman JH, Kuehn GD (1992) Boronic Acid Matrices for the Affinity Purification of Glycoproteins and Enzymes. In Kenney A, Fowell S, editors, Methods in Molecular Biology, Vol. 11: Practical Protein Chromatography. Totoway, NJ: The Humana Press Inc. pp. 45-71.

[5] Ferrier RJ, Prasad R (1965) Boric Acid Derivatives as Reagents in Carbohydrate Chemistry. 6. Phenylboronic Acid as a Protecting Group in Disaccharide Synthesis. *J. Chem. Soc.* 7429.

[6] Duggan PJ, Tyndall EM (2002) Boron acids as protective agents and catalysts in synthesis. *J. Chem. Soc., Perkin Trans. 1.* 1325-1339.

[7] Jin, S, Cheng Y, Reid S, Li M, Wang B, (2010) Carbohydrate recognition by boronolectins, small molecules and lectins. *Med. Res. Rev.* 30: 171-257.

[8] Lowry TM (1935) Optical Rotary Power. London: Longmans, Green, and Co. Reprinted (1964) New York: Dover Publishers, Inc.

[9] Böeseken J. (1913) On the storage of hydroxyl groups of polyoxide compounds in space. The configuration of saturated glycols and of alpha- and beta-glycose. *Ber. Dtsch. Chem. Ges.*, 46: 2612–2628.

[10] Lorand JP, Edwards JO (1959) Polyol Complexes and Structure of the Benzeneboronate Ion. *J. Org. Chem.* 24: 769-774.

[11] Yasuda M, Yshioka S, Yamasaki S, Somyo T, Chiba K, Baba A (2006) Cage-Shape Borate Esters with Enhanced Lewis Acidity and Catalytic Activity. *Org. Lett.* 8: 761-764.

[12] Springsteen G, Wang BH (2002) A detailed examination of the boronic acid-diol complexation. *Tetrahedron* 58: 5291-5300.

[13] Pizer RD, Tihal C (1992) Equilibria and Reaction Mechanism of the Complexation of Methylboronic Acid with Polyols. *Inorg. Chem.* 31: 3243-3247

[14] Iwatsuki S, Nakajima S, Inamo M, Takagi HD, Ishihara K (2007) Which is Reactive in Alkaline Solution, Boronate Ion of Boronic Acid? Kinetic Evidence for Reactive Trigonal Boronic Acid in an Alkaline Solution. *Inorg. Chem.* 46: 354-356.

[15] Katzin LI, Gulyas E (1966) Optical Rotatory Dispersion Studies on the Borotartrate Complexes and Remarks on the Aqueous Chemistry of Boric Acid. *J. Am. Chem. Soc.* 88: 5209-

[16] Jones B (1933) The rotatory dispersion of organic compounds. Part XXII Borotartrates and boromalates. *J. Chem. Soc.* 1933: 951-955.

[17] Kustin K, Pizer R (1969) Temperature-Jump Study of the Rate and Mechanism of the Boric Acid-Tartaric Acid Complexation. *J. Am. Chem. Soc.* 91: 317-322.

[18] Mariezcurrena RA, Rasmussen SE, (1973) Crystal Structure of Potassium Boromalate. *Acta Cryatallogr.* B29: 1035-1040.

[19] van Duin M, Peters JA, Kieboom APG, van Bekkum H (1987) Studies on Borate Esters. Part 5. The System Glucarate-Borate-Calcium(II)as studied by 'H, "B, and [13]C Nuclear Magnetic Resonance Spectroscopy. *J. Chem. Soc., Perkin Trans. 2*, 473-478.

[20] Henderson WG, How MJ, Kennedy GR, Mooney EF (1973) The Interconversion of Aqueous Boron Species and the Interaction of Borate with Diols: a ^{11}B NMR Study *Carbohydrate Res.* 28: 1-12.

[21] Collins BE, Sorey S, Hargrove AE, Shabbir SH, Lynch, VM, Anslyn, EV (2009) Probing Intramolecular B-N Interactions in Ortho-Aminomethyl Arylboronic Acids. *J. Org. Chem.* 74: 4055-4060.

[22] Johnson LL, Houston TA (2002) A Drug Targeting Motif for Glycosidase Inhibitors: An Iminosugar-Boronate Shows Unexpectedly Selective β-Galactosidase Inhibition. *Tetrahedron Lett.* 43: 8905-8908.

[23] Yoon J, Czarnik AW (1992) Fluorescent Chemosensing of Carbohydrates—A Means of Chemically Communicating the Binding of Polyols in Water by Chelation-Enhanced Quenching. *J. Am. Chem. Soc.* 114: 5874-5875.

[24] James TD, Sandanayake KRAS, Shinkai (1994) Novel Photoinduced Electron Transfer Sensor for Saccharides Based on a Boronic Acid and an Amine. *Chem. Commun.* 477-478.

[25] James TD, Shinkai S (1995) A Diboronic Acid 'Glucose Cleft' and a Biscrown Ether 'Metal Sandwich' are Allosterically Coupled. *J. Chem. Soc. Chem. Commun.* 1483-1484.

[26] Rose, M. E.; Wycherlet, D.; Preece, S. W. (1992) Negative-Ion Electrospray and Fast-Atom-Bombardment Mass Spectrometry of Esters of Boron Acids. *Org. Mass Spectrom.* 27:876-882.

[27] Penn SG, Hu H, Brown PH, Lebrilla, CB (1997) Direct Analysis of Sugar Alcohol Borate Complexes in Plant Extracts by Matrix Assisted Laser Desorption/Ionization-Fourier Transform Mass Spectrometry. *Anal. Chem.* 69: 2471-2477.

[28] Duggan PJ, Houston TA, Kiefel MJ, Levonis, SM, Smith, BD, Szydzik, ML (2008) Enhanced Fructose, Glucose and Lactose Transport Promoted by a 2-(Aminomethyl) phenylboronic Acid. *Tetrahedron.* 64: 7122-7126.

[29] Franzen S, Ni, W, Wang,B. (2003) Study of the Mechanism of Electron-Transfer Quenching by Boron-Nitrogen Adducts in Fluorescent Sensors. *J. Phys. Chem. B* 107: 12942.

[30] Zhu L, Shabbir SH, Gray M, Lynch VM, Sorey S, E. Anslyn, EV (2006) A Structural Investigation of the N-B Interaction in an *o*-(*N*,*N*-Dialkylaminomethyl)arylboronate System. *J. Am. Chem. Soc.* 128: 1222-1232.

[31] Kramp KL, DeWitt K, Flora J, Muddiman DC, Slunt KM, Houston, TA (2005) Derivatives of Pentamidine Designed to Target the *Leishmania* Lipophosphoglycan. *Tetrahedron Lett.* 46: 695-698.

[32] Mader HS, Wolfbeis OS (2008) Boronic acid based probes for microdetermination of saccharides and glycosylated biomolecules. Microchim. Acta 162: 1-34.

[33] Badugu R, Lakowicz JR, Geddes CD (2005) A glucose-sensing contact lens: from bench top to patient. *Curr. Opin. Biotechnol.*16: 100-107.

[34] James TD, Sandanayake KRAS, Shinkai S (1995) Chiral discrimination of monosaccharides using a fluorescent molecular sensor. *Nature* 374: 345-347.

[35] Gamsey S, Baxter NA, Sharrett Z, Cordes DB, Olmstead MM, Wessling RA, Singaram S (2006) The effect of boronic acid-positioning in an optical glucose-sensing ensemble. *Tetrahedron* 62: 6321-6331.

[36] Eggert H, Frederiksen J, Morin C, Norrild, JC (1999) A new glucose-selective fluorescent bisboronic acid. First report of strong alpha-furanose complexation in aqueous solution at physiological pH. *J. Org. Chem.* 64: 3846-3852.

[37] Dowlut M, Hall DG (2006) An Improved Class of Sugar-Binding Boronic Acids, Soluble and Capable of Complexing Glycosides in Neutral Water. *J. Am. Chem. Soc.* 130: 9809.

[38] Chudzinski MG, Chi Y, Taylor MS (2011) Borinic Acids: A Neglected Class of Organoboron Compounds for Recognition of Diols in Aqueous Solution. *Aust. J. Chem.* 64: 1466-1469.

[39] Mulla HR, Agard NJ, Basu A (2004) 3-Methoxycarbonyl-5-nitrophenyl boronic acid: h affinity diol recognition at neutral pH. *Bioorg. Med. Chem. Lett.* 14: 25-27.

[40] Werz H (1911) Solubility Studies. II. The Influence of d-Tartaric and dl-Tartaric Acids on the Solubility of Boric Acid. *Z. Anorg. Chem.* 70: 71-72.

[41] J. J. Lavigne, E. V. Anslyn,(1999) Teaching old indicators new tricks: A colorimetric chemosensing ensemble for tartrate/malate in beverages *Angew. Chem. Int. Ed.* 38: 3666-3669.

[42] Wiskur SL, Lavigne JJ, Metzger A, Tobey SL, Lynch V, Anslyn EV (2004) Thermodynamic Analysis of Receptors Based on Guanidinium/Boronic Acid Groups for the Complexation of Carboxylates, α-Hydroxycarboxylates, and Diols: Driving Force for Binding and Cooperativity *Chem. Eur. J.* 10: 3792-3804.

[43] Gray CW, Jr., Houston TA (2002) Boronic Acid Receptors for α-Hydroxycarboxylates: High Affinity of Shinkai's Glucose Receptor for Tartrate *J. Org. Chem.* 67: 5426-5428.

[44] Zhao, J, Fyles, TM, James, TD (2004) Chiral Binol–Bisboronic Acid as Fluorescence Sensor for Sugar Acids. *Angew. Chem. Int. Ed.* 43: 3461-3464.

[45] Houston TA, Levonis SM, Kiefel, MJ (2007) Tapping into Boron/α-Hydroxycarboxylic Acid Interactions in Sensing and Catalysis. *Aust. J. Chem.* 60: 811-815.

[46] Takeuchi M, Yamamoto M, Shinkai S (1997) Fluorescent sensing of uronic acids based on a cooperative action of boronic acid and metal chelate. *Chem. Commun.* 1731-1732.

[47] Patterson S, Smith BD, Taylor RE (1998) Tuning the affinity of a synthetic sialic acid receptor using combinatorial chemistry *Tetrahedron Lett.* 39: 3111-3114.

[48] Yang Y, Lewis PT, Escobedo, JO, St. Luce NN, Treleaven, WD, Cook RL, Strongin RM, (2004) Mild colorimetric detection of sialic acid. *Collect. Czech. Chem. Commun.* 69: 1282-1291.

[49] Levonis, SM, Kiefel MJ, Houston, TA (2009) Boronolectin with Divergent Fluorescent Response Specific for Free Sialic Acid *Chem. Commun.* 2278-2281.

[50] Ikeda M; Shinkai S; Osuka A (2000) Meso-meso-linked porphyrin dimmer as a novel scaffold for the selective binding of oligosaccharides. *Chem. Commun.* 1047-1048.

[51] Sugasaki A, Sugiyasu K, Ikeda M, Takeuchi M, Shinkai S (2001) First Successful Molecular Design of an Artificial Lewis Oligosaccharide Binding System Utilizing Positive Homotropic Allosterism. *J. Am. Chem. Soc.* 123: 10239-10244.

[52] Zhong Z, Anslyn EV (2002) A Colorimetric Sensing Ensemble for Heparin. *J. Am. Chem. Soc.* 124: 9014-9015.

[53] Wright AT, Zhong Z, Anslyn EV (2005) A Functional Assay for Heparin in Serum Using a Designed Synthetic Receptor. *Angew. Chem. Int. Ed.* 44: 5679-5682.

[54] Sun W, Bandmann H, Schrader T (2007) A Fluorescent Polymeric Heparin Sensor. *Chem. Eur. J.* 13: 7701-7707.

[55] Pal A, Bérubé M, Hall DG (2010) Design, Synthesis, and Screening of a Library of Peptidyl Bis(Boroxoles) as Oligosaccharide Receptors in Water: Identification of a Receptor for the Tumor Marker TF-Antigen Disaccharide. *Angew. Chem. Int. Ed.* 49: 1492-1495.

[56] Burnett TJ, Peebles HC, Hageman JH (1980) Synthesis of a Fluorescent Boronic Acid Which Reversibly Binds to Cell Walls and a Diboronic Acid Which Agglutinates Erythrocytes. *Biochem. Biophys. Res. Commun.* 96: 157-162.

[57] Gallop PM, Paz MA, E. Henson E (1982) Boradeption: A New Procedure for Transferring Water-Insoluble Agents Across Cell Membranes. *Science* 217: 166-169.

[58] Goldberger G, Paz MA, Torrelio BM, Okamoto Y, Gallop PM (1987) Effect of Hydroxyorganoboranes on Synthesis, Transport and N-Linked Glycosylation of Plasma Proteins. *Biochem. Biophys. Res. Commun.* 148: 493-499.

[59] Altamore TM, Duggan PJ, Krippner GY (2006) Improving the membrane permeability of sialic acid derivatives. *Bioorg. Med. Chem.* 14: 1126-1133.

[60] Yang W, Gao S, Gao X, Karnati VVR, Ni W, Wang B, Hooks WB, Carson J, Weston B (2002) Diboronic acids as fluorescent probes for cells expressing sialyl Lewis X. *Bioorg. Med. Chem. Lett.* 12: 2175-2177.

[61] Yang W, Fan H, Gao S, Gao X, Karnati VVR, Ni W, Hooks WB, Carson J, Weston B, B. Wang, B (2004) The first fluorescent diboronic acid sensor specific for hepatocellular carcinoma cells expressing sialyl Lewis X. *Chem. Biol.* 11: 439-448.

[62] Uchimura E, Otsuka H, Okano T, Sakurai Y, Kataoka K (2001) Totally synthetic polymer with lectin-like function: Induction of killer cells by the copolymer of 3-acrylamidophenylboronic acid with N,N-Dimethylacrylamide. *Biotechnol. Bioeng.* 72: 307-314.

[63] Matsumoto A, Sato N, Kataoka K, Miyahara Y (2009) Noninvasive Sialic Acid Detection at Cell Membrane by Using Phenylboronic Acid Modified Self-Assembled Monolayer Gold Electrode. *J. Am. Chem. Soc.*,132: 12022-12023.

[64] Matsumoto A, Cabral H, Sato N, Kataoka K, Miyahara Y (2010) Assessment of Tumor Metastasis by the Direct Determination of Cell-Membrane Sialic Acid Expression. *Angew. Chem. Int. Ed.* 49: 5494-5497.

[65] Houston TA (2010) Developing High-Affinity Boron-Based Receptors for Cell-Surface Carbohydrates. ChemBiochem. 11: 945-958.

Chapter 3

Conversion of Carbohydrates Under Microwave Heating

Aurore Richel and Michel Paquot

Additional information is available at the end of the chapter

1. Introduction

The non-energetic valorisation of renewable resources using efficient and eco-friendly methodologies is the central axis of the "green chemistry" concept. In particular, the chemical and chemo-enzymatical transformation of carbohydrates arising from the hydrolysis of non-edible vegetal feedstock (i.e., lignocellulosic biomass) is a widely explored thematic for the production of new high-added value materials, synthons, and platform chemicals (Bozell, 2010). The use of mono-, di- and polysaccharides for the production of new chemicals constitutes thus a subject of special relevance from both academic and industrial points of view (Lichtenthaler, 2004). Amongst the 12 principles defining this "green chemistry" concept, the development of new effective synthetic protocols, minimising wastes and energy-consumption while enhancing purity of the final product, is the corner stone (Anastas, 1998). In this regard, the use of microwaves as a non-conventional heating method has progressively gained attention due to commonly observed acceleration in reactions rates and improved (regio- and chemo-) selectivities and yields in synthetic organic transformations (Kappe, 2004; Caddick, 2009). The claimed cleaner reaction profiles of microwave-assisted processes have thus rapidly projected this kind of heating as a popular method in chemistry, which often replaces the "classical" heating ones. Numerous organic reactions are nowadays fully depicted in the peer-reviewed literature and books. They concern typical synthetic organic approaches (substitutions, alkylations, cycloadditions, esterifications, cyclisations, etc.), organometallic reactions, oxidations and reductions, or polymerisation reactions (Bogdal, 2006).

2. Scope of this contribution

Milder reaction conditions, associated to reduced run times (from several hours to a few minutes) and improved yields and selectivities, are the key advantages usually reported for microwave-promoted reactions. Even if widely employed in common organic chemistry, microwaves find however fewer applications in carbohydrate chemistry (Corsaro, 2004; Cioffi, 2008; Richel, 2011a). This fact, usually associated to a high number of unwanted and uncontrolled typical sugars reactions such as carbonisation or browning processes, has been overcame by the design of novel sophisticated microwave reactors. Advantageously, microwaves provide nowadays, with remarkable atom efficiency and yields, novel carbohydrate-based chemicals that are not easily obtainable by any other "classical" means or only using painstaking multi-step and energy-consuming protocols (Richel, 2011b). This chapter describes thus the use of microwave processes to mediate key reactions in the field of carbohydrate chemistry. Some examples of benchmark reactions (glycosylations, hydroxyl groups' protection, etc.) under microwave conditions are displayed and highlight the benefits of this microwave approach in terms of yields, selectivities and environmental impact.

3. Fundamental of microwave technology

The pioneering works in the microwave-assisted chemistry are attributed to Gedye and Giguere using household microwave ovens (Gedye, 1986; Giguere, 1986). From a theoretical point of view, microwave heating originates from interactions between a given material (reagent, solvent or catalyst) and the electric component of the electromagnetic wave through a dipolar polarisation and/or ionic conduction mechanism. Polar substances like DMSO, ethylene glycol or ethanol are thus suited candidates for microwave applications by opposition to benzene, dioxane or carbon tetrachloride which do not possess permanent dipolar moment. It is recognised that accelerations of reactions rates under microwave conditions are to correlate with a thermal/kinetic effect. Compared to analogous heating under classical conditions (using an oil bath for instance), microwave ensures a fast, selective and homogeneous heating of the reaction vessel (Fig. 1). Convection currents and temperature gradients from vessel walls to the core of the reaction medium, commonly observed using classical heating options, are thus avoided under microwave conditions.

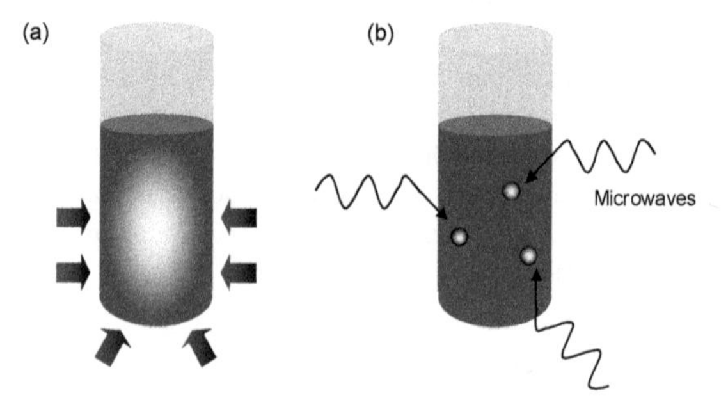

Figure 1. Difference between conventional heating (**a**) and microwave-assisted heating (**b**). In (**a**), heating is provided by thermal conduction. In (**b**), solvent and/or reagents absorb microwave heating and convert this energy into calorific components.

The ability of a given substance to convert electromagnetic energy into calorific heat, at specific frequency and temperature, is determined by the loss tangent factor (tanδ). This factor is described as the ratio between dielectric loss (ε'') and dielectric constant (ε'). Practically, substances with high tanδ values are required to ensure a convenient and fast microwave heating. By convention, substances are thus categorised into good (tanδ>0.5), medium (0.5>tanδ>0.1) and poor (tanδ<0.1) candidates for microwave (MW) applications. At the lab-scale, MW is conveniently applied to reactions performed in the liquid state in the presence of a solvent with a high tanδ. However, in order to provide a truly eco-friendly process, syntheses without solvent or in dry-media conditions are also described with reactants and/or catalysts impregnated on a solid microwave-absorbing support such as alumina or silica (Polshettiwar, 2008). In the field of carbohydrate chemistry, reactions under microwave (MW) assistance are usually achieved with saccharides in suspension or in solution in solvent and/or liquid reagents. Highly secured MW reactors, operating in batch or continuous flow conditions, are nowadays available for chemistry. They operate typically at 2.45 GHz and allow an accurate control of the temperature within the MW reaction vessel with an adjustable power, avoiding thus browning. Reactions can also be performed under "closed vessel conditions", at temperatures exceeding the boiling point of the reaction medium.

4. Chemical transformations of carbohydrates under microwave conditions

Saccharides, which contain several hydroxyl groups, are polar molecules particularly suited for MW applications. Typical MW-assisted transformations reported in the literature include either reactions involving hydroxyl groups for the production of novel entities (category 1) or dehydration reactions leading to the formation of furfural and related platform molecules (category 2).

4.1. Microwave-assisted transformations involving hydroxyl groups (category 1)

Most common reactions of the first category include protections/deprotections of hydroxyl groups, glycosylations, glycosamines formation and halogenations. These reactions are usually catalysed by chemical entities, even if enzymes are proved to offer convenient results for specific cases under microwave conditions. The several hydroxylated stereogenic centres in carbohydrates are difficult to chemically differentiate. Thus, the regioselective derivatisation of such compounds is a challenging task and requires multi-step protocols and purifications. Microwave heating has demonstrated to be a convenient solution, allowing an accurate control of the regioselectivity and the anomeric selectivity by an appropriate tuning of operating conditions. Some relevant examples are proposed herein.

4.1.1. Hydroxyls protections/deprotections

Several methods and protocols have been reported for the MW-assisted regioselective protection of hydroxyls, especially in mono- and disaccharides (Corsaro, 2004). They include mainly acetal formation with aldehydes and acylation with acetic anhydride or an acyl chloride. All these reactions are promoted using (homogeneous or heterogeneous) acids or bases (Söderberg, 2001) or enzymes (Chen, 2001) in variable solvents, reagents concentrations and operating conditions (temperature and microwave exposure time). The total or regioselective protection (and subsequent deprotection) of specific hydroxyl groups is necessary for the synthesis of building blocks suited for the production of drugs, biologically active materials, and novel high-added value materials. Particularly, acetylation reactions are often the prerequisite step in the synthesis of such complex carbohydrates (Fig. 2). The conventional per-*O*-acetylation of saccharides is mostly achieved using anhydride acetic as both reagent and solvent. Although this step is high-yielding, long reaction times (several hours) are required and noxious pyridine is mandatory to promote the reaction. Microwave dielectric heating has thus appeared as a more eco-friendly option. Indeed, replacement of pyridine by catalytic amounts of non-toxic sodium acetate or Lewis acid such as zinc chloride is found practically quantitative after only a few minutes of MW heating (Limousin, 1997). With $ZnCl_2$, an equimolar mixture of α and β pentaacetates is obtained under MW conditions, whilst a α/β ratio of 7/3 is recovered under oil bath conventional heating (Fig. 2a). Microwave heating affords also convenient results for the complete acetylation of totally *O*-unprotected mono- and disaccharides in 90 sec at 720 Watt under closed vessel conditions using indium(III) chloride catalyst. Reactions performed in acetonitrile with stoichiometric amounts of anhydride acetic are quantitative and affords predominantly the α−peracetylated form (Das, 2005). Partial acetylation of carbohydrates is usually achieved with acetyl chloride and pyridine. With 1,2:5,6-di-*O*-isopropylidene-α-D-glucofuranose as the starting material, Söderberg demonstrates that the synergic combination of microwaves and polystyrene-supported base catalysts offers a convenient strategy to obtain in 85% yield the expected mono-acetylated analogous without formation of side-products (Fig. 2b). This MW approach provides an eco-friendly and time-saving option as the catalyst can be recovered by filtration at the end of the process, thus minimising wastes and purification (Söderberg, 2001).

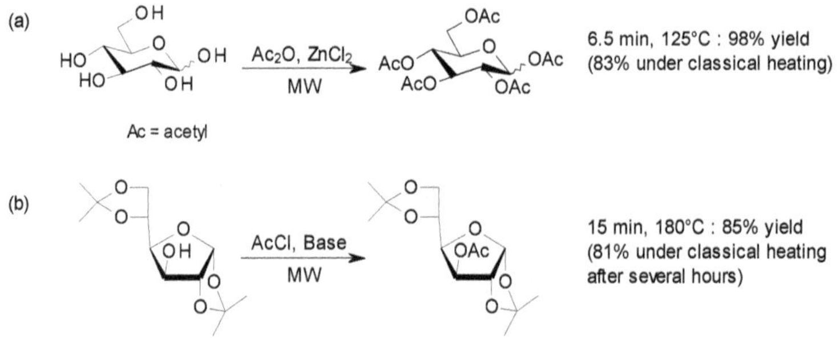

Figure 2. Peracetylation (a) and acetylation (b) of selected carbohydrates under MW conditions.

More recently, Witschi et al. describe a novel procedure for the selective acetylation of hydroxyls groups. It relies on a protecting group exchange strategy under microwave assistance starting from per-*O*-trimethylsilylated pyranosides (Fig. 3). Interestingly, whilst the 1,6-*O*-diacetate monosaccharide is formed under MW exposure, the 6-*O*-monoacetate adduct is preferentially formed under classical oil bath heating conditions. Reactions go to completion when conducted using acetic acid as a catalyst in neat anhydride acetic for 3 x 25 min under MW. For per-*O*-TMS-galactoside, a 52% yield in diacetylated adduct is recovered using 2 equivalents of acetic acid. When the galactoside is stirred for 2 days with 2 equiv of acid, the reaction shows a 50% completion with yields after column chromatography purification reaching 35% of 6-*O*-monoacetate (Witschi, 2010).

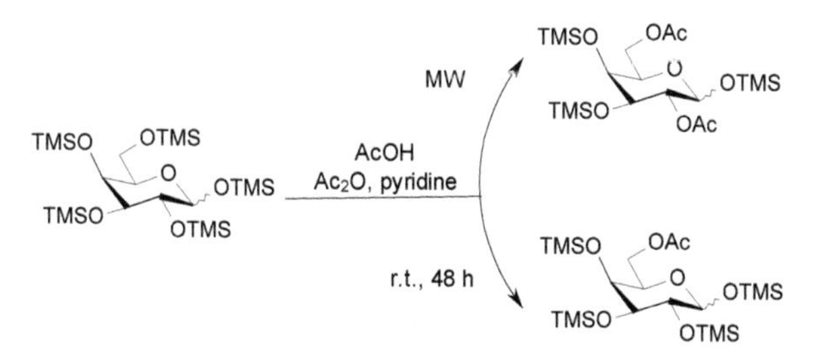

Figure 3. MW-assisted selective acetylation of per-*O*-TMS-galactoside using protecting group exchange strategy.

Regioselective benzoylation and pivaloylation of carbohydrates is another major item for free hydroxyls protection. Typically performed using benzoyl chloride and pivaloyl chloride, these reactions are only viable when using a base catalyst. Due to the steric bulk of the pivaloyl groups, the protection of hydroxyls through formation of pivaloyl esters is very slow under conventional heating method and requires several hours. Substantial reduction in reaction times (10-20 min) is however encountered under MW conditions. Pivaloylation of 1,2:5,6-di-*O*-isopropylidene-α-D-glucofuranose, catalysed by pyridine, provides thus the corresponding 3-*O*-protected target in 68% yield after 10 min at 160°C in a microwave reactor (Fig. 4a). However, presence of a side-product is recovered after MW exposure as a result of 5,6- to 3,6-acetal migration prior to acylation. The formation of this unwanted structure is repressed when using supported polystyrene-bases, i.e. *N,N*-(diisopropyl)aminoethylpolystyrene (PS-DIEA) and *N*-(methylpolystyrene)-4-(methylamino)pyridine (PS-DMAP) , as alternatives to pyridine. 79 and 88% yields are thus detected after 15-20 min at 160-180°C in the microwave cavity, whilst only 88% yield is obtained after 300 min at room temperature (Söderberg, 2001).

The regioselective benzoylation of carbohydrates affords convenient results using dibutyltin oxide mediators under microwave exposure (Herradón, 1995). Intermediate formation of dibutylstannylene acetal allows hydroxyl regioselective benzoylation after less than 10 min of microwave exposure with a 500 Watt output (Fig. 4b). This MW methodology is more advantageous than its "classical" analogous involving heating, for several hours, at reflux of a toluene suspension of saccharides and dibutyltin oxide with azeotropic removal of water. Tuning of operation conditions, including both the solvent and the MW power output, modulates the selectivity of dibutylstannylene acetal-mediated benzoylation of representative carbohydrates, such as α-D-mannopyranoside (Fig. 5).

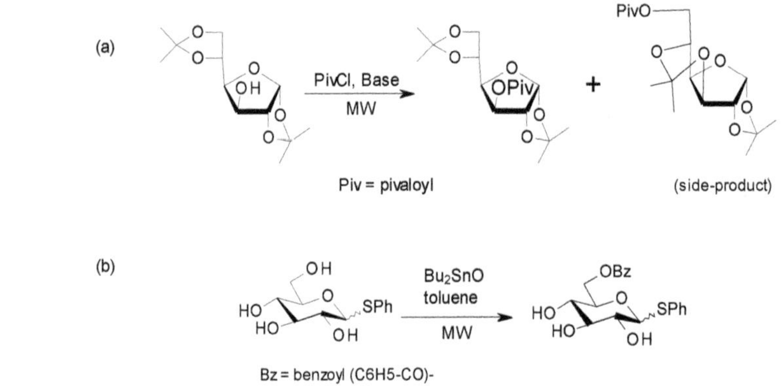

Figure 4. MW-assisted pivaloylation (a) and benzoylation (b).

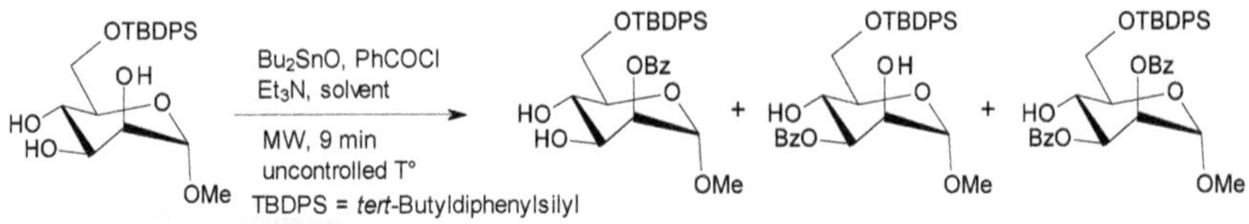

	Yield:		
Toluene, maximum MW power output	35%	25%	10%
Toluene, minimum MW power output	6%	33%	0%
Acetonitrile, minimum MW power output	0%	41%	0%

Figure 5. MW-promoted benzoylation of α-D-mannopyranoside with a tin catalyst.

For carbohydrates not stable in acid conditions, such as 4,6-*O*-benzylidene-α-D-glucopyranose, benzoylation can be achieved via transesterification with methyl benzoate using a week and non nucleophilic base promoter like potassium carbonate and a phase transfer agent (Limousin, 1998). MW permits again a tuning of the regioselectivity. Indeed, after 15 min in a domestic microwave oven, a 82% yield of di-benzoylated adduct is recovered, whilst classical oil bath heating affords preferentially the monobenzoylated structures (Fig. 6).

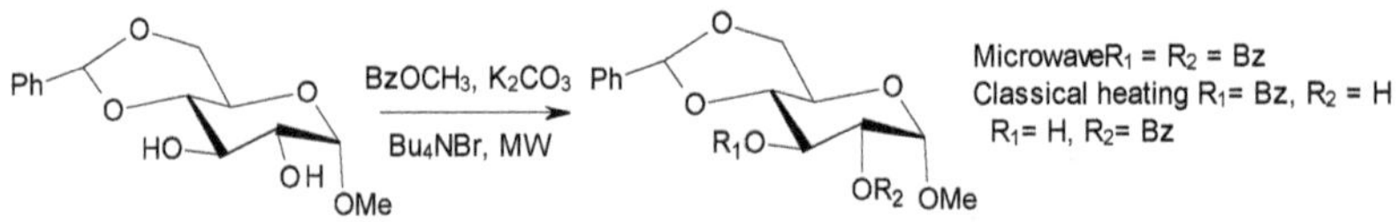

Figure 6. Competition between mono- and dibenzoylation with base catalyst.

Due to the sensitivity of some carbohydrate derivatives at elevated temperatures, microwave methodology suffers however from several limitations, notably for hydroxyls protection processes through acetal formation. As an illustration, the mono-acetalation of *O*-unprotected sucrose in the presence of *p*-toluenesulfonic acid as the catalyst is reported by Salanski and Queneau (Fig. 7). Using citral dimethylacetal, transacetalation provides, under classical heating conditions, 83% yield of a mixture of geranial and neral sucrose acetals (*E* and *Z* isomers of citral acetals) after 2 min at 100°C. When applying microwave as the heating source, yield drops to 42% after an identical runtime in open vessel conditions. Cleavage of the glycosidic linkage is denoted and leads also to the formation of unwanted side-products (17-26% yield after 2 to 10 min of MW exposure) (Salanski, 1998).

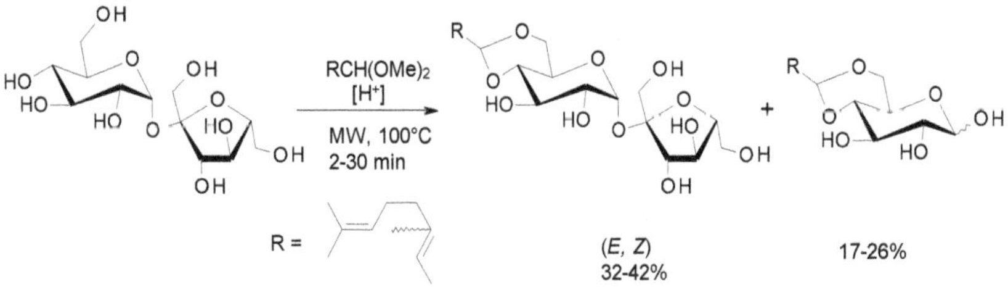

RCH(OMe)$_2$ [H$^+$] MW, 100°C 2-30 min; R = ; (*E, Z*) 32-42%; 17-26%

Figure 7. Protection of sucrose through acetal formation.

For deprotection processes, microwave heating is decisive and powerful, as a drastic diminution of reaction times is observed without affecting the initial anomeric configuration (Corsaro, 2004). An efficient and green protocol concerns the use of neutral alumina to ensure the regioselective depivaloylation of a set of carbohydrate derivatives (Fig 8a). Whilst the depivaloylation does not occur using a conventional heating source, the regioselective deprotection of primary hydroxyl functions is observed after a few minutes à 75°C under MW assistance (Ley, 1993). The fast, efficient and clean deacetalation of several carbohydrate 4,6-di-*O*-benzylidene acetals is also reported using silica supported reagents under microwave irradiation in solvent-free conditions (Fig. 8b). Yields are superior to 80% after 7 min in a domestic oven at 500 Watt, without anomerisation effect (Couri, 2005).

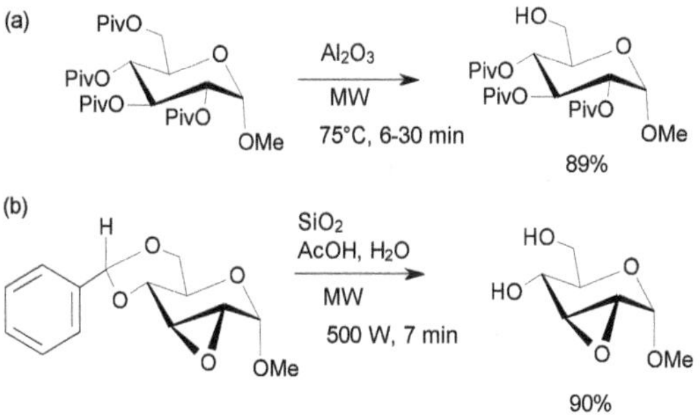

(a) Al_2O_3, MW, 75°C, 6-30 min; 89%
(b) SiO_2, AcOH, H_2O, MW, 500 W, 7 min; 90%

Figure 8. MW-assisted deprotection initiatives: deacylation (a) and deacetalation (b).

4.1.2. *O-glycosylations*

O-glycosylation consists in the specific derivatisation of the anomeric hydroxyl position in carbohydrate structures. This reaction is one of the most emblematic in carbohydrate chemistry and is extensively reviewed (Demchenko, 2008). Several parameters are found to influence the (stereoselective) formation of glycosyl linkages such as solvent, catalyst, additives, nature of both leaving groups and surrounding protecting groups. In particular, the formation of alkyl- and aryl-glycosides has find special attention in both academic and industrial worlds, due to the importance of these molecules as non-toxic and biodegradable surfactants, liquid crystals, and pharmaceutical agents (Razafindralambo 2011 & 2012). Under microwave conditions, the introduction of alkyl substituents on the OH1 position is the most extensively developed, notably for the large-scale production of surface-active agents, and relies on the reaction of an alkyl alcohol with unprotected saccharide (Fischer-type glycosylation) or with completely *O*-protected carbohydrate compounds (Helferich-type glycosylation). These glycosylations are catalysed either by chemical promoters or by enzymes. Alkylation of D-glucose by fatty alcohol using acid catalysts is the oldest reaction reported under MW conditions (Fig. 9). Fatty alcohol advantageously plays both the role of solvent and reagent, even if glucose is insoluble in most alcohols. After 10 min of MW exposure with a 200 Watt maximum output, a 15% yield in glycosylated adduct is obtained using homogeneous Brönsted (or Lewis) acids or heterogeneous promoters (zeolites, ion exchange resins, montmorillonites). No control of the temperature is provided using this domestic MW equipment. Such a low yield is to correlate with unwanted glycoside decomposition under acidic conditions during the run and to browning reactions (Limousin, 1997). This result is not completely satisfactory but remains quite noteworthy compared to other classical protocols. The versatility of this MW Fischer glycosylation is now published for various carbohydrates, including mostly easily available monosaccharides like D-mannose, D-galactose, *N*-acetyl-D-glucosamine and *N*-acetyl-D-galactosamine. Glycosylation of representative long chain alcohols is efficient with acidic ion-exchange resins between 90 and 120°C. Typically, the reaction goes to completion after less than 2 min of MW heating, compared with 4h at reflux, and affords preferentially the thermodynamic α adduct. The anomeric distribution seems quite identical under MW and conventional oil bath heating conditions (Bornaghi, 2005). A valuable scaling-up of this Fischer-type glycosylation is proposed by Nüchter up to the kg-scale with an efficient improvement of economic efficiency, using an appropriated commercial MW batch reactor (Nüchter, 2001). An accurate control of the temperature inside the reactor cavity provides a good quality result. Practically, monosaccharide (glucose, mannose or galactose) reacts with a 3-30-fold excess of alcohol (methanol, ethanol, butanol or octanol) in the presence of sulphuric acid catalyst. Nearly quantitative yields are recovered after 20-60 min of MW heating at 60-140°C, as a function of the nature of the alcohol. Further assays, using continuous flow methods, afford quantitative yields with a MW chamber heated at 110-140°C. More recent work highlights the use of Brönsted acids immobilised on silica as convenient solid catalysts with advantages in manufacturing scale synthesis (Fig. 9). The Fischer-type glycosylation of D-glucose with *n*-decanol in solvent-less conditions leads to the expected adduct in 82% yield as an equimolar mixture of α and β anomers (Richel, 2011b). This protocol offers workup, economic and environmental advantages as catalyst can be recovered by filtration and reused.

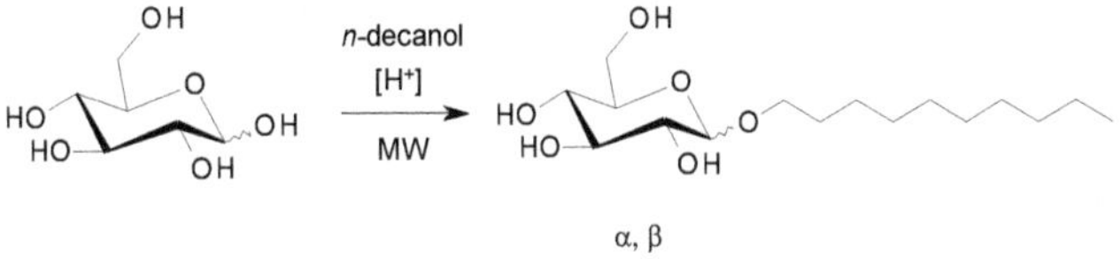

Figure 9. Fischer-type glycosylation of D-glucose with *n*-decanol under MW conditions for the production of a potent surface-active agent.

An alternative three-step protocol (complete acylation-glycosylation-deprotection) is proposed under MW. Due to the increased thermal stability of acylated compounds, MW Helferich-type reactions are expected to be more effective (Fig. 10). Indeed, addition of decanol on per-*O*-acetylated glucose offers a 74% yield (mainly the α-pyranose form) after a running time of 3 min at 113°C using zinc chloride as the catalyst. Extended MW exposure times have a detrimental effect over the yield (25% after 300 min) due to decomposition of decyl glycoside. No rationalisation between the furanose-pyranose (and anomeric) distribution and processing conditions is found at this stage (Limousin, 1997). The first elements of answer are proposed by Kovensky and Ferlin in 2008. A microwave temperature of 115°C is found as a good compromise to obtain high glycosylation rates and low decomposition products contents. Both yields and anomeric distributions are dependent on the boiling/melting points and the polarity of the selected alkyl alcohol. Polar alcohols with low boiling/melting points lead to the formation of the corresponding thermodynamic α adduct, while long chain alcohols provide preferentially the kinetic β anomer in less than 5 min. For each alcohol, an optimum reaction time, beyond which the formed alkyl glycoside is deteriorated, is estimated (Ferlin, 2008).

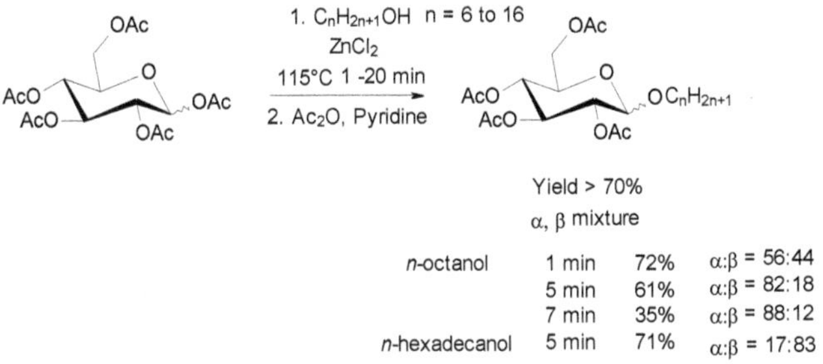

Alcohol	Time	Yield	α:β
n-octanol	1 min	72%	α:β = 56:44
	5 min	61%	α:β = 82:18
	7 min	35%	α:β = 88:12
n-hexadecanol	5 min	71%	α:β = 17:83

Figure 10. Helferich-type glycosylation of per-*O*-acetylated glucose with alkyl alcohols. Influence of the chain length on both yields and α/β ratio.

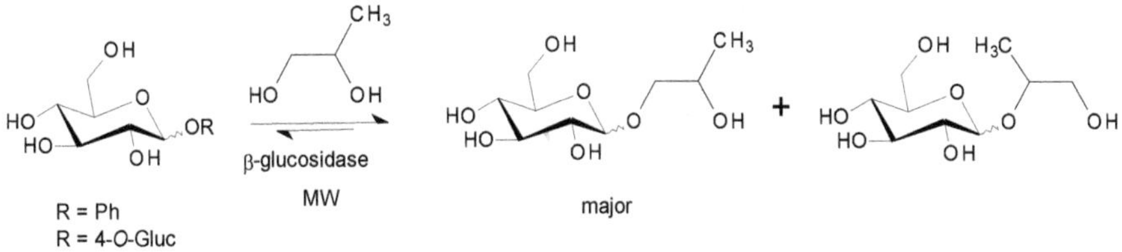

Figure 11. MW-assisted trans-glycosylation of several carbohydrates acceptors and propane-1,2-diol.

Enzymatic methods under microwave conditions are also explored as convenient option for glycosylations. Using β-glucosidase, the synthesis of several glycosides by trans-glycosylation is reported with either phenyl β-D-glucoside or cellobiose as donors and propane-1,2-diol as acceptor (Fig. 11). The process is performed in dry-media conditions, with reagents adsorbed on alumina in an open vessel system. This microwave reaction, achieved at controlled temperature (between 80 and 110°C), leads to noticeable advantages, in terms of yields and selectivity, when compared to classical heating. In particular, trans-glycosylations go to completion after less than 2h, while hydrolysis is lowered to 10% (Gelo-Pujic, 1997).

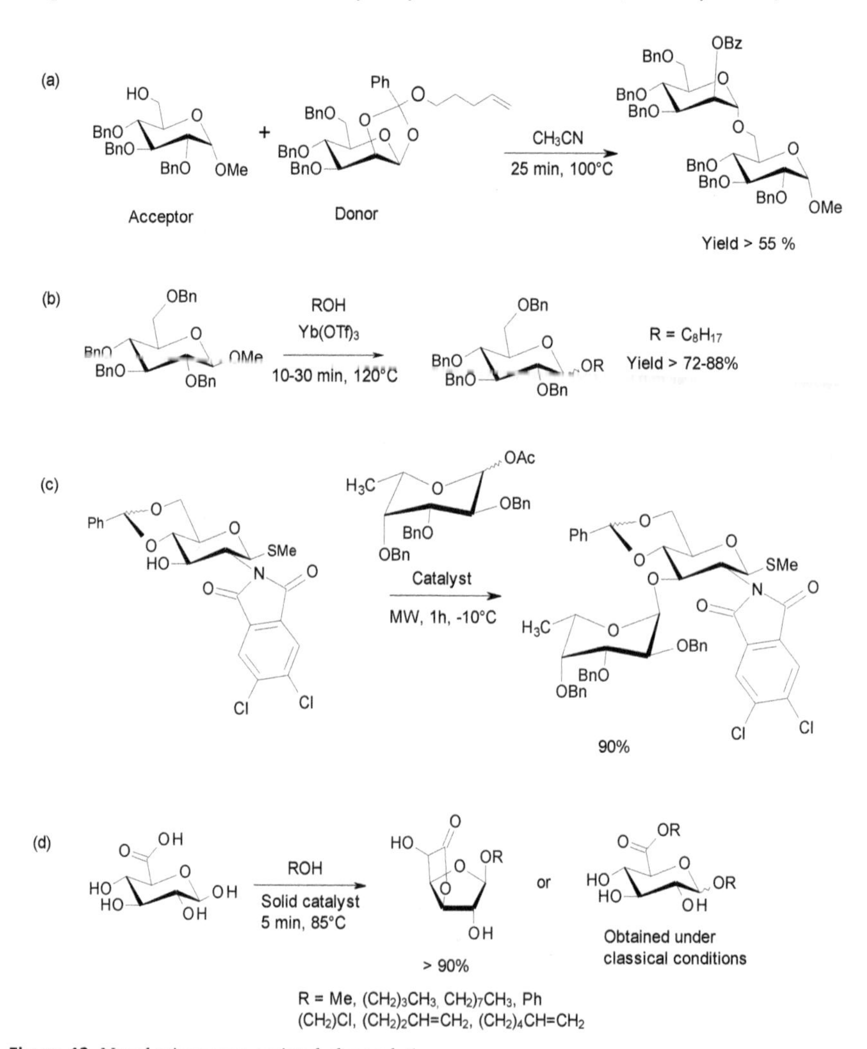

Figure 12. Novel microwave-assisted glycosylations.

Beside the benchmark Fischer- and Helferich-glycosylations, novel strategies under microwave heating are nowadays emerging in the peer-reviewed literature. Some examples are proposed in Fig. 12. Most of them concern the catalytic conjugation of alcohols (designated as acceptors) to protected 4-pentenyl glycosides or methyl glycosides (donors), notably for the production of di- and oligosaccharides. As an illustration, the synthesis of disaccharides is described in acetonitrile starting from 4-pentenyl donors and is achieved in high yields after less than 25 min at 100°C in closed vessel conditions without side-products formation (Fig. 12a) (Mathew, 2004). The use of more effective Lewis acids, like ytterbium(III) triflate, affords high yields for the anchorage of long alkyl chains on benzoylated glucose derivatives (Fig 12b). This acetal-exhange type glycosylation is performed in dichloromethane below 100°C in closed vessel conditions. This constitutes the main advantage of microwave over conventional heating techniques (Yoshimura, 2005). Application of very low temperatures (-10°C) within the MW cavity, attained by simultaneous cooling, offers a convenient approach for the synthesis of valuable saccharides and Lewis X oligosaccharides (Fig. 12c). The process involves a fucosyl acetate donor and a glucosamine acceptor in about 2h in ether. Yields culminate at about 80-90% without side-products formation, whilst the expected disaccharide is not obtained using classical protocols (Shimizu, 2008). Another strategy demonstrates the usefulness of microwave heating, in synergy with heterogeneous acid catalysts, for the regioselective functionalisation of D-glucuronic acid. Conversions are quantitative when performed in solvent-less conditions with an excess of alcohol under microwave exposure. Less than 10 min of heating at 85°C in closed vessel conditions affords selectively the corresponding β-glucurono-6,3-lactone adduct. When heating in an oil bath, only the disubstituted adducts, as a mixture of α,β-furanoses and α,β-pyranoses, are formed through competitive esterification and *O*-glycosylation processes (Fig. 12d). Comparative yields are recorded after 6 to 24 h at 85°C, evidencing that MW accelerates reactions and imparts over the selectivity. This methodology is eco-friendly as water in the only by-product and the solid catalyst can be reused for consecutive batch MW runs (Richel, 2012). These monosubstituted glucuronolactones are not easily attainable using other conventional protocols and required multi-step synthesis. These original products find special relevance in the field of surface-active agents and emulsifiers (Razafindralambo, 2011).

4.1.3. Glycosamines synthesis

Glycosamines are valuable compounds as glyco-amino acid building blocks for glycopeptides synthesis. Their preparation is traditionally achieved using a Kochetkov reductive amination procedure. Practically, a treatment at room temperature for 6 days of saccharides in aqueous solution with 50 equiv of ammonium bicarbonate affords the expected β-glycosaminated adducts in high yields. Another route involves the heating at 42°C for 36 h of mono- or disaccharides with an aqueous solution of ammonia in the presence of one equivalent of ammonium hydrogen carbonate. However, β-glycosamines can be deteriorated under aqueous solutions, even with accurate pH control. Nowadays, the preparation of these glycosamines is typically performed under microwave conditions, at 40°C for 90 min, by selective amination of reductive carbohydrates using reduced quantities of ammonium bicarbonate (5 equiv) (Fig. 13). Under closed vessels conditions, the reaction is achieved in methanol or in anhydrous dimethylsulfoxide to prevent degradation and side-products formation (Liu, 2010). Yields are of about 70-95%.

$(NH_4)_2CO_3$
DMSO or MeOH
MW
40°C, 90 min
80%, β only

Figure 13. Synthesis of 1-amino-1-deoxy-β-D-lactoside by microwave-assisted Kochetkov amination of unprotected lactose.

In summary, reactions of the first category, involving hydroxyls manipulation appear generally faster under MW conditions than their "classical" counterparts. Expected products are indeed obtained in good yields after only a few minutes. The use of eco-friendly and non-toxic (heterogeneous) and milder catalysts in synergy with microwaves, affords ideal "green chemical" conditions and minimise workup. Microwave allows also the formation of specific protected products, not easily accessible under conventional "heating" conditions.

4.2. Microwave-assisted production of platform chemicals (category 2)

The production of 5-HMF (5-hydroxymethylfurfural), a five-membered ring compounds ranging in the US Top 10 of most valuable chemicals issued from biomass, is currently extensively investigated worldwide. Most research concerns its production via thermally-induced acid-catalysed dehydration of fructose, typically in dimethylsulfoxide, in the presence of homogeneous acid catalysts (HCl, H_2SO_4, H_3PO_4, oxalic or levulinic acids) (Fig. 14). Reactions produces selectively 40-60% yields of 5-HMF with complete conversion of starting sugar materials but suffers from drawbacks in terms of acids recovery and equipment corrosion. Heterogeneous catalysts offer workup and environmental advantages associated to high 5-HMF selectivity but provide low fructose conversions (30-60% only after up to 2h). Microwave has progressively emerged as an attractive heating source allowing energy savings and improved yields and selectivity. Recent developments in the field have been reviewed by Richel in 2011 (Richel, 2011b). They include the use of heterogeneous catalysts in various media (organic solvents, water, ionic liquids or biphasic systems).

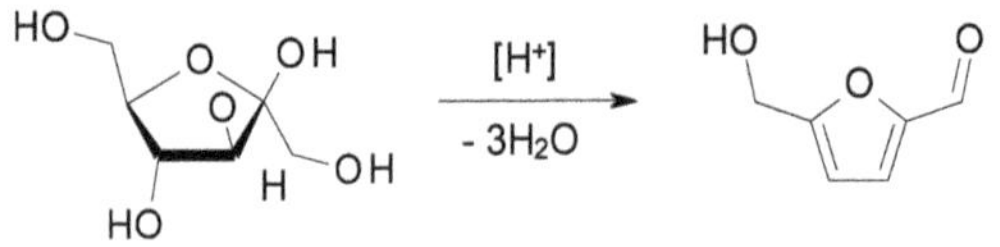

$[H^+]$
$- 3H_2O$

Figure 14. Catalytic dehydration of fructose for the production of 5-HMF.

Titanium and zirconium oxides, as sulphated zirconia, are effective catalysts for this dehydration, especially under microwave conditions. Yields in 5-HMF reach 35% for a fructose conversion of 78% after only 5 min heating at 200°C in closed vessels conditions ("microwave hot compressed water process"), while 5-HMF yield and fructose conversion are respectively of 12 and 27% after an identical runtime using a sand bath heating. Zirconium oxide demonstrates great performances for the catalytic dehydration of glucose, favouring isomerisation of glucose to fructose prior dehydration (Qi, 2009). When using an acid ion-exchange resin in an acetone/dimethylsulfoxide system, improvement in 5-HMF selectivity is encountered together with high yields (98% fructose conversion with 5-HMF selectivity of 92% for 10 min MW assisted reaction at 150°C in closed vessel conditions). Addition of low-boiling point acetone improves products separation and reduces adverse environmental impacts. The same results are published for strong ion-exchange resin in acetone-water systems, providing a truly eco-friendly process (Qi, 2008b). The common Lewis acid catalyst, aluminium(III) chloride, promotes also efficiently the dehydration of carbohydrate substrates, in water-organic solvents and in water. 5-HMF yield is higher when using an organic solvent than in pure water. Involvement of ionic liquids as convenient media for the conversion of D-glucose in 5-HMF is effective using chromium(III) chloride as a catalyst. With microwave heating at 140°C for 30 sec, a 5-HMF yield of 71% is obtained for 96% glucose in 1-butyl-3-methyl imidazolium chloride as the solvent (Qi, 2010). Lignin-based solid catalyst, produced by carbonisation and sulfonation of raw lignocellulosic materials, seems nowadays to be the most active catalyst for the dehydration of carbohydrates. In synergy with MW, 84% 5-HMF yield is obtained with 98% fructose conversion rate at 110°C for 10°C. A mixture of dimethylsulfoxide-ionic liquid has a beneficial effect over the selectivity of the process. Interestingly, the catalyst can be recovered after the reaction and reused for consecutive batch reactions (Guo, 2012).

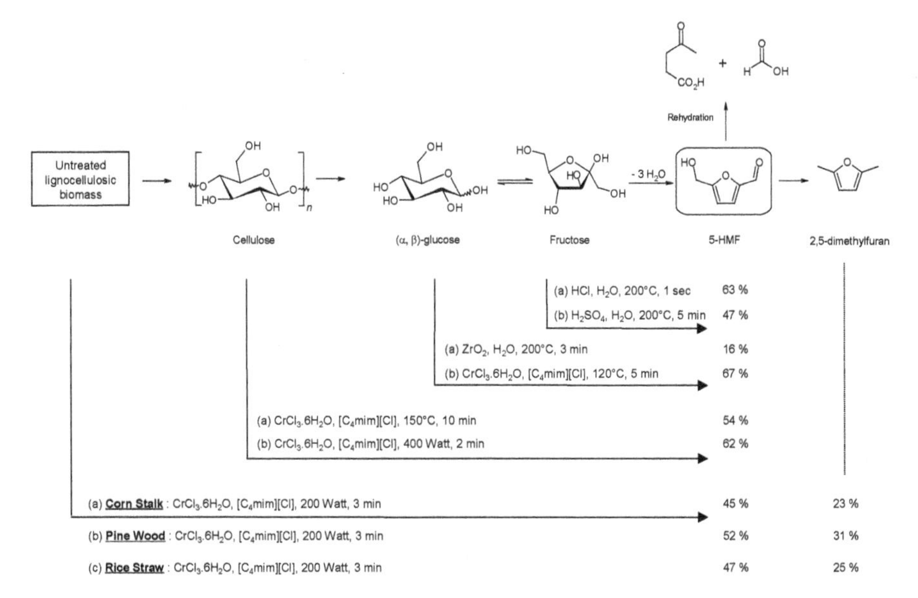

Figure 15. Overview of MW-assisted production of 5-HMF from lignocellulosic biomass materials.

Scaling up of the MW-assisted fructose dehydration process has been recently evaluated using a stop-flow MW reactor system and compared to a continuous cartridge-based reactor. Results highlight that 5-HMF yield can attain 86% yield after 2min of MW exposure (91% using a cartridge-base reactor) with a 5-HMF productivity of 0.72 g/h (versus 2.07 g/h for the continuous process). Even if quite less productive, the MW-assisted methodology is more selective and cleaner as no side-product is detected in the final mixture (Schoen, 2011).

Besides these aforementioned benchmark reactions, research has shifted toward the exploration of other carbohydrates or crude lignocellulosic feedstocks as raw materials. Some relevant illustrations are reported in Fig. 15. They concern the use of ionic liquids as solvent. For environmental concern, water is also envisioned with homogeneous Brönsted acids or metal chloride catalysts (Dutta, 2012). The production of 2-furfural, another platform chemicals, receives progressively interest. Its MW-assisted production from xylose, xylan and wheat straw is highly efficient in the presence of hydrochloric acid. A 48% furfural yield is obtained from wheat straw after 1 min of microwave heating between 140 and 190°C. (Yemis, 2011).

5. Conclusion

In the present chapter, we contribute to illustrate some relevant carbohydrate reactions under microwave heating. This heating mode appears as efficient for several key reactions like hydroxyls protection and deprotection and O-glycosylations. The production of 5-HMF from mono-/di-/polysaccharides is another special subject in MW science. Improvement of both yields and (regio- and anomeric) selectivities is generally reported together with a decrease in reaction times. MW offers also advantages as energy-saving and workup facilities in agreement with the green chemistry concept.

Author details

Aurore Richel and Michel Paquot
University of Liege, Belgium

Acknowledgement

This work was carried out in the framework of the "Technose" Excellence Programme (project number 716757) supported by the "Région Wallonne" (Belgium). The authors are grateful to the "Région Wallonne" for its financial support.

6. References

Anastas, P. T. ; Warner, J. C. (1998). Green Chemistry: Theory and Practice; Oxford University Press: Oxford

Bogdal, D. (2006). Microwave-assisted synthesis. One hundred reaction procedures. Tetrahedron Organic Chemistry Series, Vol. 25; Elsevier.

Bornaghi, L. F.; Poulsen, S. A. (2005).

Bozell, J. J.; Petersen, G. R. (2010). Technology development for the production of biobased products from biorefinery carbohydrates—the US Department of Energy's "Top 10" revisited. *Green Chem.*, 12, 539-554

Caddick, S.; Fitzmaurice, R. (2009). Microwave enhanced synthesis. *Tetrahedron*, 65, 3325-3355

Chen, S. T.; Sookkheo, B.; Phutrahul, S.; Wang, K. T. (2001). Enzymes in nonaqueous solvents: Applications in carbohydrate and peptide preparation. *Methods in Biotechnology*, 15, 373-400.

Cioffi, E. A. (2008). High-energy glycoconjugates: synthetic transformations of carbohydrates using microwave and ultrasonic energy. Curr. Top. Med. Chem., 8, 152-158

Corsaro, A.; Chiacchio, U.; Pistarà, V.; Romeo, G. (2004). Microwave-assisted chemistry of carbohydrates. *Curr. Org. Chem.*, 8, 511-538

Das, S. K.; Reddy, K. A.; Krovvidi, V. L. N. R.; Mukkanti, K. (2005). $InCl_3$ as a powerful catalyst for the acetylation of carbohydrate alcohols under microwave irradiation. *Carbohydr. Res.*, 340, 1387-1392

Demchenko, A. V. (2008). Handbook of Chemical Glycosylation; Advances in Stereoselectivity and Therapeutic Relevance, Wiley-VCH, Weinheim

Dutta, S.; De, S.; Alam, M. I. ; Abu-Omar, M. M. ; Saha, B. (2012). Direct conversion of cellulose and lignocellulosic biomass into chemicals and biofuel with metal chloride catalysts. *J. Catal.*, 288, 8-15

Ferlin, N. ; Duchet, L. ; Kovensky, J. ; Grand, E. (2008). Microwave-assisted synthesis of long-chain alkyl glucopyranosides. *Carbohydr. Res.*, 343, 2819-2821

Gedye, R. ; Smith, F.; Westaway, K.; Ali, H.; Baldisera, L.; Laberge, L.; Roussel, J. (1986). The use of microwave ovens for rapid organic synthesis. *Tetrahedron Lett.*, 27, 279-282

Gelo-Pujic, M.; Guibé-Jampel, E.; Loupy, A.; Trincone, A. (1997). Enzymatic glycosidation in dry media under microwave irradiation. *J. Chem. Soc., Perkin Trans.* 1, 1001-1002

Giguere, R. J.; Bray, T; L.; Duncan, S. M.; Majetich, G. (1986). Application of commercial microwave ovens to organic synthesis. *Tetrahedron Lett.*, 27, 4945-4948.

Herradón, B.; Morcuende, A.; Valverde, S. (1995). Microwave Accelerated Organic Transformations: Dibutylstannylene Acetal Mediated Selective Acylation of Polyols and Amino Alcohols using Catalytic Amounts of Dibutyltin Oxide. Influence of the Solvent and the Power Output on the Selectivity. *Synlett*, 455-458

Guo, F.; Fang, Z.; Zhou, T. J. (2012). Conversion of fructose and glucose with lignin-derived carbonaceous catalyst under microwave irradiation in dimethyl sulfoxide-ionic liquid mixtures. *Biores. Technol.*, 112, 313-318.

Kappe, C. O. (2004). Controlled microwave heating in modern organic synthesis. *Angew. Chem. Int. Ed.*, 43, 6250-6284

Lichtenthaler, F. W. ; Peters, S. (2004). Carbohydrates as green raw materials for the chemical industry. *C. R. Chimie*, 7, 67-90

Limousin, C.; Cléophax, J.; Petit, A.; Loupy, A. ; Lukacs, G. (1997). Solvent-Free Synthesis of Decyl D-Glycopyranosides Under Focused Microwave Irradiation. *J. Carbohydr. Chem.*, 16, 327-342

Limousin, C.; Cléophax, J.; Petit, A.; Loupy, A. (1998). Synthesis of benzoyl and dodecanoyl derivatives from protected carbohydrates under focused microwave irradiation. *Tetrahedron*, 54, 13567-13578

Liu, X.; Zhang, G.; Chan, K.; Li, J. (2010). Microwave-assisted Kochetkov amination followed by permanent charge derivatization: a facile strategy for glycomics. *Chem. Commun.*, 46, 7424-7426.

Mathew, F.; Jayaprakash, K. N.; Fraser-Reid, B.; Mathew, J.; Scicinski, J. (2004). Microwave-assisted saccharide coupling with *n*-pentenyl glycosyl donors. *Tetrahedron Lett.*, 44, 9051-9054.

Nüchter, M.; Ondruschka, B.; Lautenschlarger, W. (2001). Microwave-assisted synthesis of alkyl glycosides. *Synth. Commun.*, 31, 1277-1283

Polshettiwar, V. ; Varma, R. S. (2008). Microwave-Assisted Organic Synthesis and Transformations using Benign Reaction Media. *Acc. Chem. Res.*, 41, 629–639.

Qi, X.; Watanabe, M.; Aida, T.; Smith, T. L. (2008). Catalytical conversion of fructose and glucose into 5-hydroxymethylfurfural in hot compressed water by microwave heating. *Catal. Commun.*, 9, 2244-2249.

Qi, X.; Watanabe, M.; Aida, T.; Smith, T. L. (2009). Sulfated zirconia as a solid acid catalyst for the dehydration of fructose to 5-hydroxymethylfurfural. *Catal. Commun.*, 10, 1771-1775

Qi, X.; Watanabe, M.; Aida, T.; Smith, T. L. (2010). Fast transformation of glucose and di-/polysaccharides into 5-hydroxymethylfurfural by microwave heating in an ionic liquid/catalyst system. *ChemSusChem*, 3, 1071-1077

Razafindralambo, H.; Richel, A.; Wathelet, B.; Blecker, C.; Wathelet, J. P.; Brasseur, R.; Lins, L.; Miñones, J.; Paquot, M. (2011). Monolayer Properties of Uronic Acid Bicatenary Derivatives at the Air-Water Interface: Effect of Hydroxyl Group Stereochemistry Evidenced by Experimental and Computational Approaches. *Phys. Chem. Chem. Phys.* 13, 15291–15298

Razafindralambo, H.; Richel, A.; Paquot, M. Lins, L. ; Blecker, C. (2012). Liquid Crystalline Phases Induced by the Hydroxyl Group Stereochemistry of Amphiphilic Carbohydrate Bicatenary Derivatives. *J. Phys. Chem. B*, 116, 3998-4005

Richel, A.; Laurent, P.; Wathelet, B.; Wathelet, J. P.; Paquot, M. (2011a). Microwave-assisted conversion of carbohydrates. State of the art and outlook. *C. R. Chimie*, 14, 224-234.

Richel, A.; Laurent, P.; Wathelet, B.; Wathelet, J. P.; Paquot, M. (2011b). Current perspectives on microwave-enhanced reactions of monosaccharides promoted by heterogeneous catalysts. *Catal. Today*, 107, 141-147

Richel, A.; Laurent, P.; Wathelet, B.; Wathelet, J. P.; Paquot, M. (2012). Efficient microwave-promoted synthesis of glucuronic and galacturonic acid derivatives using sulfuric acid impregnated on silica. *Green Chem. Lett. Rev.*, 5, 179-186.

Salanski, P., Descotes, G., Bouchu, A., Queneau, Y. (1998). Monoacetalation of unprotected sucrose with citral and ionones. *J. Carbohydr. Chem.*, 17, 129-142

Schoen, M.; Schnuerch, M.; Mihovilovic, M. D. (2011). Application of continuous flow and alternative energy devices for 5-hydroxymethylfurfural production. *Mol. Divers.* 15, 639-643.

Shimizu, H.; Yoshimura, Y.; Hinou, H.; Nishimura, S.-I. (2008). A novel glycosylation method part 3: study of microwave effects at low temperatures to control reaction pathways and reduce byproducts. *Tetrahedron*, 2008, 10091-10096

Söderberg, E.; Westman, J.; Oscarson, S. (2001). Rapid carbohydrate protecting group manipulations assisted by microwave dielectric heating. *J. Carbohydr. Chem.*, 20, 397-410.

Witschi, M. A. & Gervay-Hague, J. G. (2010). Selective Acetylation of per-*O*-TMS-Protected Monosaccharides. *Org. Lett.*, 12, 4312-4315

Yemis, O.; Mazza, G. (2011). Acid-catalyzed conversion of xylose, xylan and straw into furfural by microwave-assisted reaction. *Biores. Technol.*, 102, 7371-7378.

Yoshimura, Y. ; Shimizu, H. ; Hinou, H. ; Nishimura, S.-I. (2005). A novel glycosylation concept; microwave-assisted acetal-exchange type glycosylations from methyl glycosides as donors. *Tetrahedron Lett.*, 46, 4701-4705

Investigation of Carbohydrates and Their Derivatives as Crystallization Modifiers

Josef Jampílek and Jiří Dohnal

Additional information is available at the end of the chapter

1. Introduction

Development in the field of pharmaceutical administration has resulted in the discovery of highly sophisticated drug delivery systems that allow for the maintenance of a constant drug level in an organism. Contrary to these revolution biopharmaceutical results, over the last ten years, the number of poorly soluble drugs has steadily increased. Estimates state that 40% of the drugs in the pipelines have solubility problems. Progress in high throughput screening methods leads to an even greater amount of newly discovered drugs, but a lot of them have poor water solubility. Literature states that about 60% of all drugs coming directly from synthesis are nowadays poorly soluble. Meanwhile the five key physico-chemical properties, such as pK_a, solubility, permeability, stability and lipophilicity, in early compound screening should be optimized. Compounds with insufficient solubility carry a higher risk of failure during discovery and development, since insufficient solubility may compromise other property assays, mask additional undesirable properties, influence both pharmacokinetic and pharmacodynamic properties of the compound and finally may affect the developability of the compound. Poor solubility in water correlates with poor bioavailability. If there is no way to improve drug solubility, it will not be able to be absorbed from the gastrointestinal tract into the bloodstream and reach the site of action (Junghanns & Müller, 2008; Payghan et al., 2008).

Modification/optimization of unfavourable physico-chemical properties of these drugs is possible through increasing their water solubility or improving permeability. There are many ways to solubilize certain poorly soluble drugs. But these methods are limited to drugs with certain properties in regard to their chemistry or, for example, to their molecular size or conformation.

Aqueous solubility can be increased by chemical exchange: (*i*) salts, co-crystals or solvates formation (affects also chemical stability, polymorphism, technological workability);

(*ii*) substitution by hydrophilic groups (effect of drugs with small molecules can be decreased); (*iii*) prodrug preparation (hydrolyzable amides or semiesters with polybasic acids). In general, the following structural modifications are the best way to improve permeability: (*i*) replacement of ionisable groups by non-ionizable groups; (*ii*) increase of lipophilicity; (*iii*) isosteric replacement of polar groups; (*iv*) esterification of carboxylic acid; (*v*) reduction of hydrogen bonding and polarity; (*vi*) reduction of size; (*vii*) addition of a non-polar side chain; (*viii*) preparation of prodrugs. Generally, these strategies are based on a few fundamental concepts: change of ionizability, lipophilicity, polarity or change of hydrogen bond donors or acceptors. Both approaches interact logically.

Based on these facts, pre-formulation/formulation can be another and mostly successful strategy for improving aqueous solubility and/or permeability and subsequently bioavailability. For example, selection of a suitable salt, particle size reduction (till nano size) connected with an increase of the surface area, change of polymorphic forms, selection of appropriate excipients to function as solubilizers/transporters (surfactants or pharmaceutical complexing agents, permeability enhancers) can be used for the oral dosage form (Kerns & Di, 2008).

It is well-known that crystalline materials obtain their fundamental physical properties from the molecular arrangement within the solid, and altering the placement and/or interactions between these molecules can, and usually does, have a direct impact on the properties of the particular solid. Currently, solid-state chemists call upon a variety of different strategies when attempting to alter the chemical and physical solid-state properties of active pharmaceutical ingredients (APIs), namely, the formation of salts, polymorphs, hydrates, solvates and co-crystals (Seddon & Zaworotko, 1999; Datta & Grant, 2004; Grepioni & Braga, 2007; Schultheiss & Newman, 2009).

Currently, salt formation is one of the primary solid-state approaches used to modify the physical properties of APIs, and it is estimated that over half of the medicines on the market are administered as salts (Bighley et al., 1996; Stahl & Wermuth, 2002; Gu & Grant, 2003.). However, a major limitation within this approach is that the API must possess a suitable (basic or acidic) ionizable site. In comparison, co-crystals (multi-component assemblies held together by freely reversible, non-covalent interactions) offer a different pathway, where any API regardless of acidic, basic or ionizable groups, could potentially be co-crystallized. This aspect helps to complement existing methods by reintroducing molecules that had limited pharmaceutical profiles based on their non-ionizable functional groups. Since the primary structure of a drug molecule does not change in co-crystals, their development in terms of the regulation "New Chemical Entities (NCEs)" of the U.S. Food and Drug Administration (FDA) as well as European Commission Regulation (EC) No. 258/97 carries fewer risks and takes much less time; nevertheless stability and bioequivalent studies are still necessary. In addition, the number of potential non-toxic co-crystal formers (or co-formers) that can be incorporated into a co-crystalline reaction is numerous, *e.i.* pharmaceutical excipients, amino acids, food additives, nutraceuticals, see for example the GRAS list (Generally Regarded as Safe) or the EAFUS Database (Everything Added to Food in the United States), both published by the FDA.

It should be made clear that no one particular strategy offers a solution for property enhancement of all APIs. Each API must be examined and evaluated on a case-by-case basis in terms of molecular structure and desired final properties (Schultheiss & Newman, 2009).

2. Polymorphism

The term "polymorphism" (from Greek: *polys* = many, *morfé* = form) was first used by Mitscherlich in 1822. He noticed that one compound of a certain chemical composition can crystallize into several crystal forms (Mitscherlich, 1822). Polymorphism is an ability of substances to exist in two or more crystal modifications differing from each other by structure and/or molecule conformation in the crystal lattice. The concept of polymorphism is often confounded with isomorphism or pseudopolymorphism. These concepts are interconnected but there are great differences between them.

In contrast to polymorphism when one substance able to form different crystal modifications is considered, in case of isomorphism two or more different substances that have just similar structure but form the same crystal modifications are considered. Such substances can even form so-called isomorphous series. Most often they originate as a result of co-crystallization of isomorphous substances from the mixture of their saturated solutions. A typical example is sulphates: magnesium sulphate, zinc sulphate and nickel sulphate crystallizing as heptahydrates. Also pseudopolymorphism should be distinguished from polymorphism. The concept of pseudopolymorphism is used for crystalline forms which also comprise solvent molecules as an integral part of their structure. These pseudopolymorphs are sometimes called solvates or hydrates if the solvent is water (Rodríguez-Spong et al., 2004).

The fundamental forms of solid substances can be classified into polymorphs, solvates/desolvated solvates and amorphous substances. A co-crystal can be defined as a multiple-component crystal, in which two or more molecules that are solid under ambient conditions coexist through a hydrogen bond (see Fig. 1 and Fig. 2). Polymorphism can be

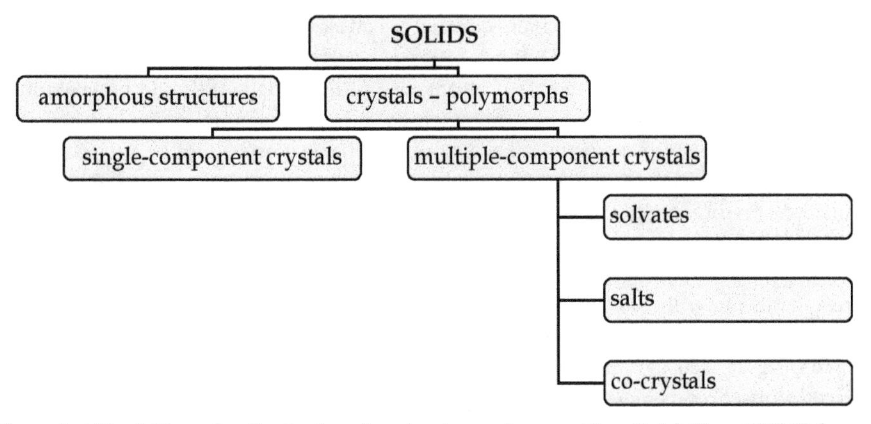

Figure 1. API solid form classification based on structure and composition. (Ref. Sekhon, 2009) (Taken and adapted with the permission of the author.)

classified depending on the fact if the molecule occurs in different conformations or is the same. If the molecule can have different conformations that crystallize differently and so the molecule is flexible, conformation polymorphism is observed. If the molecule is rigid, does not have any conformations and polymorphs differ only by their packing in the crystalline structure, this is the case of package polymorphism (Sharma et al., 2011).

Polymorphism should be distinguished from crystal morphology that represents crystallization of a substance from different solvents with a change of the form but without modification of the crystalline structure.

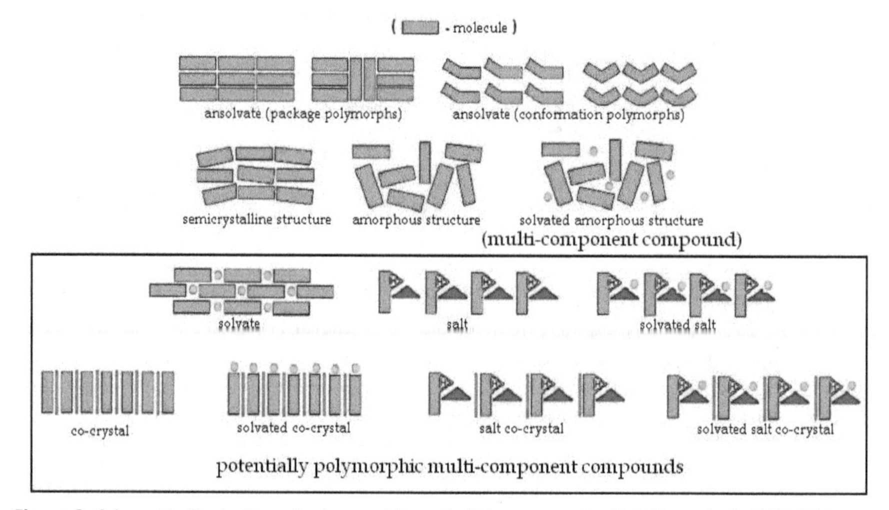

Figure 2. Schematic illustration of polymorphism of solid compounds. (Ref. Kratochvíl, 2009) (Taken and adapted from the presentation "Pharmaceutical Co-crystals" with the permission of the author.)

In the pharmaceutical practice it is conventional to collectively call all polymorphs, solvates and polyamorphs of the same API as solid forms or polymorphs in the extended meaning and designate them by, for example, Roman numerals or letters. It should be taken into consideration that designation of polymorphs is often non-uniform; it originates historically; therefore it is possible that two different polymorphs will be designated by different authors in the same way. A problem of pharmaceutical producers is polymorphic transitions in organic systems that are often hysteresis, badly defined and solvent-mediated. Polymorphic transitions can occur at all technological stages of drug production (at final API crystallization, wet granulation, micronization, tabletting or in ready tablets, for example, under the influence of auxiliaries – excipients). At drug registration, national regulation authorities demand description of all solid forms and possible phase transitions of the drug as well as a prescribed guarantee of the product polymorphic purity from pharmaceutical companies. At present approximately 85% of pharmaceutical products are solid formulations; therefore no manufacturer can afford to ignore the issue of polymorphism.

2.1. Crystallization

During crystallization a self-assembly supramolecular process takes place when randomly oriented molecules assemble into internally structured crystals (Kratochvíl, 2007). There are several ways of crystallization: (*i*) by evaporation of solvent; (*ii*) by addition of antisolvent (product precipitation); (*iii*) by cooling of solution; (*iv*) by change of pH ; (*v*) by addition of a compound that will produce the desired product by a chemical reaction; (*vi*) by lyophilisation, sublimation or cooling of melt (Kratochvíl, 2010). The first two possibilities were used in the experimental part.

The process of crystal formation is a complicated kinetic process which, if we simplify, is composed of nucleation, subsequent crystal growth around the formed crystal nuclei and growth termination. This happens at precipitation of a solid substance in the crystalline form from solution or at solidification of a substance.

The basis of nucleation is the quickest formation of the crystal nucleus. By constant sequence of molecules addition molecular aggregates, clusters, originate. The aggregates spontaneously impact and thus disintegrate but also grow. When they achieve the critical size, they become nuclei that do not disintegrate spontaneously but, to the contrary, only grow into a crystal. Nuclei can be classified into primary and secondary. Primary nucleation is further divided into homogenous and heterogeneous. Homogenous nucleation is characterized by formation of nuclei as a result of random impacts of aggregates somewhere in the solution volume without the presence of any foreign matter. By contrast, heterogeneous nucleation is assisted by foreign matters, for example, a stirrer. Secondary nucleation is also called seeded nucleation – small crystals (nuclei, seeds) of a desired crystal are added to the original solution. Seeding is successfully applied for systems where polymorphous behaviour cannot be excluded. Seeding of the system with a desired polymorph assures that an undesired polymorph will not be formed. After formation of the nucleus the system pass to the second mentioned crystallization phase. In this phase growth of crystals from the formed nucleus continues. If an API is crystallized with a co-crystallization partner, for example excipient or another API, with co-crystals formation, such process is called co-crystallization.

2.2. Properties of solid forms

For polymorphic drugs, the most thermodynamically stable polymorph is usually preferred. This polymorph normally assures reproducible bioavailability for the whole period of the pharmaceutical shelf life under different storing conditions that are possible in practice. A significant advantage is that production of such a polymorph is mostly easier controlled on an industrial scale. But in some cases a metastable crystalline form or an amorphous form is preferred due to patent and medical reasons. This happens when a higher concentration of an API in the system or faster dissolution of poorly soluble substances is required. Certain substances are used in the amorphous form also due to the fact that a crystalline form of the compound was not obtained. When an amorphous form or a metastable crystalline polymorph is used, especial attention is to be paid to safety assurance and the effectiveness of the drug for the whole shelf life period. These aspects are to be assured also for storing conditions in other climatic zones.

Though the biological effect of an API is induced by interaction of the drug molecule with a target receptor when cell natural chemistry is influenced primarily by conformation changes, it is important in what solid phase the drug is administered to the patient. That means that not only the molecular but also the crystalline structure of an API is essential. The crystalline structure influences not only chemical and physical stability but also the drug dissolution rate and so can affect markedly the drug bioavailability. On the other hand, an amorphous form may have a considerably higher dissolution rate than the most stable polymorph. Great differences in the solubility and the dissolution rate of polymorphs can be a reason of significant differences in their distribution in the organism. Low plasma concentration can theoretically cause incomplete occupation of respective membrane receptors at the site of action (they are blocked by a substrate), or biological activity can change from agonist to antagonist or vice versa.

However, the dissolution rate is not the only important parameter of the difference between polymorphs. There can be also differences in crystal size and shape that influence milling, tabletting, filtration, looseness and other important technological parameters. There are also differences in chemical reactivity, thermal stability, hygroscopicity, density, hardness, etc.

Thus it can be stated that different arrangement and conformation change of molecules in the crystal structure lead to differences in properties as follows: (*i*) mechanical (hardness, compactness, etc.); (*ii*) surface (surface energy, interfacial tension, etc.); (*iii*) kinetic (dissolution values, stability, etc.); (*iv*) spectroscopic (electronic, vibrational state transition, etc.); (*v*) thermodynamic (melting point, sublimation, enthalpy, entropy, etc.); (*vi*) packing properties (molar volume a density, hygroscopicity, etc.) (Brittain, 2009).

2.3. Multi-component solids

Usually solid crystal substances can exist as mono-component or multi-component. Mono-component systems contain, as evident from the name, only one component, and multi-component systems consist of two or more components. In pharmaceutics this is a crystal composed of an API and an additional molecule that determines the type of the multi-component system, pseudopolymorph. As illustrated in Fig. 1, the additional molecule of hydrates is water; of solvates, a solvent; of salts, an ionizable component; and of co-crystals, a co-crystallization partner (Stahly, 2007; Kratochvíl, 2010). All these systems can be polymorphic (see Fig. 1 and Fig. 2). Due to the breadth of the topic and the focus of this paper on preparation of co-crystals, only co-crystals will be described in more details.

3. Co-crystals

The term "co-crystal" (also written as cocrystal) originates from "a composite crystal" (Desiraju, 2003). The term and the definition of co-crystal is a subject of topical debate. In principle this is a multi-component system of host and guest type; both components are solid in pure state and under ambient conditions. There are a lot of definitions; for example, Stahly defines co-crystals as molecular complexes that contain two or more different molecules in the same crystal structure (Stahly, 2007). Bhogala and Nangia define co-crystals as multi-component solid-state assemblies of two or more compounds held together by any type or combination of intermolecular interactions (Bhogala & Nangia, 2008). Childs and Hardcastle define co-crystals as a crystalline material made up of two or more components, usually in a stoichiometric ratio, each component being an atom, ionic compound, or molecule (Childs & Hardcastle, 2007). The simplest definition of co-crystals was proposed by Bond: "synonym for multi-component molecular crystal" (Bond, 2007). Also the definition of Aakeröy and Salmon can be recognized. They describe co-crystal as stoichiometric structurally homogeneous multi-component crystalline material of host-guest type that contains two or more neutral building blocks (discrete neutral molecular species that are solids at ambient conditions) that are present in definite amounts whereas all solids containing ions, including complex transition-metal ions, are excluded) (Aakeröy & Salmon, 2005). Jones defines co-crystal as a crystalline complex of two or more neutral molecular constituents bound together in the crystal structure through non-covalent interactions, often including hydrogen bonding (Jones et al., 2006) and Zaworotko says that co-crystals are formed between a molecular or ionic API and a co-crystal former that is a solid under ambient conditions (Vishweshwar et al., 2006a).

The host is an API, and the guest is a co-crystallization partner (an excipient or another API). Note that according to the definition by Aakeröy and Salmon co-crystals are different from solvates by the fact that the host and the guest of co-crystals are in the solid phase, while solvates contain both a solid phase and a solvent (or its residual), *i.e.* a liquid phase. However, there is still the problem of a boundary between the salt and the co-crystal, because according to the last mentioned definition of Aakeröy and Salmon, ions as co-crystal components are excluded, so on Fig. 3 only the first example of two is a co-crystal according to the mentioned definition. The borderline between salts and co-crystals is blurred and can be distinguished by the location of the proton between an acid and a base. This state can be denoted as salt–co-crystal continuum (Childs et al., 2007).

Several types of co-crystals can be distinguished: (*i*) "simple" co-crystals; (*ii*) solvated (hydrated) co-crystals – the co-crystal contains a component that is liquid at ambient temperature; (*iii*) salt co-crystals – the host is an ionized form; (*iv*) solvated salt co-crystals; (*v*) polymorphs of all previous types of co-crystals.

3.1. Properties of co-crystals

Pharmaceutical co-crystals, that is, co-crystals that are formed between an API and pharmaceutically acceptable (GRAS) compounds (co-crystal formers, counterions) that are solid under ambient conditions, represent a new paradigm in API formulation that might address important intellectual and physical property issues in the context of drug development and delivery.

Co-crystals can be understood as "addition compounds" or "organic molecular compounds". They are attractive to the pharmaceutical industry, because they offer opportunities to modify the chemical and/or physical properties of an API without the need to make or break covalent bonds (see Fig. 4). The term "non-covalent derivatization" was coined for this approach. Co-crystals of an API with excipients become very important as a tool to tune solubility and absorption. The application of co-crystal technologies has only recently been recognised as a way to enhance solubility, stability and the intellectual property position with respect to development of APIs (Vishweshwar et al., 2006b; Sekhon, 2009; Schultheiss & Newman, 2009).

Co-crystallization with pharmaceutical excipients does not affect pharmacological activity of an API but can improve physical properties, such as solubility, hygroscopicity and compaction behaviour (Aakeröy et al., 2007; Rodríguez-Hornedo et al., 2007; Trask, 2007; Sun & Hou, 2008; Zaworotko, 2008). Co-crystals with the same active pharmaceutical ingredient will have strikingly different pharmaceutical properties (melting point, solubility, dissolution, bioavailability, moisture uptake, chemical stability, etc.), depending on the nature of the second component. Some of co-crystals formed had higher and some lower melting points as compared to their pure components, for example, succinic acid (Mp=135.3), urea (Mp=188.9), co-crystal of succinic acid-urea (Mp=149.9) (Walsh et al., 2003).

Example 1

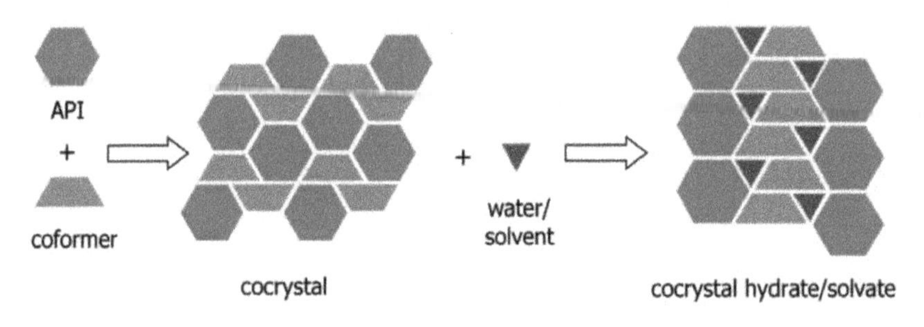

Example 2

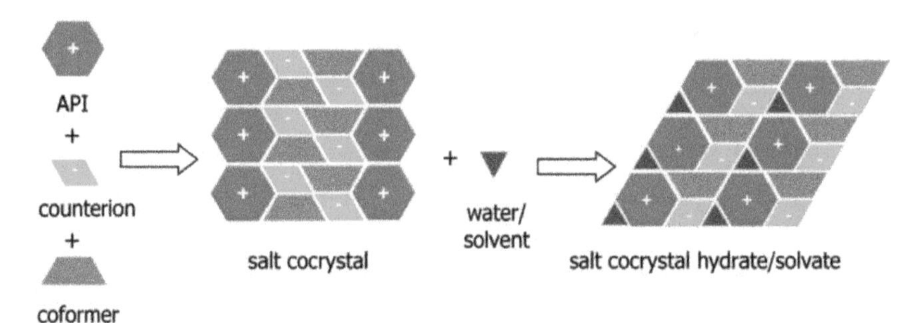

Figure 3. Possible multi-component systems: co-crystals, salt co-crystals and derived multi-component solids. (Ref. Schultheiss & Newman, 2009; reprinted and adapted with permission from Schultheiss N., Newman A. Pharmaceutical co-crystals and their physicochemical properties. Crystal Growth & Design 2009, 9(6): 2950–2967. Copyright 2009 American Chemical Society.)

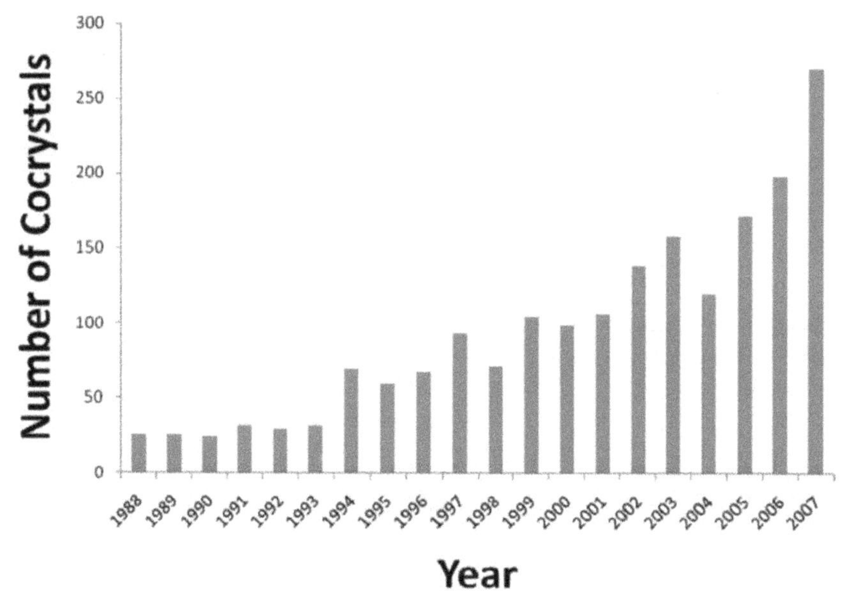

Figure 4. Frequency of occurrence of organic molecular co-crystals in the Cambridge Structural Database from 1988 to 2007. For the purposes of this graph, co-crystals are distinct from solvates, hydrates and simple salts. (Ref. Childs & Zaworotko, 2009; reprinted with permission from Childs S.L., Zaworotko M.J. The reemergence of cocrystals: The crystal clear writing is on the wall introduction to virtual special issue on pharmaceutical cocrystals. Crystal Growth&Design 2009, 9(10): 4208-4211. Copyright 2009 American Chemical Society.)

Scientists showed that modifying the physical properties of a pharmaceutical compound through pharmaceutical co-crystal formation improved the performance of a drug known to have poor solubility (Vishweshwar et al., 2006a). Pharmaceutical co-crystallization is a reliable method to modify physical and technical properties of drugs such as solubility increase/decrease, change of dissolution rate, stability, hygroscopicity and compressibility without alternating their pharmacological behaviour (Ranganathan, 1999; Fleischman et al., 2003; Almarsson & Zaworotko, 2004; Childs et al., 2004; Peterson et al.; 2006; Hickey et al., 2007; Žegarać et al., 2007; Schultheiss & Newman, 2009).

The use of co-crystals in drug design and delivery and as functional materials with potential applications as pharmaceuticals has recently attracted considerable interest (Fleischman et al., 2003; Jones et al., 2006; McNamara et al., 2006; Peterson et al., 2006; Vishweshwar et al., 2006a). Pharmaceutical co-crystals have been described for many drugs such as acetaminophen, aspirin, ibuprofen, flurbiprofen, sulfadimidine, etc. (Vishweshwar et al., 2005; Vishweshwar et al., 2006b; Caira, 2007; Rodríguez-Hornedo et al., 2007; Babu et al., 2008; Sarma et al., 2008,). Co-crystals of antitubercular drugs with dicarboxylic acids were reported using carboxylic acid-pyridine synthon as a reliable tool (Vishweshwar et al., 2003). The co-crystal of piracetam-L-tartaric acid showed improved hygroscopic properties

(Viertelhaus et al., 2009). Co-crystal forming abilities of two anti-HIV drugs (lamivudine and zidovudine) were studied to investigate the general applicability (Bhatt et al., 2009). Trimer co-crystals of *cis*-itraconazole-succinic acid were prepared and characterized by the possibility of achieving the higher oral bioavailability normally observed for amorphous forms of water-insoluble drugs (Remenar et al., 2003). The novel pharmaceutical co-crystal norfloxacin saccharinate dihydrate and its co-crystal, norfloxacin saccharinate–saccharin dihydrate were reported (Velaga et al., 2008).

3.2. Design of co-crystals

Co-crystallization is a result of competing molecular associations between similar molecules, or homomers, and different molecules, or heteromers. Instead, both components (host and guest) utilise prominent intermolecular non-covalent interactions such as hydrogen bonding, van der Waals forces and π-π stacking interactions to combine and yield a uniform crystalline material. Hydrogen bonds are the basis of molecular recognition phenomena in pharmaceutical systems and are responsible for the generation of families of molecular networks with the same molecular components (single-component crystals and their polymorphs) or with different molecular components (multiple-component crystals or co-crystals) in the crystalline state (Jayasankar et al., 2006; Sekhon, 2009). The components in a co-crystal exist in simple definite stoichiometric ratios, e.g., 1:1, 1:2, 2:1, etc. Co-crystals have different crystal structures than the pure components, contain different intermolecular packing patterns and as such they often exhibit different physical properties than the pure components. Unlike salt formation, co-crystallisation does not rely on ionisation of the API and the counterion to make a solid. Co-crystals are an alternative to salts when these do not have the appropriate solid state properties or cannot be formed due to the absence of ionization sites in the API (Aakeröy & Salmon, 2005; Miroshnyk et al., 2009).

The formation of a salt or a co-crystal can be predicted from pK_a value of an acid (A) and a base (B). Salt formation generally requires a difference of about 3 pK_a units between the conjugate base and the conjugate acid (A) *i.e.* [pK_a (B) – pK_a (A) ≥ 3] (Etter, 1990; Whitesides & Wong, 2006; Sekhon, 2009). In cases when $\Delta pK_a = pK_a (B) - pK_a (A) = 0\text{-}3$, the transfer of proton is ambiguous, and we can talk about the salt–co-crystal continuum (Childs et al., 2007).

Co-crystals can be prepared from two molecules of any shape or size having complementary hydrogen bond functionalities. The ability of an API to form a co-crystal is dependent on a range of variables, including the types of co-former, the API co-former ratio, the solvents, the temperature, the pressure, the crystallization technique, etc. Common functional groups, such as carboxylic or amino acids, amides, alcohols and carbohydrates are typically found to interact with one another in co-crystals (see Fig. 5) (Miroshnyk et al., 2009, Sarma et al., 2011; Qiao et al., 2011). Etter has studied hydrogen bonds in co-crystals and uses them as design elements. The hydrogen bond general rules are the following: (*i*) all good proton donors and acceptors are used in hydrogen bonding, (*ii*) if six-membered ring intramolecular hydrogen bonds can form, they will usually do so in preference to forming intermolecular hydrogen bonds, and (*iii*) the best proton donors and acceptors remaining after intramolecular hydrogen-bond formation form intermolecular hydrogen bonds with one another. In addition, the selectivity of hydrogen bonding in co-crystals was demonstrated by using pyridines (Etter, 1991; Sarma et al., 2011).

Based on the above mentioned facts co-crystal prediction includes the following steps: (*i*) determining whether a given set of two or more molecular components will undergo co-crystallization; (*ii*) identifying the primary intermolecular interactions, e.g., hydrogen-bond motifs that will exist within a particular co-crystal structure; and (*iii*) envisioning the overall packing arrangement in the resulting co-crystal structure (Trask, 2007).

Design and preparation of pharmaceutical co-crystals is a multi-stage process that can be schematically described in the following steps: (*i*) selection and research of APIs; (*ii*) selection of co-crystal formers; (*iii*) empirical and theoretical guidance; (*iv*) co-crystal screening; (*v*) co-crystal characterisation; (*vi*) co-crystal performance (Miroshnyk et al., 2009; Qiao et al., 2011).

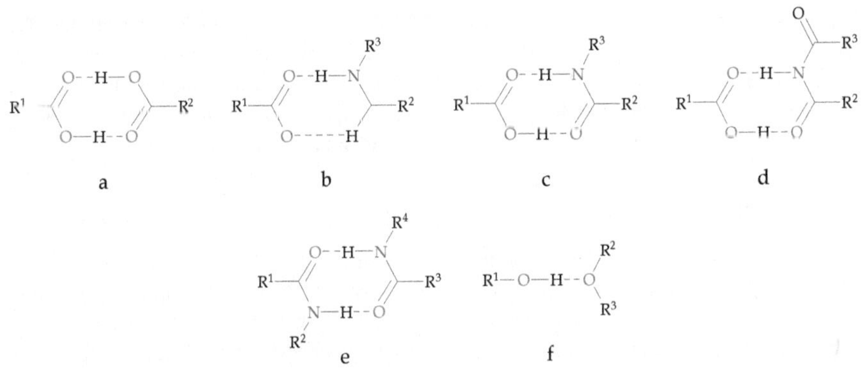

Figure 5. Possible formation of hydrogen bonds between synthons used in crystal engineering: acid-acid (a), acid-amine (b), acid-amide (c), acid-imide (d), amide-amide (e), and alcohol-ether (f).

3.3. Synthesis of co-crystals

A pharmaceutical co-crystal is a single-crystalline solid that incorporates two neutral molecules, one being an API and the other a co-crystal former (Vishweshwar et al., 2006). Co-crystal former may be an excipient or another drug (Rodríguez-Hornedo et al., 2007). Pharmaceutical co-crystal technology is used to identify and develop new proprietary forms of widely prescribed drugs and offers a chance to increase the number of forms of an API.

Crystalline forms can be generated by means of kinetically or thermodynamically controlled crystallization processes. Synthesis/processing of co-crystals can be accomplished via a number of methods, including slow solvent evaporation crystallization from solution, solvent-reduced (e.g. slurrying, solvent-drop grinding) and solvent-free (e.g. grinding, melt), high throughput crystallization and co-sublimation techniques (Shekunov & York,

2000; Morissette et al., 2004; Trask & Jones, 2005; Trask et al., 2005a; Trask et al., 2005b; Trask et al., 2006; Stahly, 2007; Berry et al., 2008; Takata et al., 2008; Friščić & Jones, 2009; Schultheiss & Newman, 2009; Qiao et al., 2011). Typically co-crystals are prepared by slow solvent evaporation that is only viable if compatible solubility in a given solvent exists between the components comprising the potential co-crystal. The potential benefits, disadvantages and methods of preparation of co-crystals were reported (Blagden et al., 2007). Solvent drop grinding has been reported to be a cost-effective, green, and reliable method for discovery of new co-crystals as well as for preparation of existing co-crystals. A slurry crystallization technique was used in co-crystal screening of two non-ionizable pharmaceutical host compounds, stanolone and mestanolone, with 11 pharmaceutically acceptable guest acids, and the results demonstrated the importance not only of hydrogen bonding but also of geometric fit in co-crystal formation (Takata et al., 2008). In addition to these classical techniques of co-crystal synthesis, also more sophisticated methods can be mentioned such as electrochemical crystallization, gel crystallization, vapour-diffusion crystallization, cryogenic grinding (cryomilling), microporous membranes crystallization, supercritical fluids crystallization and sonocrystallization (Dahlin et al., 2011; Hsu et al., 2002; Forsythe et al., 2002; He & Lavernia, 2001; Di Profio et al., 2007; Tong et al., 2001; Ruecroft et al., 2005). The combinations and variations of the above techniques may be used to cause co-crystal formation (McMahon et al., 2007). It is evident that many of the reported co-crystals appear to be the result of serendipity, although several groups have successfully exploited crystal engineering principles for design and synthesis of co-crystals.

4. Characterization of co-crystals

Co-crystal characterisation is an important constituent part of co-crystal research. The basic techniques of analysis of co-crystals involve especially solid state analysis methods (Zakrzewski & Zakrzewski, 2006; Dohnal et al., 2010), *i.e.* vibration-rotation spectroscopy, solid state NMR, thermal analysis, microscopy techniques and X-ray diffraction. The most often used solid-state analytical techniques will be discussed below.

4.1. Spectroscopy of vibration-rotation transitions

According to the used part of infrared (IR) spectra (energy), near infrared spectroscopy (NIR), middle infrared spectroscopy (MIR), Raman scattering and terahertz (THz) spectroscopy are distinguished.

4.1.1. NIR Spectroscopy

Photons in NIR region have the highest energy and can therefore vibrationally excite molecules into even higher excited vibrational states than the first level, *i.e.* the second, the third and others. These transitions are called overtons. Absorption of radiation in the NIR region is usually based on higher energy transitions between vibrational levels of molecules, namely combination transitions and overtones and not fundamental transitions, which are dominant in mid infrared region (MIR). NIR spectrometers are not so demanding of applied materials, as are the instruments working in the mid infrared region, due to different radiation frequencies used in the NIR spectrometry. Acquisition of NIR spectra requires several seconds and can be performed during manufacturing process. Therefore, the NIR spectroscopy is still more often applied as a tool of process analytical technology. In co-crystal analysis NIR spectroscopy is commonly used as a screening method of the first choice (Pekárek & Jampílek, 2010).

4.1.2. MIR spectroscopy and Raman scattering

The aim is not only to present the basic principles of both methods, but also to compare them. Raman spectroscopy and mid-infrared spectroscopy are both widely used in the pharmaceutical industry for the solid-state characterization because of their specificity. MIR region is the most important region in terms of analytical application. It is a region, where the majority of the so-called fundamental vibrations appear, *i.e.* the vibrational transitions from the basic to the first excited vibrational state. Vibrational spectroscopy provides key information about the structure of molecules. Positions and intensities of bands in a vibrational spectrum can be used to determine the structure of a molecule or to determine the chemical identity of a sample. With sufficient experience it is possible to identify chemical compounds or monitor intermolecular interactions by evaluating changes in positions and intensities of Raman bands which is extremely useful for the above mentioned purposes.

The mid-infrared and Raman spectra (with the exception of optical isomers) of a drug substance or any chemical compound are unique. Raman spectroscopy is a vibrational spectroscopy method which complements mainly the mid-IR spectroscopy. The intensity of bands in IR spectra is proportional to the dipole moment change occurring during the given type of the vibrational motion. Modes with a large change in the dipole moment having intensive bands in the IR spectra generally provide low-intensity bands in Raman spectra. Conversely, vibrations of non-polar functional groups provide intense bands in Raman spectra and weak bands in infrared spectra. (Zakrzewski & Zakrzewski, 2006; Pekárek & Jampílek, 2010). An important advantage of Raman spectroscopy over IR spectroscopy is the possibility to measure aqueous solutions. This advantage can be used in identification of APIs in aqueous solutions, emulsions or suspensions.

4.1.3. THz spectroscopy

This technique covers a wide interval, which can be roughly defined, for example, by frequencies of 100 GHz and 3 THz, which corresponds to wavelengths between 3 and 0.1 mm, *i.e.* wave numbers from 3 to 100 cm^{-1}. Hence THz region partially overlaps with far-infrared region. The characteristic frequency of 1 THz can be equivalently expressed in other spectroscopic units. The decisive factor for expansion of this method was the development of optical femtosecond lasers being an integral part of the most current laboratory and commercial THz spectrometers. Attractiveness of the THz field lies in the scale of the specific options for application of electromagnetic radiation of these frequencies. Application options in pharmaceutical analysis are mainly spectroscopy and imaging.

Solid substances often exhibit specific interactions in the THz region. While some crystals are transparent in this range, many others show low-frequency oscillations of the crystal structure (called phonon bands) with characteristic frequencies being determined by short-range inter-atom interactions and even arrangement of atoms at long distances (Kadlec & Kadlec, 2012). Similar to NIR and mid-IR spectroscopy, in THz spectroscopy reflectance (less often) and transmittance measurements are used.

It is known that vibrational modes of amorphous and crystalline substances are very different. This was experimentally proven, for example, in THz spectra of samples of crystalline and amorphous saccharides (Walther et al., 2003). The absence of sharp vibrational modes that can be observed in the crystalline form was later also confirmed in THz spectra of indomethacin (Strachan et al., 2004).

Determination of tablet coating thickness and determination of particle size belong among special applications of THz spectroscopy. Radiation in THz region has the lowest energy, which can cause changes mainly in rotational energy of molecules therefore it has the lowest (destructive) influence on measured co-crystals.

4.2. Solid state nuclear magnetic resonance

Solid state NMR (ssNMR) spectroscopy is not used for routine analyses in pharmacy because of its demands, but its role in pharmaceutical development is indispensable. It has very wide application. The most important applications can be divided into the following groups (Havlíček, 2010): (*i*) API structural analysis; (*ii*) polymorphism; (*iii*) co-crystal analysis; (*iv*) dosage form analysis; (*v*) solvate analysis; (*vi*) salt analysis.

Among the most widely used ssNMR techniques CP/MAS NMR belongs. It has three modifications: cross polarization (CP), magic angle spinning (MAS) and high-power heteronuclear decoupling. With this arrangement the sensitivity and line broadening problems were overcome, and high-resolution ssNMR was brought in practical use.

Solid-state NMR is capable of providing detailed structural information about organic and pharmaceutical co-crystals and complexes. ssNMR non-destructively analyzes small amounts of powdered material and generally yields data with higher information content than vibrational spectroscopy and powder X-ray diffraction methods. Particularly, its ability to prove or disprove molecular association and possibility to observe structural features (such as hydrogen bonding) are great advantages of this method. These advantages can be utilized in the analysis of pharmaceutical co-crystals, which are often initially produced using solvent drop grinding techniques that do not lend themselves to single-crystal growth for X-ray diffraction studies (Vogt et al., 2009).

4.3. Thermal analysis

Thermal analysis is a broad term referring to methods (see Table 1), which measure physical and chemical properties of a substance, a mixture of substances or also of a reaction mixture as a function of temperature or time during a controlled temperature programme. Most of these methods monitor corresponding system properties (mass, energy, size, conductivity, etc.) as dynamic functions of temperature.

For co-crystal analysis mainly differential scanning calorimetry (DSC) and its modifications, differential thermal analysis (DTA) and thermogravimetry analysis (TGA), are of great importance. For special cases a very useful method is thermally stimulated current (TSC). DSC and DTA are the most applied methods of thermal analysis in pharmaceutical development. In DSC, the sample is subjected to linear (or modulated) heating, and the heat flow rate in the sample is proportional to the actual specific heat and is continuously measured. Very interesting and useful modifications of DSC are hyperDSC, microDSC and modulated DSC (Krumbholcová & Dohnal, 2010).

Method name	Tracked value
Differential thermal analysis (DTA)	temperature difference between the studied and the standard sample
Differential scanning calorimetry (DSC)	thermal energy provided for compensation of temperature between the studied and the standard sample
Thermogravimetry analysis (TGA)	weight change
Thermomechanical analysis (TMA)	change in a mechanical property (module, hardness)
Dilatometry	change in volume
Effluent gas analysis	volume of the studied gas
Pyrolysis	pyrolysis products
Thermal luminescence analysis	emission of light
Electric conductivity analysis	change in electric conductivity

Table 1. List of the most common methods of thermal analysis. (Ref. Krumbholcová & Dohnal, 2010).

4.3.1. Thermally stimulated current

This special thermal method uses a so-called molecular mobility, which provides information about the structure of substances, about dynamic parameters ΔH and ΔS and the relaxation (release) time τ. In TSC the substance is heated to a temperature T_P in an electric field with intensity E_P for a period of time t_P, which is sufficient for differently moving particles of the studied substance to reach identical orientation in the electric field. In this state, the sample is rapidly cooled to a temperature T_0, which ensures zero motion of the particles. The effect of the electric field is then also deactivated, and the substance is then kept at temperature T_0 for time t_0. The temperature then linearly increases, and the substance returns to the previous balance, and depolarization current I_D is recorded as a function of temperature. Each depolarization peak is characterized by temperature T_{max} with intensity

I_{max}. This technique is particularly suitable for substances with polar character. It can distinguish polymorphs or co-crystals, which cannot be determined by classical DSC (Krumbholcová & Dohnal, 2010).

4.4. X-Ray diffraction

X-Ray diffraction is one of the basic solid state analytical techniques labelled as the "gold standard". It is used not only for characterization and identification of crystalline substances but also for their discrimination.

Two basic methods are discerned according to the type of the analysed sample. They are:

1. Diffraction on single crystal (single-crystal X-ray diffraction, SCXRD),
2. Powder diffraction (X-ray powder diffraction, XRPD).

Name	**single-crystal X-ray diffraction**	**X-ray powder diffraction**
Abbreviation	SCXRD	XRPD
Sample – type	Single crystal	Powder
Sample – amount	1 single crystal with size of 0.1 – 1 mm	100 – 500 mg
Sample – preparation	Selection of a suitable single crystal and its fixation onto the goniometric head.	Filling of the holder cavity or possibly milling.
Sample – consumption	Non-destructive	Non-destructive
Measurement time	Hours/days	Minutes/hours
Analysis of obtained data	Complete information about the molecule, conformation, bond lengths, chirality, spatial arrangement of molecules in the crystal, interactions between molecules, determination and location of solvents.	„Fingerprint" of the crystal structure
Principle of the method	Monochromatic light falls on a single crystal, and because only one lattice diffracts, it is necessary to rotate the crystal to obtain the desired set of reflections (the principle of four-circle diffractometer).	Monochromatic radiation impacts a polycrystalline material and all lattices that meet the diffraction condition diffract at the same time.
Practical application	Crystallographic studies: determination of the complete structure of an API, a protein, a complex, etc.	Mainly industrial, screening and control method for characterization of crystalline materials.

Table 2. Comparison of basic diffraction methods. (Ref. Brusová, 2010).

Their basic characteristics, advantages and disadvantages are summarized in Table 2. The X-ray structural analysis methods, which use single-crystal samples, often allow a complete determination of crystallographic characteristics. Diffraction on a single-crystal yields intensities of diffractions on individual configurations of the crystallography planes, individually for each of their orientation, and the number of the detected maxima is very high, in orders of $10^2 - 10^4$. In the powder diffraction, on a debyeogram or a diffractogram there at most 100 lines can be distinguished. In addition, each diffraction line is a superposition of diffractions of all inequivalent and equivalent planes (Hušák et al., 2007; Kratochvíl et al., 2008).

5. Carbohydrates and their derivatives as crystallization modifiers

As discussed above, pharmaceutical co-crystals have rapidly emerged as a new class of API solids demonstrating great promise and numerous advantages. Various co-crystals of APIs were prepared by reason of intellectual property protection, for example, imatimib (see Fig. 6, structure **I**) with co-crystal carbohydrate formers such as α-D-glucopyranose, D-fructofuranose and *N*-methyl-D-glucamine (meglumine) (Král et al., 2010); agomelatine (see Fig. 6, structure **II**) with carbohydrate counterions such as D-sorbitol, aspartame, phenyl-β-D-glucopyranoside, D-glucoheptono-1,4-lactone, D-(+)-trehalose, lactose, α-D-glucopyranose, saccharose, *N*-methyl-D-glucamine and D-(+)-glucosamine hydrochloride (Ferencova et al., 2012); or to improve permeability, for example, alendronate (see Fig. 6, structure **III**), ibandronate (see Fig. 6, structure **IV**) or risedronate (see Fig. 6, structure **V**) with a number of carbohydrates as co-crystal formers (Jampílek et al., 2009; Haroková, 2010; Havelková, 2010; Hrušková, 2010; Jampílek et al., 2010; Kos, 2010; Oktábec et. al, 2010; Kos et al., 2011; Ťažká 2011; Havelková, 2012; Oktábec, 2012).

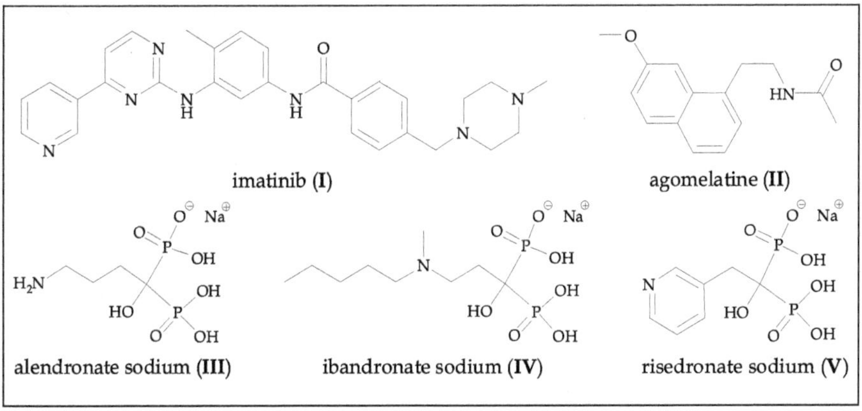

imatinib (**I**) agomelatine (**II**)

alendronate sodium (**III**) ibandronate sodium (**IV**) risedronate sodium (**V**)

Figure 6. APIs used as host of carbohydrate formers.

The present study deals with the design and an effort to prepare co-crystals/new entities, generally new solid phases, of the above discussed bisphosphonates **III-V**. These compounds were investigated in detail, and a lot of valuable knowledge concerning generation of solids was obtained. Bisphosphonates (BPs) can be denoted as top-selling APIs. BPs are the most widely used and the most effective bone resorption inhibitors currently available for treatment of Paget's disease, tumour-associated bone disease and osteoporosis. All BPs have high affinity for bone mineral as a consequence of their P-C-P backbone structure, which allows chelation of calcium ions (Ebetino et al., 1998). Following release from bone mineral during acidification by osteoclasts, BPs appear to be internalized specifically by osteoclasts, but not other bone cells. The intracellular accumulation of BP leads to inhibition of osteoclast function due to changes in the cytoskeleton, loss of the ruffled border (Carano et al., 1990; Sato et al., 1991) and apoptosis (Hughes et al., 1995; Selander et al., 1996; Ito et al., 1999; Reszka et al., 1999). The ability of BPs to inhibit bone resorption depends on the presence of two phosphonate groups in the P-C-P structure, which appears to be required for interaction with a molecular target in the osteoclast as well as for binding bone mineral (Rogers et al., 1995; van Beek et al., 1998; Rogers et al., 2000). BPs as pyrophosphate analogues are a group of drugs that are widely used in practice. There are several injectable bisphosphonates: etidronate, pamidronate and zoledronate, which may be administered every three months or yearly. Peroral BPs alendronate and risedronate are taken daily, weekly or monthly, and ibandronate is approved to be taken monthly. Oral bioavailability of these BPs is very low (their gastrointestinal absorption is about 1%) due to their high hydrophilicity (Ezra & Golomb 2000).

A number of various patented solid forms of each API from this group can be found, which, for example, complicates their utilization for generic formulation from the intellectual property point of view. Eiermann et al. prepared crystalline forms of ibandronate (**IV**) B and A (Eiermann et al., 2006a; Eiermann et al., 2006b). Lifshitz-Liron et al. obtained forms C, D, E, F, G, H, J, K, K2, K3, Q, Q1, Q2, Q3, Q4, Q5, Q6, QQ, R, S and T (Lifshitz-Liron et al., 2006). Muddasani et al. prepared polymorphs I and II (Muddasani et al., 2007), and Devaraconda et al. generated ibandronate forms III-XXXI (Devarakonda et al., 2010). Ten different polymorphic and pseudo-polymorphic forms of sodium risedronate (**V**) identified as A, B, B1, BB, D, E, F, G and H and a semi-crystalline form were described (Cazer et al., 2001; Aronhime et al., 2003a; Aronhime et al., 2003b; Richter et al., 2007). The crystal structures of four different hydrates (monohydrate, dihydrate, hemipentahydrate and variable hydrate) and an anhydrate of sodium risedronate (**V**) have been elucidated and discussed by Redman-Furey (Redman-Furey et al., 2005) and Gossman (Gossman et al., 2003). Recently three new phases were found and named J, K and M (Bruning et al., 2011).

This paper deals with investigation of various types of carbohydrates and their derivatives as crystallization modifiers applied to crystal study concerning the BP family. Carbohydrates were used due to their hydroxyl moieties, which are able to interact with a phosphoric group and/or a nitrogen atom in the alkyl chain or heterocycle. Carbohydrates also provided a unique excellent system of hydroxyl moieties in different stereochemical modifications. These hydroxyl groups can be straightly modified, for example, by alkylation/arylation, and the structure of the carbohydrate molecule obtains absolutely different three-dimensional/space properties.

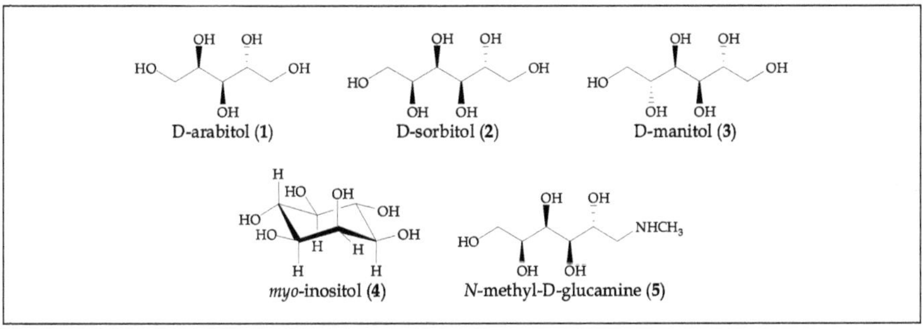

D-arabitol (1) D-sorbitol (2) D-manitol (3)

myo-inositol (4) *N*-methyl-D-glucamine (5)

Figure 7. Structures of sugar alcohols used as potential co-crystal/crystallization formers.

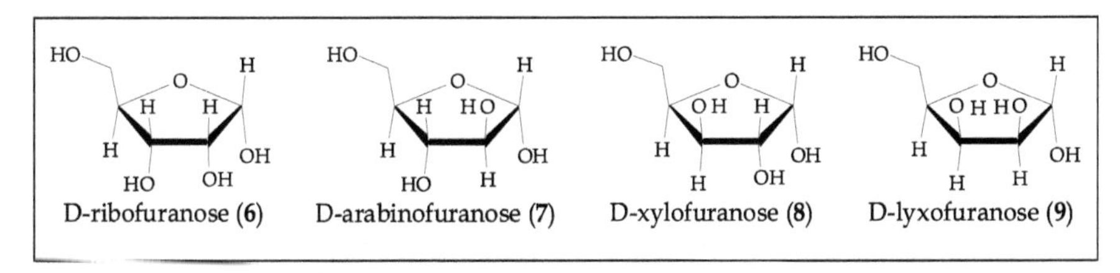

Figure 8. Structures of furanoses used as potential co-crystal/crystallization formers.

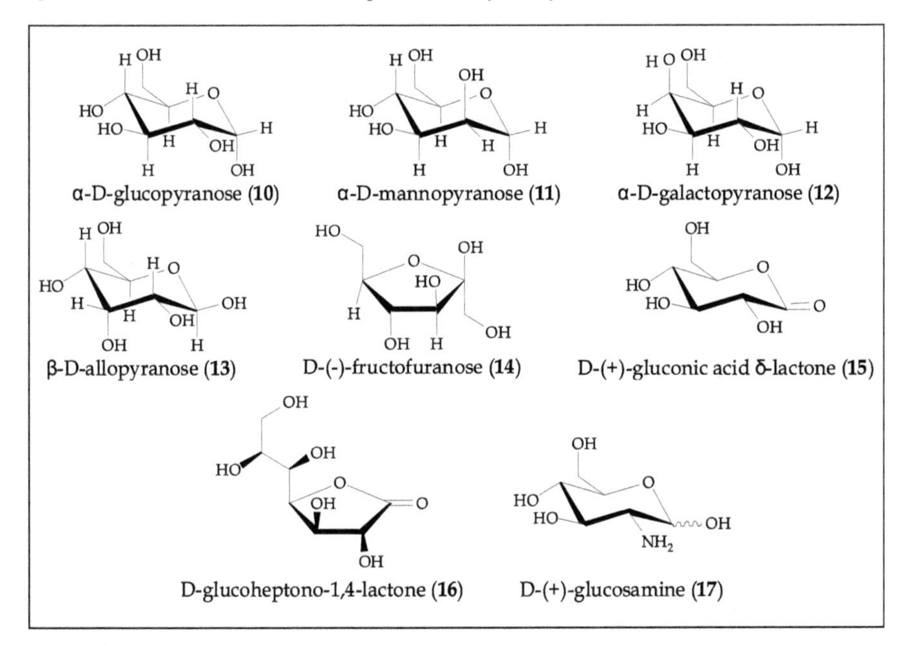

Figure 9. Structures of hexoses used as potential co-crystal/crystallization formers.

Thus mixtures of BPs with various sugar alcohols, furanoses, pyranoses and gluco-, manno- and galactopyranoside derivatives, some amino carbohydrates and disaccharides (see Figs. 7-11) as counterions were designed in an effort to prepare new crystalline forms or co-crystals/new entities. Mixtures of BPs and carbohydrates in different ratios and under various conditions were prepared. All the prepared mixtures (solid compounds) were characterized using some of the above mentioned solid state analytical techniques (Haroková, 2010; Havelková, 2010; Hrušková, 2010; Kos, 2010; Oktábec et. al, 2010; Kos et al., 2011; Ťažká 2011; Havelková, 2012; Oktábec, 2012).

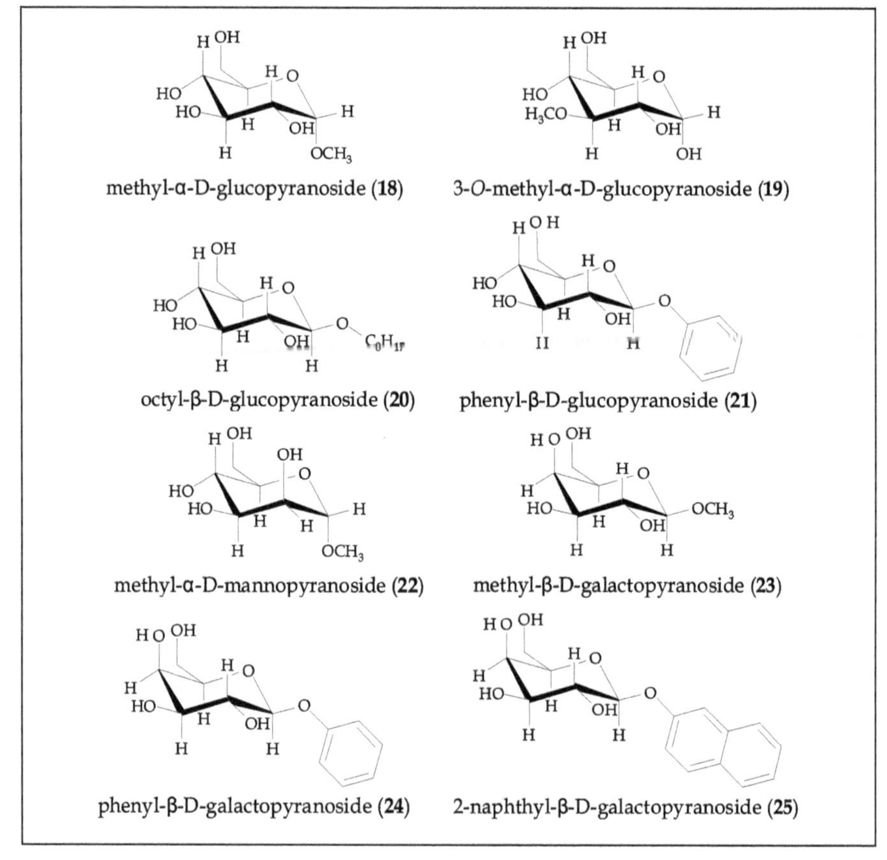

Figure 10. Structures of pyranosides used as potential co-crystal/crystallization formers.

5.1. Generation of samples

All the evaluated samples with ratios 1:1 (**A**), 1:2 (**B**) and 1:3 (**C**) were prepared by means of dissolution of bisphosphonate monosodium salt and the excipient in water, subsequently mixed (1 h) and slowly evaporated at ambient temperature. To some samples with ratios 1:2

and 1:3 methanol was slowly added dropwise as an anti-solvent. The solid precipitated compound was filtered and dried at ambient temperature, samples 1:2 (**D**) and 1:3 (**E**), and the remaining liquid part was slowly evaporated at ambient temperature, samples 1:2 (**F**) and 1:3 (**G**). All generated solid compounds were subsequently screened by means of FT-NIR and FT-Raman spectroscopy. If a sample differing from the starting materials was found, it was additionally characterized by the below mentioned methods (Jampílek et al., 2009; Haroková, 2010; Havelková, 2010; Hrušková, 2010; Jampílek et al., 2010; Kos, 2010; Oktábec et. al, 2010; Kos et al., 2011; Ťažká 2011; Havelková, 2012; Oktábec, 2012).

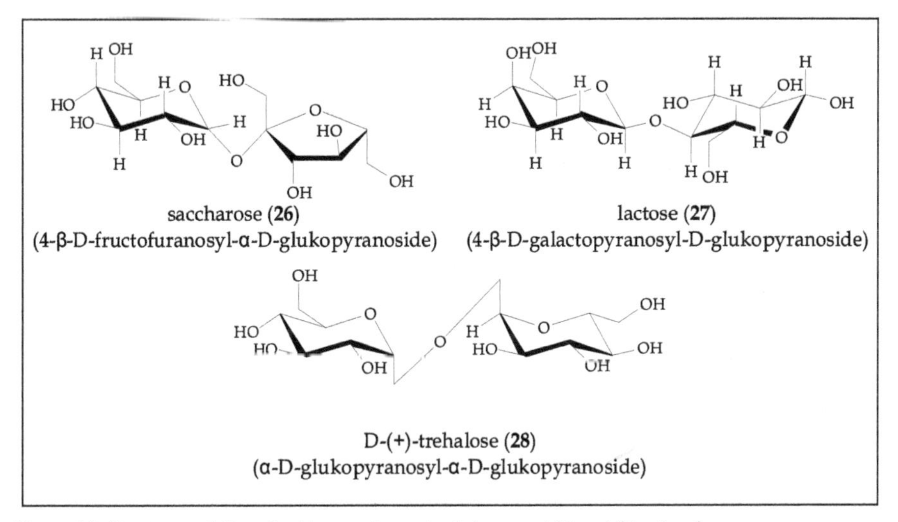

Figure 11. Structures of disaccharides used as potential co-crystal/crystallization formers.

5.2. Used solid-state analytical techniques

Near infrared spectra were recorded using a Smart Near-IR UpDrift™, Nicolet™ 6700 FT-IR Spectrometer (Thermo Scientific, USA). The spectra were obtained by accumulation of 128 scans with 4 cm^{-1} resolution in the region of 12800–4000 cm^{-1}. FT-Raman spectra were accumulated by an FT-Raman spectrometer RFS 100/S (Karlsruhe, Bruker, Germany). The spectra were obtained by accumulation of 256 scans with 4 cm^{-1} resolution in the back scattering geometry with the laser wavelength of 1064 nm. ^{31}P CP/MAS NMR Spectra were recorded on a Bruker AVANCE 500 MHz spectrometer (Karlsruhe, Bruker, Germany). The ^{31}P CP/MAS spectra were measured in 4 mm rotor at 10 kHz with 2 ms contact time. ^{31}P chemical shift of $NH_4H_2PO_4$ (0 ppm) was used as an external reference for ^{31}P chemical shift. The ^{13}C CP/MAS spectra were measured in 4 mm rotor at 13 kHz with 2 ms contact time. Carbon chemical shifts were referenced to the signal for TMS via a replacement sample of glycine (176 ppm for the carbonyl group signal). The XRPD patterns were obtained on a PANalytical X'PERT PRO MPD diffractometer with Cu Kα radiation (45 kV, 40 mA). The powder samples were measured on Silica plate holder. Data were recorded in the range

2-40° 2θ, with 0.01° 2θ step size and 50 s/step scan speed. For the measurement of differential scanning calorimetry (DSC) curve an instrument DSC Pyris 1 (PerkinElmer, USA) was used. Maximum sample weight was 3.5 mg, and the standard Al sample pan was used. The record of the DSC curve was in the range of 50–300 °C at the rate of 10.0 °C/min under a nitrogen atmosphere.

5.3. Crystal products of carbohydrates with alendronate and ibandronate

Although a number of carbohydrates were tested as potential co-crystal/crystallization formers, no change in the crystal structure of the used alendronate sodium salt (**III**) was obtained. The starting material was always obtained.

Mixtures of ibandronate monosodium salt (**IV**) with twenty-eight carbohydrates were generated by means of thermodynamically and/or kinetically controlled crystallization processes. Polymorph B of ibandronate monosodium salt monohydrate (sodium hydrogen {1-hydroxy-3-[methyl(pentyl)amino]-1-phosphonopropyl}phosphonate, **IV**) was used as a starting material (Eiermann, 2006a), which is the most common in pharmaceutical formulations. It is a white powder, freely soluble in water and practically insoluble in organic solvents.

From all the tested agents, only phenyl-β-D-galactopyranoside (see Fig. 10, structure **24**) yielded noteworthy products with BP **IV**. The rest of the tested carbohydrates, except phenyl-β-D-glucopyranoside (see Fig. 10, structure **21**) and 2-naphthyl-β-D-galactopyranoside (see Fig. 10, structure **25**), generated again the starting polymorph B. The mentioned two carbohydrates **21** and **25** provided mixtures of polymorphs A+B, as shown in Table 3. Samples of **IV-24** in ratios 1:1 (**A**), 1:2 (**B**) and 1:3 (**C**) were prepared by mixing and subsequent evaporation at ambient temperature. In all three samples a change in the NIR spectra can be observed in the range of 5,300-4,800 cm^{-1}. The spectra of samples **IV-24/B** and **IV-24/C** are very similar, only slightly different from sample **IV-24/A**, probably due to the lower crystallinity of sample **IV-24/A**, which causes broader bands in the spectrum. As samples **IV-24/A-C** were prepared in the same way, it can be concluded that increasing concentration of compound **24** influences the sample crystallinity.

Comp.	A	B	C	D	E	F	G
IV-21	B	B	B	B	B	A+B	A+B
IV-24	new	new	new	B	B	new	new
IV-25	B	B	B	B	B	A+B	A+B

Table 3. Samples of ibandronate (**IV**) and used carbohydrates **21, 24** and **25** in ratios 1:1, 1:2 and 1:3 prepared by evaporation at ambient temperature (**A, B, C**), samples in ratios 1:2 and 1:3 prepared by methanol precipitation (**D, E**) and samples in ratios 1:2 and 1:3 prepared by addition of methanol and evaporation of liquid part at ambient temperature (**F, G**). (Ref. Oktábec et al., 2010; Havelková, 2012).

Based on the NIR spectra of samples of **IV-24/D** and **IV-24/E** in ratios 1:2 and 1:3 precipitated by methanol and filtered, it can be concluded that both samples contain only form B of compound **IV**. The same characteristic bands in the range of 5,300-4,800 cm^{-1} as for samples **IV-24/A-C** can be observed for samples **IV-24/F** and **IV-24/G** in ratios 1:2 and 1:3 after addition of methanol, filtration of the obtained precipitate and evaporation at ambient temperature. Based on this fact it can be concluded that addition of methanol does not influence generation of a new unknown solid phase, because the same products were yielded with and without methanol addition. Slow evaporation seems to be important, *i.e.* thermodynamically controlled crystal modification is probable. The presence of carbohydrate **24** is fundamental for generation of a new entity. The samples **IV-24/A-C** and **IV-24/F**, **IV-24/G** were also characterized by means of the FT-Raman spectrometry and ^{31}P CP/MAS NMR spectroscopy for verification of this hypothesis. Both methods confirmed the presence of new solid phases.

From the above mentioned results (Table 3) it is evident that ibandronate (**IV**) provided a new solid phase only with phenyl-β-D-galactopyranoside (**24**). These samples, **IV-24/A-C** and **IV-24/F**, **IV-24/G**, were generated under the thermodynamic conditions (slow evaporation at ambient temperature) without or with methanol. It can be concluded that the presence of the co-crystal/crystallization former and the thermodynamic conditions were essential for generation of the new solid phase. Note that only β-D-pyranosides with substituted hydroxyl moiety in $C_{(1)}$, position 2 of the tetrahydropyran ring, *e.i.* in the equatorial position, showed interactions with BP **IV**. Although phenyl-α-D-pyranosides were not evaluated, it is possible to suppose that α-D-pyranosides possess probably a disadvantageous configuration of $C_{(1)}$ hydroxyl moiety.

Based on the above discussed results it can be also stated that substitution of $C_{(1)}$ hydroxyl with the aromatic group is necessary for interactions, because e.g. methyl-β-D-galactopyranoside (**23**) contrary to phenyl- (**24**) or naphthyl-β-D-galactopyranoside (**25**) provided no modification of the starting polymorph of compound **IV**. This hypothesis is supported by the fact that phenyl-β-D-glucopyranoside (**21**) afforded also the change of polymorph B to form A of BP **IV**. It can be assumed that naphthyl is too bulky compared to the phenyl ring. Nevertheless from all the evaluated substituted pyranosides only phenyl-β-D-galactopyranoside (**24**) yielded the new solid phase of compound **IV**. As non-covalent interactions are important for generation of crystal forms, the space configuration of all the hydroxyl moieties is probably essential for interactions between BP **IV** and carbohydrate **24**. As illustrated in Fig. 10, where basic spatial configuration of substituted pyranosides is shown, different interactions between phenyl-β-D-galactopyranoside (**24**) and phenyl-β-D-glucopyranoside (**21**) are probably caused by opposite orientation of hydroxyl moiety in $C_{(4)}$ in position 5 of the tetrahydropyran ring. In compound **21** this hydroxyl moiety is *trans*-oriented (in equatorial configuration) to the hydroxymethyl group in $C_{(5)}$ in position 6 of the tetrahydropyran ring, while in compound **24** it is *cis*-oriented (in axial configuration), which at the same time guarantees space proximity to the pyran oxygen. Pyranoside **24** possesses *cis*-orientation of hydroxyl moieties in $C_{(3)}$ and $C_{(4)}$ in positions 4 and 5 of the tetrahydropyran ring together with the phenoxy moiety in $C_{(1)}$ in position 2 of the tetrahydropyran ring. This fact together with *cis*-orientation of the hydroxymethyl group in $C_{(5)}$ in position 6 probably results in essential three-point interaction of neighbouring hydroxyl moieties between compound **24** and BP **IV** that is completed by interactions of the adjacent phenoxy moiety and pyran oxygen in comparison with carbohydrate **21**, see Fig. 10.

5.4. Crystal products of carbohydrates with risedronate

The semi-crystalline risedronate monosodium salt sodium 1-hydroxy-1-phosphono-2-(pyridin-3-yl-ethyl)phosphonate, (**V**) was used as a starting material (Richter et al., 2007). It is a white powder, freely soluble in water and practically insoluble in organic solvents. The sodium hemipentahydrate, which is the marketed form A, is the most stable of all these forms at ambient conditions (298 K, 50% room humidity) (Cazer et al., 2001).

From all the tested agents only phenyl-β-D-galactopyranoside (**24**) with risedronate (**V**) yielded noteworthy products. Other tested carbohydrates yielded either risedronate form A (in most cases), form H (in the case of the samples with *myo*-inositol (see Fig. 7, structure **4**), D-lyxofuranose (see Fig. 8, structure **9**), phenyl-β-D-glucopyranoside (**21**) and naphthyl-β-D-galactopyranoside (**25**) prepared by addition of methanol and evaporation of the liquid part at ambient temperature) or impure form B in the case of the sample with β-D-allopyranose (see Fig. 9, structure **13**) precipitated by methanol. A rapid change in solubility equilibrium and fast precipitation (kinetically controlled crystallization process) caused generation of different form B (samples **V-13/D** and **V-13/E**), while slow evaporation, *i.e.* thermodynamically controlled crystallization, led to preparation of stable polymorphs A or H. β-D-Allopyranose (**13**) modifies the environment from which compound **V** was crystallized, but this carbohydrate was not detectable in the final crystalline form. Based on this fact it can be concluded that the addition of methanol as an anti-solvent is crucial for generation of this different/uncommon solid form.

Different interactions of BP **V** with carbohydrate **13** are probably caused by the opposite orientation of hydroxyl moieties in $C_{(1)}$ and $C_{(3)}$ in positions 2 and 4 of the tetrahydropyran ring in comparison with α-D-gluco-, α-D-manno- and α-D-galactopyranose (**10-12**). The β-position of the hydroxyl moiety in $C_{(1)}$ of β-D-allopyranose (**13**) possesses also *cis*-orientation (axial configuration, see Fig. 9) with respect to the pyran oxygen in position 1 of the tetrahydropyran ring. As bonds influencing generation of crystalline forms are formed by non-binding interactions (e.g. by *H*-bonds, ionic bonds, van der Waals forces (dispersion attractions, dipole-dipole, dipole-induced dipole interactions) and hydrophobic interactions), the steric arrangement of hydroxyl moieties on pyranose skeletons is important for formation of interactions, as discussed above.

Samples of **V-24/A-C** were prepared by mixing saturated aqueous solutions and subsequent evaporation of water at ambient temperature. All three samples contained polymorph A of risedronate (**V**) (the most thermodynamically stable form). Samples **V-24/D** and **V-24/E** precipitated by methanol and filtered yielded again polymorph A of compound **V**. Samples **V-24/F** and **V-24/G** were generated by addition of methanol and filtration of the obtained precipitate with following evaporation at ambient temperature. Samples **V-24/F** and **V-24/G** were absolutely different from all the above mentioned samples. A change in the NIR spectra of samples **V-24/F** and **V-24/G** was observed in the range of 7,100–4,900 cm^{-1}. Both samples were also characterized by means of FT-Raman spectrometry and ^{31}P and ^{13}C CP/MAS NMR spectroscopy for verification of the above mentioned hypothesis. Sample **V-24/G** was a mixture of polymorph A and the amorphous form of **V**, but sample **V-24/F** was confirmed as a new crystalline form of BP **V**. Therefore solid **V-24/F** was additionally characterized by means of XRPD (Fig. 12) and also by DSC. A XRPD pattern corresponds to a crystalline sample. Visual comparison of the measured pattern with those published previously (Aronhime et al., 2003; Bruning et al., 2011) revealed that a new solid phase of BP **V** was formed. It is also supported by the absence of peaks of co-crystal former **24**. It can be concluded that the presence of compound **24** and slow evaporation, *i.e.* thermodynamically controlled crystallization process, with a small amount of methanol as anti-solvent provided risedronate (**V**) in an unknown form that was named as polymorph P.

Sugar alcohols did not provide any different forms or co-crystals with risedronate (**V**). The polyols used are acyclic compounds, or probably important heterocyclic oxygen is not present in the ring. In the case of *myo*-inositol (**4**), where only the different polymorph H of compound **V** was detected, *cis*-oriented hydroxyl moieties are in $C_{(1)}$, $C_{(2)}$, $C_{(3)}$ and $C_{(5)}$ or conversely oriented hydroxyl moieties are in $C_{(4)}$ and $C_{(6)}$, see Fig. 7. Contrary to the rest of the tested unsubstituted carbohydrates, only β-D-allopyranose (**13**) shows *cis*-orientation of hydroxyl moieties in $C_{(1)}$ and $C_{(5/6)}$ in positions 2 and 6 of the tetrahydropyran ring together with the pyran oxygen in position 1 and *cis*-orientation of hydroxyl moieties in $C_{(2)}$, $C_{(3)}$ and $C_{(4)}$ in positions 3, 4 and 5 of the tetrahydropyran ring, *i.e. cis*-orientation of three sequential hydroxyl moieties. These facts are probably essential for interactions between BP **V** and carbohydrate **13**. For example, α-D-galactopyranose (**12**) possesses 1, 4, 5, 6 *cis*-oriented pyran oxygen together with hydroxyl moieties; α-D-glucopyranose (**10**) possesses 1, 4, 6 *cis*-oriented pyran oxygen together with hydroxyl moiety; and α-D-mannopyranose (**11**) possesses 3, 4, 6 *cis*-oriented hydroxyl moieties together with pyran oxygen in position 1.

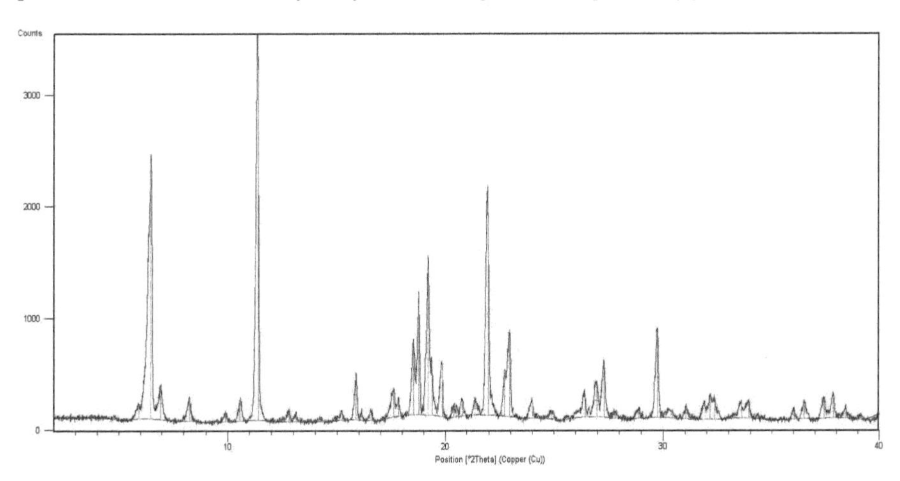

Figure 12. XRPD patterns of new solid phase **V-24/F** named as polymorph P.

According to the above mentioned hypothesis, interactions with BP **V** should be predicted only for D-lyxofuranose (Fig. 8, structure **9**) from the furanose family. D-Lyxofuranose (**9**) shows *cis*-orientation of hydroxyl moieties in $C_{(2)}$ and $C_{(3)}$ in positions 3 and 4 of the tetrahydrofuran ring together with the furan oxygen in position 1. Nevertheless, this three-point interaction of carbohydrate **9** with BP **V**, which however is not completed by other additional/secondary interactions, is not sufficient for generation of a different form or co-crystal of BP **V**, when only form H was generated using compound **9**. Probably the conformation of the tetrahydrofuran ring different from the pyranose chair conformation is also important.

Different interactions of risedronate monosodium salt (**V**) with phenyl-β-D-galactopyranoside (**24**) compared to other evaluated *O*-substituted pyranosides are probably caused by the opposite orientation of the hydroxyl moiety in $C_{(4)}$ in position 5 of the tetrahydropyran ring, as it is shown in Fig. 10 and discussed in Section 5.3. β-Position of the hydroxyl moiety in $C_{(1)}$ of β-D-gluco- and β-D-galactopyranoside (as well as in β-D-allopyranose) possessing also *cis*-orientation to the pyran oxygen together with phenyl substitution of this hydroxyl moiety seems also to be an important assumption for interactions. For example, methyl-β-D-galactopyranoside (**23**) did not show any interactions with compound **V**, whereas phenyl-β-D-glucopyranoside (**21**) and naphtyl-β-D-galactopyranoside (**25**) generated polymorph H of compound **V**, and phenyl-β-D-galactopyranoside (**24**) provided a new solid phase. Aliphatic alkoxy moieties (methoxy, octyloxy) show absolutely different physico-chemical properties, *i.e.* non-binding interactions compared with the aromatic phenyl nucleus. On the other hand, a naphthyl moiety, which is comparable with a phenyl ring, does not meet steric requirements to generate a new solid phase with compound **V**. Contrary to the rest of the tested *O*-substituted pyranosides, carbohydrate **24** shows *cis*-orientation of hydroxyl moieties in $C_{(3)}$, $C_{(4)}$ and $C_{(5\text{-}6)}$ in positions 4, 5 and 6 of the tetrahydropyran ring, *i.e.* three sequential hydroxyl moieties that possess *cis*-orientation with the phenoxy moiety in $C_{(1)}$ in position 2 of the tetrahydropyran ring together with the pyran oxygen in position 1, as was mentioned in Section 5.3. This configuration of all the hydroxyl moieties, as illustrated in Fig. 10, is probably essential for interactions between risedronate (**V**) and phenyl-β-D-galactopyranoside (**24**).

6. Future research

Samples **IV-24/A-B** and **IV-24/F**, **IV-24/G** should be characterized by XRPD. Also some of the above mentioned samples as well as sample **V-24/F** (polymorph P) should be analyzed using single-crystal X-ray diffraction (SCXRD), *i.e.* monocrystals should be prepared for absolute characterization of their structure. Based on the above described primary screening, other carbohydrate derivatives should be synthesized, especially various substituted β-D-pyranosides, e.g. *O*-arylated or *O*-alkylated. For example, based on the screening of these carbohydrates, it was confirmed that the structure derived from β-D-galactopyranose could be a successful candidate for modification of a crystalline form of BPs.

In the recent past, molecular modelling and/or molecular dynamics simulates started playing an increasingly important role in detection of molecule interactions. These computational techniques allow investigating possible binding modes of compounds. Therefore the priorities are advanced simulation of interactions between carbohydrate derivatives and various APIs and modelling/computing of physico-chemical properties of these new potential solids by means of various advanced simulating software products and subsequent transfer of the results of this systematic virtual screening to practice.

7. Conclusion

Twenty-eight carbohydrate derivatives were evaluated as formers during crystallization process of monosodium salts of alendronate (**III**), ibandronate (**IV**) and risedronate (**V**). All prepared samples were screened by FT-NIR and FT-Raman spectroscopy, and some new entities were checked by ^{31}P and ^{13}C CP/MAS NMR spectroscopy, XRPD and DSC. In the present study the relationships between the chemical structures of bisphosphonates and carbohydrates required for crystalline form change are investigated and discussed. It can be concluded that in general carbohydrates can be used as crystallization modifiers, although none of carbohydrates afforded a new solid phase with alendronate (**III**). Ibandronate (**IV**) and risedronate (**V**) generated new solid phases with phenyl-β-D-galactopyranoside (**24**). It is worth to note that both BPs **IV** and **V** contain trisubstituted nitrogen in contrast to alendronate (**III**), which can be an important factor that can cause different interactions. In case of BP **IV** it can be speculated about co-crystal generation; in case of BP **V** the new crystal form, polymorph P, is explicitly characterized. In both cases thermodynamically controlled crystallization was successful. The fundamental steric requirements to carbohydrate formers for generation of a new solid phase is β-orientation of hydroxyl moiety in $C_{(1)}$ in position 2 of the tetrahydropyran ring together with *cis*-orientation of at least three vicinal hydroxyl moieties and the pyran oxygen. β-Hydroxyl moiety in $C_{(1)}$ have to be substituted by an aromatic or heteroaromatic ring. It is also important to note that all the used carbohydrates can chelate the sodium cation in monosodium salts of BPs, and thus the sodium cation can contribute to the binding of BPs and carbohydrate with convenient conformation. The sodium cation can make the complex energetically favourable and help to retain the proper topology of the binding phosphates of BPs and the proper orientation of the hydroxyl moieties of a carbohydrate.

Author details

Josef Jampílek and Jiří Dohnal
Faculty of Pharmacy, University of Veterinary and Pharmaceutical Sciences Brno,
Research Institute for Pharmacy and Biochemistry (VUFB, s.r.o.),
Czech Republic

Acknowledgement

This study was partly supported by the Grant Agency of the Czech Republic (Czech Science Foundation), project number GACR P304/11/2246. Thanks to all our colleagues and numerous graduate and undergraduate students who participated in the synthesis and analytical evaluation of the discussed solid forms. The authors would like to thank Prof. Bohumil Kratochvíl for his constructive comments that helped us to improve the manuscript. The authors also thank Natalia Jampílková for her valuable assistance with the English proofreading of the manuscript.

8. References

Almarsson, Ö. & Zaworotko, M. J. (2004). Crystal engineering of the composition of pharmaceutical phases. Do pharmaceutical co-crystals represent a new path to improved medicines? *Chemical Communications*, Vol.2004, No.17, pp.1889-1896, ISSN 1364-548X

Aakeröy, C. B. & Salmon, D. J. (2005). Building co-crystals with molecular sense and supramolecular sensibility. *CrystEngComm*, Vol.7, No.72, pp.439-448, ISSN 1466-8033

Aakeröy, C. B.; Fasulo, M. E. & Desper, J. (2007). Cocrystal or salt: Does it really matter? *Molecular Pharmaceutics*, Vol.4, No.3, pp.317–322, ISSN 1543-8384

Aronhime, J.; Lifshitz-Liron, R.; Kovalevski-Ishai, E. & Lidor-Hadas, R. (TEVA Pharmaceutical Industries, Ltd.), (2003a). *Novel polymorphs and pseudopolymorphs of risedronate sodium*. CA 2480764

Aronhime, J.; Lifshitz-Liron, R.; Kovalevski-Ishai, E. & Lidor-Hadas, R. (TEVA Pharmaceutical Industries, Ltd.), (2003b). *Novel polymorphs and pseudopolymorphs of risedronate sodium*. WO 03086355

Babu, N. J.; Reddy, L. S.; Aitipamula, S. & Nangia, A. (2008). Polymorphs and polymorphic cocrystals of temozolomide. *Chemistry – An Asian Journal*, Vol.3, No.7, pp.1122-1133, ISSN 1861-471X

Berry, D. J.; Seaton, C. C.; Clegg, W.; Harrington, R. W.; Coles, S. J.; Horton, P. N.; Hursthouse, M. B.; Storey, R.; Jones, W.; Friščić, T. & Blagden, N. (2008). Applying hot-stage microscopy to co-crystal screening: A study of nicotinamide with seven active pharmaceutical ingredients. *Crystal Growth & Design*, Vol.8, No.5, pp.1697-1712, ISSN 1528-7483

Bhatt, P. M.; Azim, Y.; Thakur, T. S. & Desiraju, G. R. (2009). Co-crystals of the anti-HIV drugs lamivudine and zidovudine. *Crystal Growth & Design*, Vol.9, No.2, pp.951-957, ISSN 1528-7483

Bhogala, B. R. & Nangia, A. (2008). Ternary and quaternary co-crystals of 1,3-*cis*,5-*cis*-cyclohexanetricarboxylic acid and 4,4'-bipyridines. *New Journal of Chemistry*, Vol.32, No.5, pp.800-807, ISSN 1144-0546

Bighley, L. D.; Berge, S. M. & Monkhouse, D. C. (1996). Salt forms of drugs and absorption, In: *Encyclopedia of Pharmaceutical Technology*, Vol.13, Swarbick, J. & Boylan, J. C. (Eds.), pp.453-499, Marcel Dekker, ISBN 978-0-8247-2814-9, New York, NY, USA

Blagden, N.; de Matas, M.; Gavan, P. T. & York, P. (2007). Crystal engineering of active pharmaceutical ingredients to improve solubility and dissolution rates. *Advanced Drug Delivery Reviews*, Vol.59, No.7, pp.617-630, ISSN 0169-409X

Bond, A. B. (2007). What is a co-crystal? *CrystEngComm*, Vol.9, No.9, pp.833–834, ISSN 1466-8033

Brittain, H. (Ed.). (2009). *Polymorphism in Pharmaceutical Solids*. Informa Healthcare, ISBN 978-1-4200-7321-8, New York, NY, USA

Bruning, J.; Petereit, A. C.; Alig, E.; Bolte, M.; Dressman, J. B. & Schmidt, M. U. (2011). Characterization of a new solvate of risedronate. *Journal of Pharmaceutical Sciences*, Vol.100, No.3, pp.863–873, ISSN 1520-6017

Brusová, H. (2010). X-Ray diffraction, In: *Modern Approaches to Pharmaceutical Analysis*, Dohnal, J.; Jampílek, J.; Král, V. & Řezáčová, A. (Eds.). pp.433-477, Faculty of Pharmacy, University of Veterinary and Pharmaceutical Sciences Brno, ISBN 978-80-7305-085-6, Prague, Czech Republic

Caira, M. R. (2007). Sulfa drugs as model co-crystal formers. *Molecular Pharmaceutics*, Vol.4, No.3, pp.310–316, ISSN 1543-8384

Carano, A.; Teitelbaum, S. L.; Konsek, J. D.; Schlesinger, P. H. & Blair, H. C. (1990). Bisphosphonates directly inhibit the bone resorption activity of isolated avian osteoclasts *in vitro*. *Journal of Clinical Investigation*, Vol.85, No.2, pp.456–461, ISSN 0021-9738

Cazer, F. D.; Perry, G. E.; Billings, D. M. & Redman-Furey, N. L. (Warner Chilcott Company, LLC.), (2001). *Selective crystallization of 3-pyridyl-1-hydroxy-ethylidene-1,1-bisphosphonic acid sodium as the hemipentahydrate or monohydrate*. EP 1252170

Childs, S. L.; Chyall, L. J.; Dunlap, J. T.; Smolenskaya, V. N.; Stahly, B. C. & Stahly, G. P. (2004). Crystal engineering approach to forming cocrystals of amine hydrochlorides with organic acids. Molecular complexes of fluoxetine hydrochloride with benzoic, succinic, and fumaric acids. *Journal of American Chemical Society*, Vol.126, No.41, pp.13335–13342, ISSN 0002-7863

Childs, S. L. & Hardcastle, K. I. (2007). Cocrystals of piroxicam with carboxylic acids. *Crystal Growth & Design*, Vol.7, No.7, pp.1291–1304, ISSN 1528-7483

Childs, S. L.; Stahly, G. P. & Park, A. (2007). The salt-cocrystal continuum: The influence of crystal structure on ionization state. *Molecular Pharmaceutics*, Vol.4, No.3, pp.323-338, ISSN 1543-8384

Childs, S. L. & Zaworotko, M. J. (2009). The reemergence of cocrystals: The crystal clear writing is on the wall introduction to virtual special issue on pharmaceutical cocrystals. *Crystal Growth & Design*, Vol.9, No.10, pp.4208-4211, ISSN 1528-7483

Dahlin, A. B.; Sannomiya, T.; Zahn, R.; Sotiriou, G. A. & Vőrős, J. (2011). Electrochemical crystallization of plasmonic nanostructures. *Nano Letters*, Vol.11, No.3, pp.1337-1343, ISSN 1530-6984

Datta, S. & Grant, D. J. W. (2004). Crystal structures of drugs: Advances in determination, prediction and engineering. *Nature Reviews Drug Discovery*, Vol.3, No.1, pp.42–57, ISSN 1474-1776.

Desiraju, G. R. (2003). Crystal and co-crystal. *CrystEngComm*, Vol.5, No.82, pp.466–467, ISSN 1466-8033

Devarakonda, S. N.; Thaimattam, R.; Raghupati, B.; Asnani, M.; Vasamsetti, S. K. Rangineni, S. & Muppidi, V. K. (Dr. Reddy's Laboratories, Ltd.), (2010). *Ibandronate sodium polymorphs*. US 20100125149

Di Profio, G.; Tucci, S.; Curcio, E. & Drioli, E. (2007). Selective glycine polymorph crystallization by using microporous membranes. *Crystal Growth & Design*, Vol.7, No.3, pp.526-530, ISSN 1528-7483

Dohnal, J.; Jampílek, J.; Král, V. & Řezáčová, A. (Eds.). (2010). *Modern Approaches to Pharmaceutical Analysis*. Faculty of Pharmacy, University of Veterinary and Pharmaceutical Sciences Brno, ISBN 978-80-7305-085-6, Prague, Czech Republic

Ebetino, F. H.; Francis, M. D.; Rogers, M. J.; Russell, R. G. G. (1998). Mechanisms of action of etidronate and other bisphosphonates. *Reviews in Contemporary Pharmacotherapy*, Vol.9, No.4, pp.233–243, ISSN 0954-8602

Eiermann, U.; Junghans, B.; Knipp, B. & Sattelkau, T. (Hoffmann-La Roche, Inc.), (2006a). *Ibandronate polymorph B*. WO 2006081962

Eiermann, U.; Junghans, B.; Knipp, B. & Sattelkau, T. (Hoffmann-La Roche, Inc.), (2006b). *Ibandronate polymorph A*. WO 2006081963

Etter, M. C. (1990) Encoding and decoding hydrogen bonds patterns of organic compounds. *Accounts of Chemical Research*, Vol.23, No.4, pp.120–126, ISSN 0001-4842

Etter, M. C. (1991). Hydrogen bonds as design elements in organic chemistry. *Journal of Physical Chemistry*, Vol.95, No.12, pp.4601-4610, ISSN 0022-365

Ezra, A. & Golomb, G. (2000). Administration routes and delivery systems of bisphosphonates for the treatment of bone resorption. *Advanced Drug Delivery Reviews*, Vol.42, No.3, pp.175–195, ISSN 0169-409X

Ferencová, Z. (2012). *Physico-Chemical Properties Modification of Biologically Active Compounds I*, Diploma Thesis, Faculty of Pharmacy, University of Veterinary and Pharmaceutical Sciences Brno

Fleischman, S. G.; Kuduva, S. S.; McMahon, J. A.; Moulton, B.; Walsh, R. B.; Rodríguez-Hornedo, N. & Zaworotko, M. J. (2003). Crystal engineering of the composition of pharmaceutical phases. Multiple component crystalline solids involving carbamazepine. *Crystal Growth & Design*, Vol.3, No.6, pp.909-919, ISSN 1528-7483

Forsythe, E.L.; Maxwell, D.L. & Pusey, M. (2002). Vapor diffusion, nucleation rates and the reservoir to crystallization volume ratio. *Acta Crystallographica Section D*, Vol.58, No.10-1, pp.1601-1605, ISSN 0907-4449

Friščić, T. & Jones, W. (2009). Recent advances in understanding the mechanism of cocrystal formation via grinding. *Crystal Growth & Design*, Vol.9, No.3, pp.1621-1637, ISSN 1528-7483

Gossman, W. L.; Wilson, S. R. & Oldfield, E. (2003). Three hydrates of the bisphosphonate risedronate, consisting of one molecular and two ionic structures. *Acta Crystallographica Section C*, Vol.C59, No.2, pp.m33–m36, ISSN 0108-2701

Grepioni, F. & Braga, D. (Eds.). (2007). *Making Crystals by Design - from Molecules to Molecular Materials, Methods, Techniques, Applications*, Wiley-VCH, ISBN 978-3-527-31506-2, Weinheim, Germany.

Gu, C. H. & Grant, D. J. W. (2003). In: *Handbook of Experimental Pharmacology: Stereochemical Aspects of Drug Action and Disposition*; Eichelbaum, M.; Testa, B. & Somogyi, A. (Eds.), pp.113–137, Springer, ISBN 978-3-5404-1593-0, Berlin, Germany

Haroková, P. (2010). *Study of Polymorphism of Biologically Active Compounds I*, Diploma Thesis, Faculty of Pharmacy, University of Veterinary and Pharmaceutical Sciences Brno

Havelková, L. (2010). *Study of Drug Polymorphism I*, Diploma Thesis, Faculty of Pharmacy, University of Veterinary and Pharmaceutical Sciences Brno

Havelková, L. (2012). *Crystallization Products of Ibandronate with Gluco- and Galactopyranoside Derivatives*, Rigorous Thesis, Faculty of Pharmacy, University of Veterinary and Pharmaceutical Sciences Brno

Havlíček, J. (2010). NMR Spectroscopy, In: *Modern Approaches to Pharmaceutical Analysis*, Dohnal, J.; Jampílek, J.; Král, V. & Řezáčová, A. (Eds.). pp.261-313, Faculty of Pharmacy, University of Veterinary and Pharmaceutical Sciences Brno, ISBN 978-80-7305-085-6, Prague, Czech Republic

He, J. H. & Lavernia, E. J. (2001). Development of nanocrystalline structure during cryomilling of Inconel 625. *Journal of Materials Research*, Vol.16, No.9, pp.2724-2732, ISSN 0884-2914

Hickey, M. B.; Peterson, M. L.; Scoppettuolo, L. A.; Morrisette, S. L.; Vetter, A.; Guzmán, H.; Remenar, J. F.; Zhang, Z.; Tawa, M. D.; Haley, S.; Zaworotko, M. J. & Almarsson, Ö. (2007). Performance comparison of a co-crystal of carbamazepine with marketed product. *European Journal of Pharmaceutics and Biopharmaceutics*, Vol.67, No.1, pp.112–119, ISSN 0939-6411

Hilfiker, R. (Ed.). (2006). *Polymorphism in the Pharmaceutical Industry*. Wiley-VCH, ISBN 978-3-527-31146-0, Weinheim, Germany

Hrušková, J. (2010). *Study of Polymorphism of Biologically Active Compounds II*, Diploma Thesis, Faculty of Pharmacy, University of Veterinary and Pharmaceutical Sciences Brno

Hsu, W. P.; Koo, K. K. & Myerson, A. S. (2002). The gel-crystallization of 1-phenylalanine and aspartame from aqueous solutions. *Chemical Engineering Communications*, Vol.189, No.8, pp.1079-1090, ISSN 0098-6445

Hughes, D. E.; Wright, K. R.; Uy, H. L.; Sasaki, A.; Yoneda, T.; Roodman, G. D.; Mundy, G. R. & Boyce, B.F. (1995). Bisphosphonates promote apoptosis in murine osteoclasts *in vitro* and *in vivo*. *Journal of Bone and Mineral Research*, Vol.10, No.10, pp.1478–1487, ISSN 0884-0431

Hušák, M.; Rohlíček, J.; Čejka, J. & Kratochvíl, B. (2007). Structure determination from powder diffraction data - it is unrealizable dream or daily use? *Chemické Listy*, Vol.101, No.9, pp. 697–705, ISSN 0009-2770

Ito, M.; Amizuka, N.; Nakajima, T. & Ozawa, H. Ultrastructural and cytochemical studies on cell death of osteoclasts induced by bisphosphonate treatment. *Bone*, Vol.25, No.4, pp.447–452, ISSN 8756-3282

Jampílek, J.; Oktábec, Z.; Řezáčová, A.; Plaček, L.; Kos, J.; Havelková, L.; Dohnal, J. & Král V. (2009). Preparation and Properties of New Co-crystals of Ibandronate. *Proceedings of the 13th International Electronic Conference on Synthetic Organic Chemistry (ECSOC-13)*, ISBN 3-906980-23-5, Basel, Switzerland, November, 2009, Available from: <http://www.sciforum.net/presentation/201>

Jampílek, J.; Kos, J.; Oktábec, Z.; Mandelová, Z.; Pekárek, T.; Tkadlecová, M.; Havlíček, J.; Dohnal, J. & Král V. (2010). Co-crystal Screening Study of Risedronate and Unsubstituted Hexoses. *Proceedings of the 14th International Electronic Conference on Synthetic Organic Chemistry (ECSOC-14)*, ISBN 3-906980-24-3, Basel, Switzerland, November, 2010, Available from: <http://www.sciforum.net/presentation/421>

Jayasankar, A.; Somwangthanaroj, A.; Shao, Z. J. & Rodríguez-Hornedo, N. (2006). Co-crystal formation during cogrinding and storage is mediated by amorphous phase. *Pharmaceutical Research*, Vol.23, No.10, pp.2381-2392, ISSN 0724-8741

Jones, W.; Motherwell, W. D. S. & Trask, A. V. (2006). Pharmaceutical co-crystals: An emerging approach to physical property enhancement. *Materials Research Society Bulletin*, Vol.31, No.11, pp.875-879, ISSN 0883-7694

Junghanns, J. U. A. H. & Müller, R. H (2008). Nanocrystal technology, drug delivery and clinical applications. *International Journal of Nanomedicine*, Vol.3, No.3, pp.295–309, ISSN 1178-2013.

Kadlec, F. & Kadlec, Ch. (2012). Terahertz spectroscopy in time-domain, In: *Modern Pharmaceutical Analysis of Solid State*. Dohnal, J.; Jampílek, J. & Čulen, M. (Eds.), in press, Research Institute for Pharmacy and Biochemistry (VUFB), ISBN 978-80-905220-0-8, Brno, Czech Republic

Kerns, E. H. & Di, L. (2008). *Drug-like Properties: Concepts, Structure, Design and Methods: From ADME to Toxicity Optimization*. Elsevier/Academic Press, ISBN 978-0-1236-9520-8, San Diego, CA, USA.

Kos, J. (2010). *Study of Drug Polymorphism II*, Diploma Thesis, Faculty of Pharmacy, University of Veterinary and Pharmaceutical Sciences Brno

Kos, J.; Pěntáková, M.; Oktábec, Z.; Krejčík, L.; Mandelová, Z.; Haroková, P.; Hrušková, J.; Pekárek, T.; Dammer, O.; Tkadlecová, M.; Havlíček, J.; Král, V.; Vinšová, J.; Dohnal, J. & Jampílek, J. (2011). Crystallization products of risedronate with carbohydrates and their substituted derivatives. *Molecules*, Vol.16, No.5, pp.3740-3760, ISSN 1420-3049

Král, V.; Jampílek, J.; Havlíček, J., Brusová, H. & Pekárek, T. (Zentiva, k.s.), (2010). *Dosage forms of tyrosine kinase inhibitors*. WO 2010081443

Kratochvíl, B. (2007). Crystallization of pharmaceutical substances. *Chemické Listy*, Vol.101, No.1, pp. 3–12, ISSN 0009-2770

Kratochvíl, B.; Hušák, M.; Brynda, J. & Sedláček, J. (2008). What can the current x-ray structure analysis offer? *Chemické Listy*, Vol.102, No.10, pp. 889–901, ISSN 0009-2770

Kratochvíl, B. (2009). Pharmaceutical cocrystals. *Proceedings of the XXXVIII Conference Drug Synthesis and Analysis*, ISBN 978-80-7305-078-8, Hradec Králové, Czech Republic, September, 2009

Kratochvíl, B. (2010). Cocrystals and their expected pharmaceutical applications. *Chemické Listy*, Vol.104, No.9, pp.823-830, ISSN 0009-2770

Krumbholcová, L. & Dohnal, J. (2010). Thermal analysis, In: *Modern Approaches to Pharmaceutical Analysis*, Dohnal, J.; Jampílek, J.; Král, V. & Řezáčová, A. (Eds.). pp.347-381, Faculty of Pharmacy, University of Veterinary and Pharmaceutical Sciences Brno, ISBN 978-80-7305-085-6, Prague, Czech Republic

Lifshitz-Liron, R.; Bayer, T.; Aronhime, J. (TEVA Pharmaceutical), (2006). *Solid and crystalline ibandronate sodium and processes for preparation thereof*. WO 2006024024

McMahon, J.; Peterson, M.; Zaworotko, M. J.; Shattock, T. & Magali, B. H. (Trans Form Pharmaceuticals, Inc.), (2007). *Pharmaceutical co-crystal compositions and related methods of use*. US 20070299033

McNamara, D. P.; Childs, S. L.; Giordano, J.; Iarriccio, A.; Cassidy, J.; Shet, M. S.; Mannion, R.; O'Donnell, E & Park, A. (2006). Use of a glutaric acid cocrystal to improve oral bioavailability of a low solubility API. *Pharmaceutical Research*, Vol.23, No.8, pp.1888-1897, ISSN 0724-8741

Miroshnyk, I.; Mirza, S. & Sandler, N. (2009). Pharmaceutical co-crystals – An opportunity for drug product enhancement. *Expert Opinion Drug Delivery*, Vol.6, No.4, pp.333-341, ISSN 1742-5247

Mitscherlich, E. (1822). Über das Verhältnis der Krystalform zu den chemischen Proportionen. *Annales de Chimie et de Physique*, Vol.19, pp.350–419, ISSN 0003-4169

Morissette, S.L.; Almarsson, Ö.; Peterson, M. L.; Remenar, J. F.; Read, M. J.; Lemmo, A. V.; Ellis, S.; Cima, M. J. & Gardner, C. R. (2004). High-throughput crystallization: polymorphs, salts, co-crystals and solvates of pharmaceutical solids. *Advanced Drug Delivery Reviews*, Vol.56, No.3, pp.275-300, ISSN 0169-409X

Muddasani, P. R.; Vattikuti, U. & Nannapaneni, V. C. (NATCO Pharma, Ltd.), (2007). *Novel polymorphic forms of ibandronate*. WO 2007074475

Oktábec, Z.; Kos, J.; Mandelová, Z.; Havelková, L.; Pekárek, T.; Řezáčová, A.; Plaček, L.; Tkadlecová, M.; Havlíček, J.; Dohnal, J. & Jampílek, J. (2010). Preparation and properties of new co-crystals of ibandronate with gluco- or galactopyranoside derivatives. *Molecules*, Vol.15, No.12, pp.8973-8987, ISSN 1420-3049

Oktábec, Z. (2012). *Modification of Physico-Chemical Properties of Biologically Active Compounds*, Ph.D. Thesis, Faculty of Pharmacy in Hradec Králové, Charles University in Prague

Payghan, S. A.; Bhat, M.; Savla, A.; Toppo, E. & Purohit, S. (2008). Solubility of active pharmaceutical ingredients (API) has always been a concern for formulators, since inadequate aqueous solubility may hamper development of parenteral products and limit bioavailability of oral products. [cited 2012 February 28], Retrieved from <http://www.pharmainfo.net/reviews/potential-solubility-drug-discovery-and-development>

Pekárek, T. & Jampílek, J. (2010). Infrared and Raman spectroscopy, In: *Modern Approaches to Pharmaceutical Analysis*, Dohnal, J.; Jampílek, J.; Král, V. & Řezáčová, A. (Eds.). pp.222-260, Faculty of Pharmacy, University of Veterinary and Pharmaceutical Sciences Brno, ISBN 978-80-7305-085-6, Prague, Czech Republic

Peterson, M. L.; Hickey, M. B.; Zaworotko, M. J. & Almarsson Ö. (2006). Expanding the scope of crystal form evaluation in pharmaceutical science. *Journal of Pharmacy & Pharmaceutical Sciences*, Vol.9, No.3, pp.317-326, ISSN 1482-1826

Qiao, N.; Li, M. Z.; Schlindwein, W.; Malek, N.; Davies, A. & Trappitt, G. (2011). Pharmaceutical cocrystals: An overview. *International Journal of Pharmaceutics*, Vol.419, No.1-2, pp.1-11, ISSN 0378-5173

Ranganathan, A. (1999). Hydrothermal synthesis of organic channel structures: 1:1 Hydrogen-bonded adducts of melanine with cyanuric amd trithiocyanuric acids. *Journal of American Chemical Society*, Vol.121, No.8, pp.1752–1753, ISSN 0002-7863

Redman-Furey, N. L.; Ficka, M.; Bigalow-Kern, A.; Cambron, R. T.; Lubey, G.; Lester, C. & Vaughn, D. (2005). Structural and analytical characterization of three hydrates and an anhydrate form. *Journal of Pharmaceutical Sciences*, Vol.94, No.4, pp.893–911, ISSN 1520-6017

Remenar, J. F.; Morissette, S. L.; Peterson, M. L.; Moulton, B.; MacPhee, J. M.; Guzmán, H. & Almarsson, Ö. (2003). Crystal engineering of novel co-crystals of a triazole drug with

1,4-dicarboxylic acids. *Journal of American Chemical Society*, Vol.125, No.28, pp.8456-8457, ISSN 0002-7863

Reszka, A. A.; Halasy-Nagy, J. M.; Masarachia, P. J. & Rodan, G. A. (1999). Bisphosphonates act directly on the osteoclast to induce caspase cleavage of Mst1 kinase during apoptosis. A link between inhibition of the mevalonate pathway and regulation of an apoptosis-promoting kinase. *Journal of Biological Chemistry*, Vol.274, No.49, pp.34967–34973, ISSN 0021-9258

Richter, J. & Jirman, J. (Zentiva, a.s.), (2007). *Crystalline form of the sodium salt of 3-pyridyl-1-hydroxyethylidene-1,1-bisphosphonic acid*. US 7276604

Rodríguez-Hornedo, N.; Nehm, S. J. & Jayasankar, A. (2007). Cocrystals: Design, properties and formation mechanisms, In: *Encyclopedia of Pharmaceutical Technology*, 3rd ed.; Swarbrick, J. (Ed.), pp 615–635, Taylor & Francis, ISBN 978-1841848198, London, UK

Rodríguez-Spong, B.; Price, C. P.; Jayasankar, A.; Matzger, A. J. & Rodríguez-Hornedo, N. (2004). General principles of pharmaceutical solid polymorphism: A supramolecular perspective. *Advanced Drug Delivery Reviews*, Vol.56, No.3, pp.241-274, ISSN 0169-409X

Rogers, M. J.; Xiong, X.; Brown, R. J.; Watts, D. J.; Russell, R. G.; Bayless, A. V. & Ebetino, F. H. (1995). Structure-activity relationships of new heterocycle-containing bisphosphonates as inhibitors of bone resorption and as inhibitors of growth of *Dictyostelium discoideum* amoebae. *Molecular Pharmacology*, Vol.47, No.2, pp.398–402, ISSN 0026-895X

Rogers, M. J.; Gordon, S.; Benford, H. L.; Coxon, F. P.; Luckman, S. P.; Monkkonen, J. & Frith, J. C. (2000). Cellular and molecular mechanisms of action of bisphosphonates. *Cancer*, Vol.88, No.12, pp.2961–2978, ISSN 1097-0142

Ruecroft, G.; Hipkiss, D.; Ly, T.; Maxted, N. & Cains, P. W. (2005). Sonocrystallization: The use of ultrasound for improved industrial crystallization. *Organic Process Research & Development*, Vol.9, No.6, pp.923–932, ISSN 1083-6160

Sarma, B.; Reddy, L. S. & Nangia, A. (2008). The role of π-stacking in the composition of phloroglucinol and phenazine cocrystals. *Crystal Growth & Design*, Vol.8, No.12, pp.4546-4552, ISSN 1528-7483

Sarma, B.; Chen, J.; Hsi H.-Y. & Myerson, A. S. Solid forms of pharmaceuticals: Polymophs, salts and cocrystals. (2011). *Korean Journal of Chemical Engineering*, Vol.28, No.2, pp.315-322, ISSN 0256-1115

Sato, M.; Grasser, W.; Endo, N.; Akins, R.; Simmons, H.; Thompson, D. D.; Golub, E. & Rodan, G. A. (1991). Bisphosphonate action. Alendronate localization in rat bone and effects on osteoclast ultrastructure. *Journal of Clinical Investigation*, Vol.88, No.6, pp.2095–2105, ISSN 0021-9738

Schultheiss, N. & Newman, A. (2009). Pharmaceutical co-crystals and their physicochemical properties. *Crystal Growth & Design*, Vol.9, No.6, pp.2950–2967, ISSN 1528-7483

Seddon, K. R. & Zaworotko, M. J. (Eds.). (1999). *Crystal Engineering: The Design and Application of Functional Solids*, NATO-ASI Series, Vol. 539, Kluwer Academic Publisher, ISBN 978-0-7923-5905-4, Dordrecht, the Netherlands

Sekhon, B. S. Pharmaceutical co-crystals - A review. (2009). *Ars Pharmaceutica*, Vol.50, No.3, pp.99-117, ISSN 0004-2927

Selander, K. S.; Monkkonen, J.; Karhukorpi, E. K.; Harkonen, P.; Hannuniemi, R. & Vaananen, H. K. (1996). Characteristics of clodronate-induced apoptosis in osteoclasts and macrophages. *Molecular Pharmacology*, Vol.50, No.5, pp.1127–1138, ISSN 0026-895X

Shan, N. & Zaworotko, M. J. (2008). The role of cocrystals in pharmaceutical science. *Drug Discovery Today*, Vol.13, No.9-10, pp.440-446, ISSN 1359-6446

Shekunov, B. Y. & York, P. (2000). Crystallization processes in pharmaceutical technology and drug delivery design. *Journal of Crystal Growth*, Vol.211, No.1-4, pp.122-136.

Stahl, P. H. & Wermuth, C. G. (Eds.). (2008). *Handbook of Pharmaceutical Salts: Properties, Selection and Use*, Verlag Helvetica Chimica Acta, ISBN 978-3-906390-58-1, Zürich, Switzerland

Stahly, G. P. (2007). Diversity in single-and multiple-component crystals. The search for and prevalence of polymorphs and co-crystals. *Crystal Growth & Design*, Vol.7, No.6, pp.1007-1026, ISSN 1528-7483

Strachan, C. J.; Rades, T.; Newnham, D. A.; Gordon, K. C.; Proper, M. & Taday P. F. (2004). Using terahertz pulsed spectroscopy to study crystallinity of pharmaceutical materials. *Chemical Physics Letters*, Vol.390, No.1-3, pp. 20–24, ISSN 0009-2614

Sun, C. C. & Hou, H. (2008). Improving mechanical properties of caffeine and methyl gallate crystals by co-crystallization. *Crystal Growth & Design*, Vol.8, No.5, pp.1575-1570, ISSN 1528-7483

Takata, N.; Shiraki, K.; Takano, R.; Hayashi, Y. & Terada, Y. (2008). Co-crystal screening of stanolone and mestanolone using slurry crystallization. *Crystal Growth & Design*, Vol.8, No.9, pp.3032-3037, ISSN 1528-7483

Ťažká, A. (2011). *Study of Polymorphism of Biologically Active Compounds II*, Diploma Thesis, Faculty of Pharmacy, University of Veterinary and Pharmaceutical Sciences Brno

Tong, H. H.; Shekunov, B.Y.; York, P. & Chow, A. H. (2001). Characterization of two polymorphs of salmeterol xinafoate crystallized from supercritical fluids. *Pharmaceutical Research*, Vol.18, No.6, pp.852-858, ISSN 0724-8741

Trask, A. V. & Jones, W. (2005). Crystal engineering of organic co-crystals by the solid-state grinding approach. *Topics in Current Chemistry*, Vol.254, pp.41-70, ISSN 0340-1022

Trask, A. V.; van de Streek, J.; Motherwell, W. D. S. & Jones, W. (2005a). Achieving polymorphic and stoichiometric diversity in co-crystal formation: Importance of solid-state grinding, powder X-ray structure determination and seeding. *Crystal Growth & Design*, Vol.5, No.6, pp.2233-2241, ISSN 1528-7483

Trask, A. V.; Motherwell, W. D. S. & Jones, W. Pharmaceutical co-crystallization: Engineering a remedy for caffeine hydration. (2005b). *Crystal Growth & Design*, Vol.5, No.3, pp.1013-1021, ISSN 1528-7483

Trask, A. V.; Haynes, D. A.; Motherwell, W. D. S. & Jones, W. (2006). Screening for crystalline salts via mechanochemistry. *Chemical Communications*, Vol.2006, No.1, pp.51-53, ISSN 1364-548X

Trask, A. V. (2007). An overview of pharmaceutical co-crystals as intellectual property. *Molecular Pharmaceutics* Vol.4, No.3, pp.301-309, ISSN 1543-8384 van Beek, E. R.; Lowik, C. W.; Ebetino, F. H. & Papapoulos, S. E. (1998). Binding and antiresorptive properties of heterocycle-containing bisphosphonate analogs: Structure-activity relationships. *Bone*, Vol.23, No.5, pp.437–442, ISSN 8756-3282

Velaga, S. P.; Basavoju, S. & Boström D. (2008). Norfloxacin saccharinate–saccharin dihydrate cocrystal – A new pharmaceutical co-crystal with an organic counter ion. *Journal of Molecular Structure*, Vol.889, No.1-3, pp.150-153, ISSN 0022-2860

Viertelhaus, M.; Hilfiker, R. & Blatter, F. (2009). Piracetam co-crystals with OH-group functionalized carboxylic acids. *Crystal Growth & Design*, Vol.9, No.7, pp.2220–2228, ISSN 1528-7483

Vishweshwar, P.; Nangia, A. & Lynch, V. M. (2003). Molecular complexes of homologous alkanedicarboxylic acids with isonicotinamide: X-Ray crystal structures, hydrogen bond synthons, and melting point alternation. *Crystal Growth & Design*, Vol.3, No.5, pp.783-790, ISSN 1528-7483

Vishweshwar, P.; McMahon, J. A.; Peterson, M. L.; Hickey, M. B.; Shattock, T. R. & Zaworotko, M. J. (2005). Crystal engineering of pharmaceutical co-crystals from polymorphic active pharmaceutical ingredients. *Chemical Communications*, Vol.2005, No.36, pp.4601-4603, ISSN 1364-548X

Vishweshwar, P.; McMahon, J. A., Bis, J. A. & Zaworotko, M. J. (2006a). Pharmaceutical co-crystals. *Journal of Pharmaceutical Sciences*, Vol.95, No.3 , pp.449-516, ISSN 1520-6017

Vishweshwar, P.; McMahon, J. A. & Zaworotko, M. J. (2006b). Crystal engineering of pharmaceutical co-crystals, In: *Frontiers in Crystal Engineering*, Tiekink, E. R. T. & Vittal, J. J. (Eds.)., pp.25-50, John Wiley & Sons, ISBN 978-0-470-02258-0, Chichester, UK

Vogt, F. G; Clawson, J. S.; Strohmeier, M.; Edwards, A. J.; Pham, T. N. & Watson, S. A. (2009). Solid-state NMR analysis of organic cocrystals and complexes. *Crystal Growth & Design*, Vol.9, No.2, pp. 921–937, ISSN 1528-7483

Walsh, R. D. B.; Bradner, M. W.; Fleischman, S.; Morales, L. A.; Moulton, B.; Rodríguez-Hornedo, N. & Zaworotko, M. J. (2003). Crystal engineering of the composition of pharmaceutical phases. *Chemical Communications*, Vol.2003, No.2, pp.186-187, ISSN 1364-548X

Walther, M.; Fischer, B. M. & Jepsen, P. U. (2003). Noncovalent intermolecular forces in polycrystalline and amorphous saccharides in the far infrared. *Chemical Physics*, Vol.288, No.2-3, pp. 261–288, ISSN 0301-0104

Whitesides, G. M. & Wong, A. P. (2006). The intersection of biology and materials science. *Materials Research Society Bulletin*, Vol. 31, No.1, pp.19-27, ISSN 0883-7694

Zakrzewski, A. & Zakrzewski, M. (Eds.) (2006). *Solid State Characterization of Pharmaceuticals*. Assa, ISBN 83-920584-5-3, Danbury, CT, USA

Zaworotko, M. (2008). Crystal engineering of co-crystals and their relevance to pharmaceuticals and solid-state chemistry. *Acta Crystallographica Section A*, Vol.A64, No.Supplement, pp.C11-C12, ISSN 0108-7673

Žegarać, M.; Meštrović, E.; Dumbović, A.; Devčić, M.; Tudja, P. (Pliva Hrvatska D.o.o.), (2007). *Pharmaceutically acceptable co-crystalline forms of sildenafil*. WO 2007080362

Chapter 5

Liquid Chromatography – Mass Spectrometry of Carbohydrates Derivatized with Biotinamidocaproyl Hydrazide

Stephanie Bank and Petra Kapková

Additional information is available at the end of the chapter

1. Introduction

Glycans are involved in a number of cell processes. The glycocomponents of a protein encompass structural and modulatory functions, trigger receptor/ligand binding and many other processes which include mechanisms of interest to immunology, hematology, neurobiology and others (Marino *et al.*, 2010; Raman *et al.*, 2005). Because they are associated with both the physiological and the pathological processes, there is constant interest paid to analysis of protein glycosylation and its changes (North *et al.*, 2009; Leymarie & Zaia, 2012).

In order to get a full picture about the structure and function of a carbohydrate, combination of different analytical methods, detection and derivatization techniques is usually applied (Bindila & Peter-Katalinic, 2009; Geyer & Geyer, 2006; Harvey, 2011). Derivatization of sugars for the purpose of their structural or functional analysis bears various advantages (Rosenfeld, 2011). Most of the labels used for the derivatization of carbohydrates possess a chromophor or a flurophor, enabling a sensitive detection of these analytes by means of spectroscopic methods. Apart from an increase in detection sensitivity, chromatographic properties of derivatized sugars may improve in comparison with their underivatized counterparts (Melmer *et al.*, 2011; Pabst *et al.*, 2009). Furthermore, mass spectrometric signal intensity showed to profit from the derivatization step, as an effective ionization was achieved through derivatization. Moreover, depending on the type of the carbohydrate considered, mass spectrometric fragmentation showed to be influenced by the kind of label introduced, and provided additional information about the structure of sugars (Lattová *et al.*, 2005).

There are many possible labelling reactions for the generation of respective derivatives. The most applied derivatization method is the reductive amination via Schiff-base in the presence of the sodium cyanoborohydride as reductant. The greatest success and a broad area of usage over the last decades experienced the arylamine tag such as 2-aminobenzamide (2-AB) and is still extensively used in mass spectrometric studies of glycans (Bigge *et al.*, 1995). Other common tags from this group are 2-aminobenzoic acid (2-AA), 2-aminopyridine (2-AP) and 2-aminoacridone (2-AMAC). Another labelling approach which has been widely used is the formation of hydrazones, where the hydrazine label of the tag reacts with the aldehyde group of the sugar. The mentioned approaches and other derivatizations which have shown their use in analytics of glycans were reviewed recently (Harvey, 2011; Ruhaak *et al.*, 2010).

Generally, selection criteria for a certain derivatization approach depend on many different factors, including the nature of the sample, the compatibility of the label with the analytical protocol being applied and very importantly, the objective of the analysis. Here, very often, not only the structure of a glycan is a matter of a question but also its functionality, which is an essential feature, that often wants to be examined. Consequently, a derivatization method, which would allow for both the structural and functional studies of sugars can be of interest in the characterization of the glycan species. Such a possibility is given e.g. by biotinylated reagents (Grün *et al.*, 2006; Hsu *et al.*, 2006; Leteux *et al.*, 1998; Ridley *et al.*, 1997; Rothenberg *et al.*, 1993; Shinohara *et al.*, 1996; Toomre & Varki, 1994).

In our recent study, biotinamidocaproyl hydrazide (BACH) was presented as a label for carbohydrates analyzed by mass spectrometry (Kapková, 2009). This bifunctional label combines two features: the sugar reactivity and the bioaffinity. In this way, beside the structural characterization of carbohydrates, it enables also for performing interaction studies with carbohydrate-binding proteins. The derivatization was performed under non-reducing conditions via hydrazone formation and showed an increase in the ion abundance of small sugars and *N*-linked glycans.

Here, chip-based electrospray tandem mass spectrometric structural analysis of different glycan-derivatives generated via BACH-labeling using triple quadrupole instrument was performed. The methodology has been applied to glycans released from glycoproteins (ovalbumin, ribonuclease B) including different types of *N*-glycans e.g. high-mannose, complex and hybride glycans. Further, an analysis of small BACH-labeled carbohydrates was conducted as these represent important building unit of the large glycan entities. Reversed liquid chromatography and hydrophilic interaction chromatography (HILIC) coupled with ion-trap mass spectrometry (LC-ESI-IT-MS) of BACH-derivatized monosaccharides were carried out. Advantageous properties of HILIC in terms of separation of non-reducing trisaccharides and BACH-derivatized isomeric disaccharides are presented. Here, derivatization helped to obtain a favorable separation profile or a higher signal intensity. The isomers were compared in their fragmentation pattern, and a distinction of their linkage or anomeric configuration was sought.

2. Experimental

2.1. Materials

Hen ovalbumin, ribonuclease B from bovine pancreas, melezitose, raffinose and biotinamidocaproyl hydrazide were purchased from Sigma-Aldrich (Taufkirchen,

Germany). The BACH-label was used without further purification. Monosaccharides (glucose, galactose, mannose, fucose, xylose, *N*-acetylglucosamine) and disaccharides (maltose, cellobiose) were kindly provided by the Division of Chemicals (Department of Pharmacy, University Würzburg). The standard *N*-glycans MAN5, MAN6, NA2, NGA2 as well as PNGase F and the GlycoClean™ S Cartridges were from Europa Bioproducts Ltd. (Cambridge, UK). The water was from a Milli-Q system.

2.2. Enzymatic digestion of glycoproteins

About 500μg of the protein were dissolved in water (40μl) and reaction buffer (10μl) and denatured by boiling in a water-bath (100°C, 10min). After cooling, the samples were incubated with PNGase F (2μl) at 37°C for 18 hours. By adding 200μl of cold ethanol and keeping in ice for 2h, the deglycosylated protein was precipitated and centrifuged down to a pellet. The supernatant, containing the glycans, was dried in vacuum and then the residue was derivatized.

2.3. Derivatization with BACH

The released glycans from glycoproteins were mixed with 100μl of a 5mmol solution of BACH in 30% acetonitrile and subsequently evaporated. For the derivatization of standard *N*-glycans (0.5μg) the same protocol was followed. According to the evaporation, the mixture was dissolved in 30μl of methanol/water (95/5), and incubated at 90°C for 1h. Then the BACH conjugates were further purified using GlycoClean™ S cartridges according to the accompanying protocol.

2.4. Derivatization with 2-Aminobenzamid (2-AB)

50μl of the AB-derivatization solution, containing 2-AB (5mg), $NaBH_3CN$ (7.5mg), DMSO (500μl) and acetic acid (200μl), were added to the glycans or small sugars. After the incubation at 65°C for 2.5h, the AB-conjugates were purified with GlycoClean™ S cartridges according to the manufacturer's protocol.

2.5. LC/ESI-MS

2.5.1. Nano LC/ESI-MS of N-glycans

The mass spectra were measured on the triple quadrupole mass spectrometer 6460 (Agilent Technologies, Waldbronn, Germany) coupled with an Agilent 1200 liquid chromatography nanoflow system with a ChipCube interface. The HPLC-Chip used for the separation contained a 0.075 × 43mm porous graphitized carbon column and an integrated 40nl enrichment column. This column was filled with identical media, e.g. the same phase was used for both the separation and trapping. Solvents for the capillary and the nanoflow pump were A: 0.1% formic acid in water and B: 0.1% formic acid in methanol. The chromatographic conditions were as follows: capillary pump flow rate (1μl/min), nanoflow pump (0.4μl/min for standard glycans and 0.5μl/min for released glycans); 4μl of the samples were loaded. Gradient was set from 30% to 90%B in 50min. MS conditions were: positive ionization, drying gas flow 4l/min at a drying temperature of 325°C, capillary voltage 1800V, and a scan range from 300 to 2200 m/z.

The MS/MS experiments were performed based on the collision-induced dissociations (CID). In order to selectively monitor the glycans in the triple quadrupole instrument, precursor ion scanning was performed. Characteristic fragment ions used in this mode were the oxonium ions of hexose (HexNAc) at m/z 163 and of *N*-acetylhexosamine (HexNAc) at m/z 204 and the larger oxonium ion at m/z 366 (HexHexNAc). In this approach, quadrupole 1 works as a scanner across the range of interest in order to determine the mass of injected glycans, the second quadrupole acts as a collision cell and the third quadrupole serves to analyze the fragment ions generated in the collision cell. Upon the detection of precursor ions which are losing a specific fragment ion of interest (oxonium ion) the precursor ions of glycans were analyzed in the product ion scan mode which gave the fragment spectrum of the glycan.

2.5.2. LC/ESI-MS of mono-, di- and trisaccharides

Analyses of small sugars were performed on the LC-MSD ion-trap instrument (Agilent) operated with Agilent HPLC 1100 or infusion syringe pump. The separation was conducted on the C18 reversed-phase and HILIC-phase (Kinetex™, Phenomenex, 100 × 2.1 mm, 100Å, 2.6 μm). The chromatographic conditions of C18 were: A: 100mM ammonium acetate (pH 5.8), B: MeOH (0.1% formic acid), flow: 0.3 mL/min, gradient: 0 min, 11% B; 17 min, 11% B; 30 min, 50% B. The chromatographic conditions of HILIC: A: 100mM ammonium acetate (pH 5.8), B: acetonitril (0.1% formic acid), flow: 0.3 mL/min, gradient: 0 min, 99% B; 12 min, 99% B, 16min, 92% B; 30min, 60% B.

3. Results and discussion

3.1. Derivatization with BACH

Labeling reagent biotinamidocaproyl hydrazide (BACH) is a bifunctional label which possesses a hydrazide group for coupling with the reducing end of sugars and a biotin group in order to interact with a solid support which possesses affinity to biotin. This chemical process may be run under reducing or non-reducing conditions depending on the purpose of the analysis. Because the reaction with hydrazines gives much more stable products than the easily reversible Schiff-base generated by the reaction of an amine with an aldehyde, carbohydrate-hydrazone bond formed *via* hydrazide chemistry does not necessarily require a reducing agent (Hermannson, 2008).

In this way, the resultant product is a glycosylhydrazide (Fig. 1), which preserves the pyranose ring closed, i.e. it preserves its near-native conformation. This is a favorable feature in many functional studies where the closed pyranose ring is essential, in order to be recognized by a carbohydrate-binding protein. Differences in binding activity of certain sugars have been shown (Grün *et al.*, 2006), when reacting with lectins or other carbohydrate binding proteins. Hence, whether a biological interaction is affected by the reduction or not, depends on the type of the sugar and on the kind of the carbohydrate binding protein.

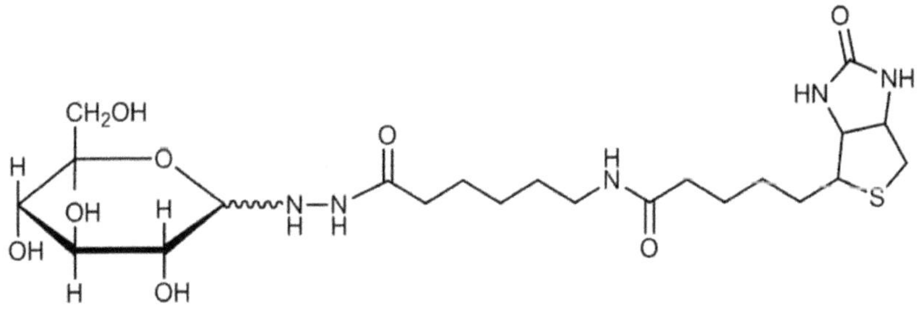

Figure 1. Scheme of glycosylhydrazide formation of carbohydrate with biotinamidocaproyl hydrazide

3.2. Structural characterization of BACH-derivatized *N*-linked glycans

In general, the structures of *N*-linked glycans fall within three main types, namely, high-mannose, complex and hybrid. These all share a common chitobiose core and differ in branching patterns mainly at the non-reducing end.

Here, different standard asparagine-linked glycans and glycans released from model glycoproteins such as ovalbumin and ribonuclease B (RNAse B) were analyzed by means of nano-LC-chip/electrospray mass spectrometry in order to cover the main types of *N*-linked glycans. The *N*-linked glycans were released by treating the glycoproteins with peptide-*N*-deglycanase F (PNGaseF). Released glycans were labeled with the biotinamidocaproyl hydrazide and analyzed by nanoLC- triple quadrupole mass spectrometry using the precursor ion scan mode. The oxonium ions chosen for the scanning mode were m/z 163 (Hex+H), m/z 325 (Hex-Hex+H) and m/z 204.1 (GlcNAc+H). Fragmentation was carried out at three different collision energies (15V, 25V and 50V). The best fragmentation spectra resulted from the application of the collision energy of 15V and this was used throughout. Tandem mass spectra of the respective glycans were recorded in the product ion scan mode by the collision induced dissociation.

RNase B has one glycosylation site occupied by five high-mannose glycans differing in one mannose residue. Because of the high-mannose nature of these glycans, precursor ion scanning with the oxonium ion of hexose (m/z 163) was performed and the resultant mass spectra showed high abundance.

Fig. 2 shows the mass spectrum of the BACH-derivatized high-mannose glycans present in RNaseB and features the characteristic spacing pattern of its glycoforms [$(Man)_nGlcNAc)_2$ with n = 5,6,7,8,9]. The signals were registered as doubly charged hydrogenated species. Hence, the spectrum exhibits the differences of 81 Da, which account for half of the mass of one mannose unit. The most abundant glycoforms were MAN6 and MAN5. This proportionality in pattern was observed also in studies on high-mannose glycopeptides (Alley *et al.*, 2009).

In Fig. 3 the tandem mass spectrum of the doubly charged BACH-labeled MAN6 glycan is shown, which is dominated by the glycosidic cleavages, which correspond with losses of mannose residues. The prominent signals were the doubly charged signals related to consecutive losses of 81Da directly from the molecular ion $[M+2H]^{2+}$. A series of weaker singly charged ions corresponded to losses of mannose residues (162Da) from the chitobiose cleavage between the two GlcNAc-moieties (B_4-ion). In this way, the collision-induced dissociation of BACH-derivatized ribonuclease B glycans provided fast and informative spectra in order to characterize the identity of the analyzed glycans. Fragments have been assigned using the systematic nomenclature for carbohydrate fragmentation proposed by Domon and Costello (Domon & Costello, 1988).

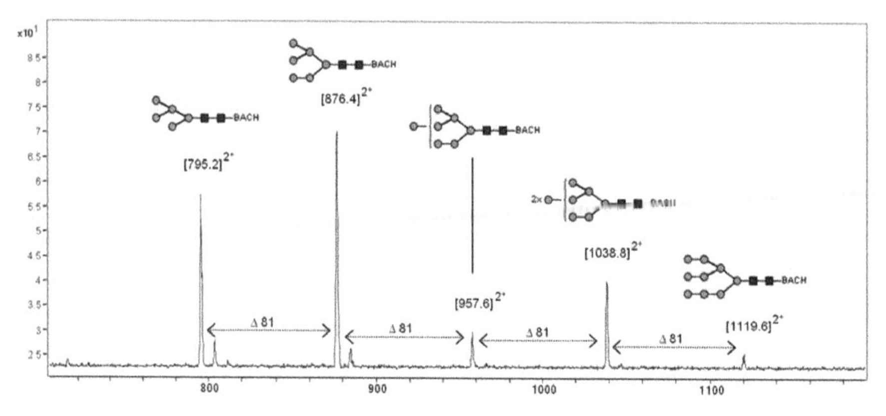

Figure 2. Precursor ion scan mass spectrum of BACH-biotinylated high-mannose glycans from ribonuclease B for oxonium ion of hexose at m/z 163. Doubly charged ions show the differences of 81Da.

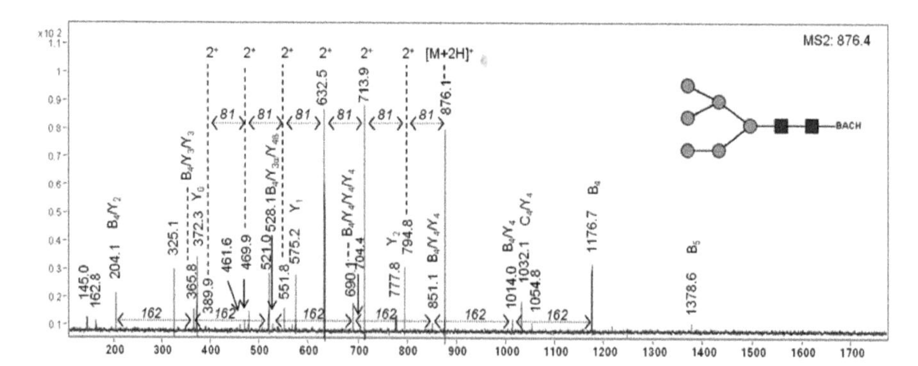

Figure 3. Product ion mass spectrum of the high-mannose glycan $(GlcNAc)_2(Man)_6$ prepared by BACH-derivatization. Spectrum was derived by fragmenting the $[M+2H]^{2+}$ glycan precursor ion of the m/z 876.4.

Ovalbumin, a major glycoprotein in hen egg white, contains one glycosylation site with a series of asparagine-linked oligosaccharides of mannose- and hybrid-type; some rare complex-type oligosaccharides are also known. Due to its structural diversity, this protein is an optimal source of a library of *N*-linked glycans and a good model glycoprotein. The mixture of glycans released from ovalbumin was analyzed by precursor ion scanning in the triple quadrupole. The oxonium ion which has been scanned for was the protonated *N*-acetylglucosamine (GlcNAc) at m/z 204. The mass spectrum of the identified glycans can be seen in the Fig. 4. Found oligosaccharides cover the glycan forms typically present in this glycoprotein (Fig. 4, Table 1). Under the conditions described in the experimental part, the glycan ions appeared as doubly or triply charged ions. Doubly charged ions were fragmented by product ion scanning and their structure was elucidated. Elucidation of the fragment spectra of triply charged ions was a cumbersome task, as the spectra contained ions with different charge states. The fragment ion with m/z 204, corresponding to the oxonium ion $[HexNAc+H]^+$, was frequently observed in the MS/MS spectra of ovalbumin glycans. Generally, this type of ion is highly diagnostic of the presence of glycans that terminate in one or more *N*-acetylglucosamine residues.

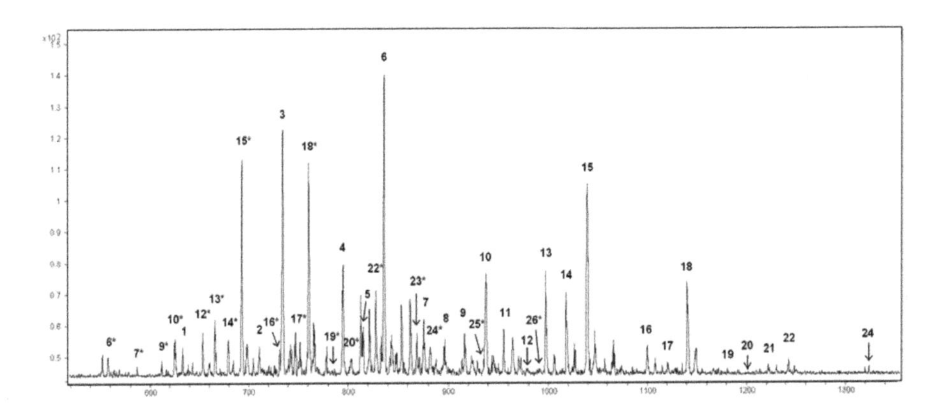

Figure 4. Precursor ion scan for oxonium ion of *N*-acetylglucosamine at m/z 204 of BACH-biotinylated glycans from ovalbumin. Doubly and triply (*) charged ions are depicted. For the structure see the Table 1.

Compd.[1]	Comp.[2]	Observed *m/z* BACH-glycans, $[M+2H]^{2+}$/$[M+3H]^{3+}$*	Structure
1	H_3N_2	632.9	
2	H_4N_2	714.0	
3	H_3N_3	734.5	
4	H_5N_2	795.4	
5	H_4N_3	815.4	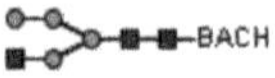
6	H_3N_4	836.4/557.4*	
7	H_6N_2	876.1/584.5*	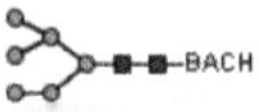
8	H_5N_3	896.7	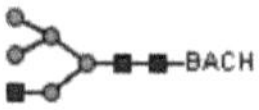
9	H_4N_4	916.5/611.9*	
10	H_3N_5	937.8/625.6*	
11	H_7N_2	956.6	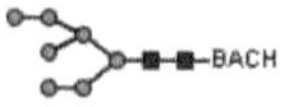
12	H_6N_3	978.0/652.3*	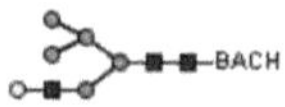
13	H_5N_4	997.6/665.4*	
14	H_4N_5	1018.0/679.8*	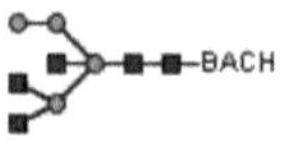
15	H_3N_6	1039.1/692.8*	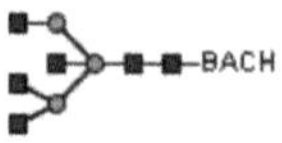
16	H_5N_5	1099.8/733.6*	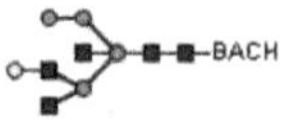

Compd.[1]	Comp.[2]	Observed m/z BACH-glycans, $[M+2H]^{2+}/[M+3H]^{3+*}$	Structure
17	H_4N_6	1120.0/747.3*	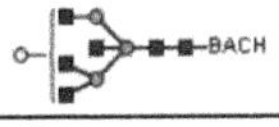 BACH
18	H_3N_7	1140.0/760.5*	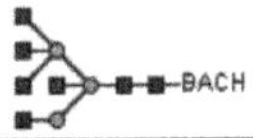 BACH
19	H_6N_5	1181.0/786.8*	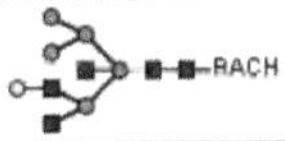 BACH
20	H_5N_6	1201.9/801.8*	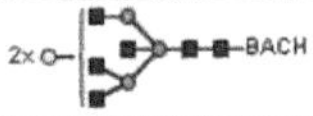 2x BACH
21	H_4N_7	1221.9	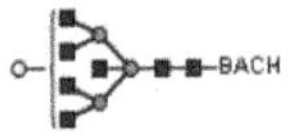 BACH
22	H_3N_8	1242.5/828.2*	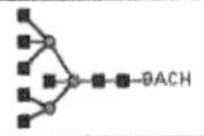 BACH
23	H_5N_7	869.2*	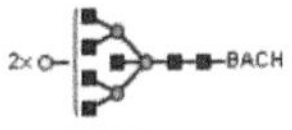 2x BACH
24	H_4N_8	1323.5/882.6*	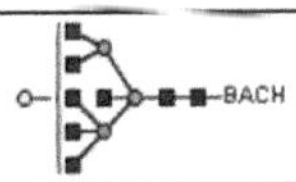 BACH
25	H_5N_8	936.8*	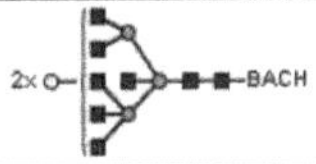 2x BACH
26	H_6N_8	990.9*	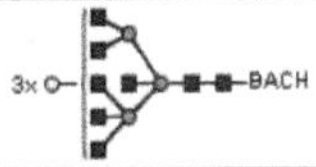 3x BACH

Table 1. Composition and m/z of the doubly and triply charged (*) glycans released by PNGase from the hen ovalbumin observed by precursor ion scanning experiment (m/z 204). For the spectrum see the Fig. 4. [1]Compound number, [2]Composition of *N*-glycans (H = hexose, N = GlcNAc)

Several diagnostic ions were present in the ESI-MS/MS spectra of complex and hybride glycans from ovalbumin. Fig. 5 shows the tandem mass spectrum of one of the complex glycans of ovalbumin. The MS/MS fragmentation of this BACH-derivatized glycan was characterized by sequential losses of sugar residues from its non-reducing end. Arrows with continuous lines indicate differences in the number of mannose, and the dashed lines in the number of *N*-acetylglucosamines.

The Fig. 6 represents the product ion spectrum of the galactosylated standard glycan NA2. Several diagnostic ions were present in the MS/MS spectrum. The loss of the protonated fragment with m/z 366 was the preferred route in the breaking up of this galactosylated species. The oxonium ion is diagnostic of the loss of a fragment containing a hexose residue attached to *N*-acetylhexosamine. The sequencial losses of 365 (i.a. Hex-HexNAc) from the precursor ion (m/z 1630 and m/z 1264) show that this glycan accommodates two galactosyl-*N*-acetylglucosamine residues. The appearance of the fragment at m/z 1264 distinguishes this molecule from the core-bisected species.

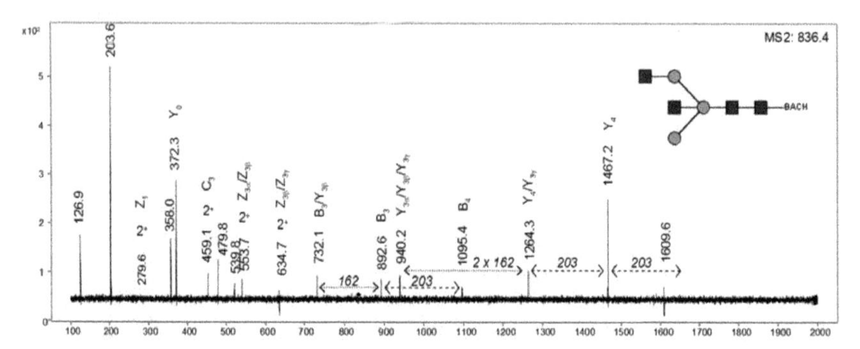

Figure 5. Product ion scan of BACH-derivatized ovalbumin glycan GlcNAc4Man3 at m/z 836.4.

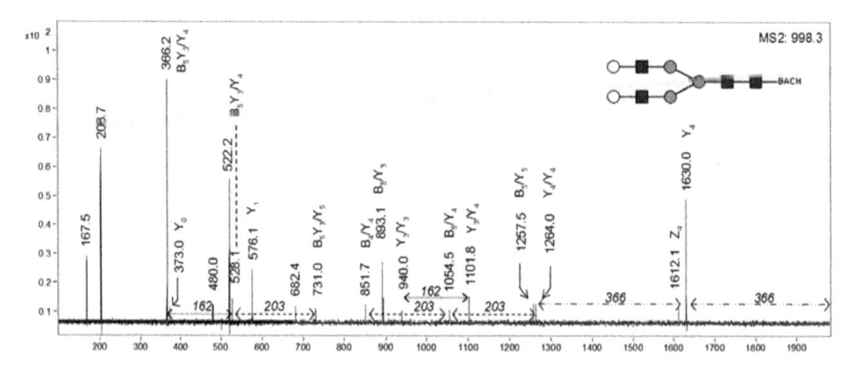

Figure 6. Product ion mass spectrum of the galactosylated NA2 glycan derivatized with BACH. The fragmentation was performed by collision-induced dissociation of the $[M+2H]^{2+}$-ion at the m/z 998.3

When compared with literature of frequently used 2-AB derivatization, tandem mass spectra of BACH-derivatives showed similar fragmentation as 2-AB-derivatives encompassing dominant glycosidic cleavages from reducing and non-reducing end (Harvey, 2000; Morelle & Michalski, 2004; Wuhrer *et al.*, 2004). These simple spectra make the MS/MS elucidation a straightforward task. Moreover, one should keep in mind that with BACH-derivatives not only the structure of the glycans could be explored, but also the function (Kapková, 2009). The bioaffinity of the BACH-label makes these derivatives suitable for the experiments exploring the interaction capabilities of a glycan (Wuhrer *et al.*, 2010).

Usually, the complete sugar analysis encompasses the characterization of the composition, sequence, branching and linkage of the monosaccharide residues constituting the oligosaccharides. The fragments arising from glycosidic cleavages are useful for the elucidation of the sugar composition, sequence and branching pattern of the glycan. Using this methodology alone, the type of linkage between the various sugar residues was not obtained. However, combining the product ion scanning with other methods e.g. enzymatic exoglycosidase digestion studies would provide some of this information.

3.3. LC-MS of BACH-derivatized monosaccharides

Mixture of carbohydrates (*N*-acetylglucosamine (GlcNAc), galactose (Gal), mannose (Man), Fucose (Fuc) and Xylose (Xyl)) which are typically present in mammalian proteins was analyzed by ESI-MS in the ion trap. Comparative examinations of pre-column labeled BACH-monosaccharides were performed by means of reversed phase liquid chromatography (RPLC) and hydrophilic interaction liquid chromatography (HILIC) as these methods shown to be highly complementary in resolution of mixtures of carbohydrates (Melmer *et al.*, 2011; Schlichtherle-Cerny *et al.*, 2003; Strege *et al.*, 2000). HILIC is based on the hydrophilic character of the analyte which accounts for the interaction of the molecule with the stationary phase (Wuhrer *et al.*, 2009). In our study, diol-modified silica phase was used. Chemically bonded diol groups were shown in the past to give similar functions as an amino-bonded silica. Moreover, they possess better stability than amino-sorbent and provide a good separation of structurally related polar molecules, e.g. sugars (Brons & Olieman, 1983; Churms, 1996; Ikegami *et al.*, 2008).

The BACH-derivatized monosaccharides, non-labeled and 2-AB-labeled counterparts (as 2-AB-labeling is one of the most widely used methods) were compared in terms of their chromatographic behavior and signal intensity of the mass spectrometric measurement.

Under the conditions used, underivatized sugars did not provide sufficient retention and separation on either of the phases (C18 and HILIC) and eluted almost at the same time (t_R = 1 min, C18; t_R = 2-5 min, HILIC) with a poor sensitivity (data not shown), especially for glucose, xylose and fucose (Intens. = 0.5×10^6). In order to seek an enhanced detection and better separation, derivatization of sugars was performed. After the derivatization with BACH or 2-AB, mass spectrometric signal intensity significantly enhanced about an order of a magnitude (BACH-derivatives Ints. = 0.5-4 $\times 10^7$, 2-AB-derivatives 0.5-3 $\times 10^7$). The extent of the intensity enhancement depended on the kind of the sugar analyzed (Fig. 7).

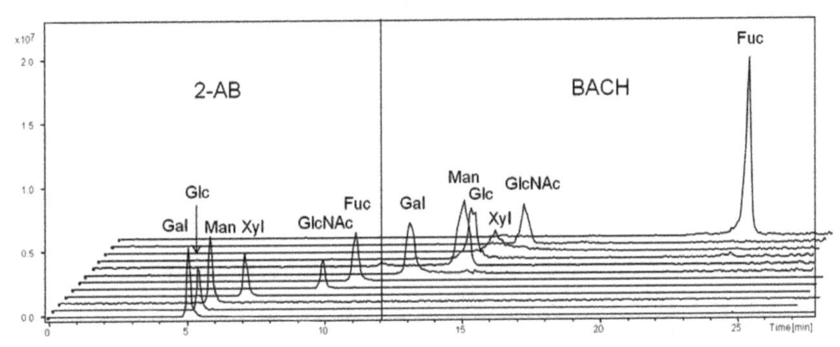

Figure 7. Total ion chromatogram (TIC) of RPLC-MS of 2-AB and BACH-derivatized monosaccharides

The BACH-derivatives interacted with the C18-phase ~8-15 min longer than the respective 2-AB-derivatives. The elution order of the sugars on the C-18-phase remained similar except for the triplet of the sugars galactose, glucose and mannose (Gal-2-AB, Glc-2AB, Man-2AB and Gal-BACH, Man-BACH, Glc-BACH). Galactose and mannose separation was achieved with both labels under the named conditions on the RP (Fig. 7), whereas the BACH-derivatives provided a clear baseline resolution.

Glucose is rarely present in glycoproteins. It can however originate from hydrolysis of contaminating polysaccharides from the environment. This problem was often encountered in reversed phase chromatography of monosaccharides (Toomre & Varki, 1994; Kwon & Kim, 1993) where glucose (at least partially) overlaps with the peak of galactose. Therefore, glucose contamination in carbohydrate samples should be addressed and examined.

In our experiments on C18 and HILIC-phase, 2-AB-derivatized glucose and galactose eluted very close to each other (Fig. 7 and Fig. 8, respectively). Through BACH-derivatization, on the other hand, glucose and galactose could have been separated on C18 (Fig. 7). However, the separation of glucose from mannose was still insufficient with BACH-labeling.

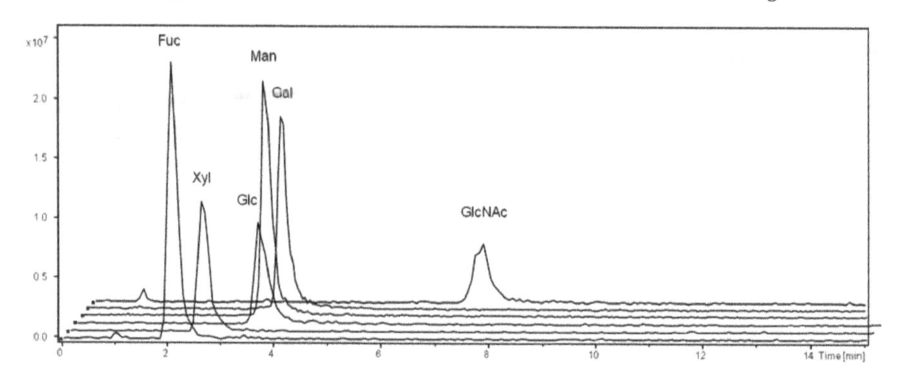

Figure 8. Hydrophilic interaction LC/MS of 2-AB-derivatized monosaccharides

A complete separation of glucose from galactose and mannose was reached only after the application of BACH-sugars to the HILIC-phase as shown in Fig. 9. However, the monosaccharides galactose and mannose could have been separated from each other only by means of RP chromatography (Fig.7). In this regard, the two phases (RP and HILIC) provided complementary features and their interchange may help to solve some analytical obstacles.

The BACH-label was used in the original form without any purification after its purchase. The presence of an isomeric byproduct was evident when the chromatography of the native label was performed. Two peaks (t_R: 16.8 min and 24 min, C18; 2 min and 6 min, HILIC) with the same m/z at 372 were present when chromatographically examined (data not shown). An example of this fact can be seen later in Fig.11 with the disaccharides, where the unreacted BACH shows two peaks. Due to this fact, we obtained after derivatization of saccharides a main product with a byproduct of the same m/z (Fig. 9). When necessary, the byproduct can be easily purified away with the excess BACH-label. Detailed structure of this byproduct is under investigation. Our further measurements concentrated only on the main product. As the BACH-tag eluted on the reversed phase between 16- 17 min with other sugars such as Man and Glc, a simultaneous clean-up of this label with the method on the C18 would not be feasible. However, when applied to HILIC-phase, the BACH-reagent eluted earlier than all the sugars and could be separated away when desired. As 2-Aminobenzamid label eluted before the mixture of monosaccharides on both stationary phases RP (6 min.) and HILIC (1.5 min.), a clean-up of the 2AB-label by this methodology would be very straightforward by both chromatographic methods.

The peaks of the investigated monosaccharides on the HILIC phase show broadening or splitting (Fig. 9). This is due to mutarotation of the sugars and consequently the presence of its anomers. The broadening is more evident with sugars, which mutarotate fast e.g. glucose and mannose. This phenomenon was observed with small reducing sugars on the HILIC-phase already in the past (Churms, 1996; Moriyasu *et al.*, 1984).

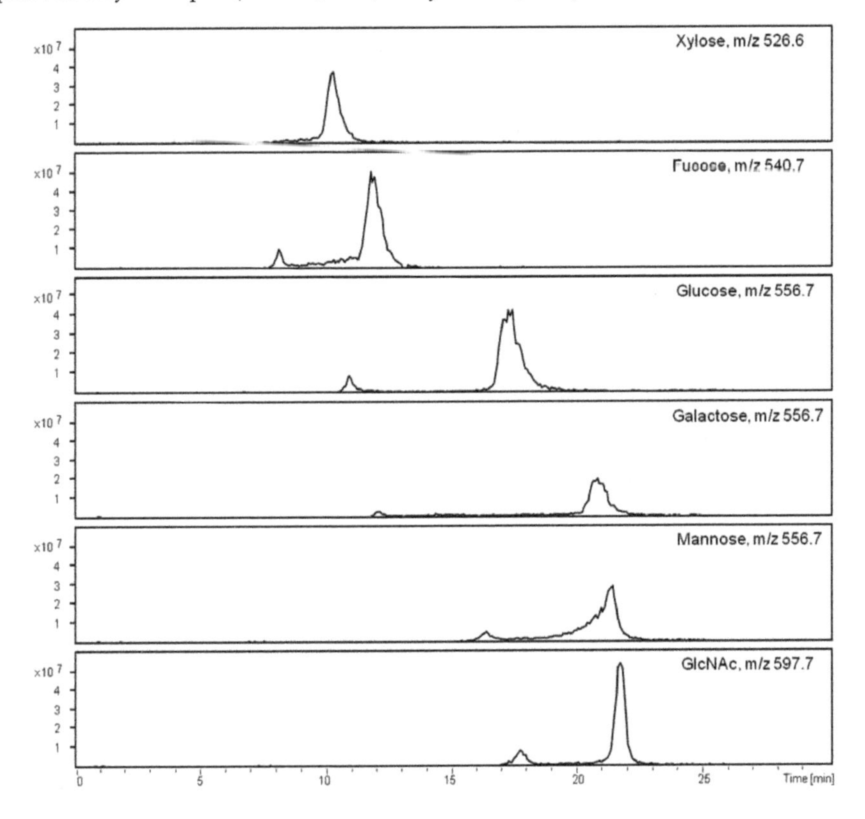

Figure 9. Extracted ion chromatograms (EIC) of BACH-derivatized monosaccharides separated by hydrophilic interaction liquid chromatography.

If the mutarotation should be suppressed, chromatography on the reversed phase can be followed and usage of a solvent with a higher content of organic phase might be preferable. A full separation of such isomeric structures (anomers) was achieved hitherto by means of chiral HPLC (Lopes & Gaspar, 2008), ion-exchange chromatography or amino-columns (Schumacher & Kroh, 1995; Nishikawa *et al.*, 1996; Moriyasu *et al.*, 1984). More recently, separation of metal adducts of anomeric methyl glycoside isomers was achieved by ion mobility mass spectrometry (Dwivedi *et al.*, 2007).

3.3.1. MS/MS of BACH-derivatized monosaccharides

The BACH-derivatized monosaccharides were subjected to tandem mass spectrometry. Analysis by ESI-ion trap showed that sodium adducts were the most abundant ions. Less abundant $[M+H]^+$ ions were detected in positive ion mode as well. Negative ion mode MS yielded $[M-H]^-$ ions of derivatized monosaccharides. All three types of ions were subjected to MS/MS in order to examine their fragmentation (Fig. 10).

MS/MS of sodium adducts $[M+Na]^+$ of neutral monosaccharides provided X-ring-cleavage in position 0-2 (m/z 436, 10% rel. intensity (RI)) and cleavage between two nitrogen atoms of the hydrazide group with m/z 379 (100% RI). With the sodiated ion of BACH-derivatized aminosugar *N*-acetylglucosamine, both of the cleavages were abundantly present in the spectrum (^{0,2}X at m/z 477, 100% RI; m/z 379, ~60% RI) and were accompanied by minor water losses from the molecular ion.

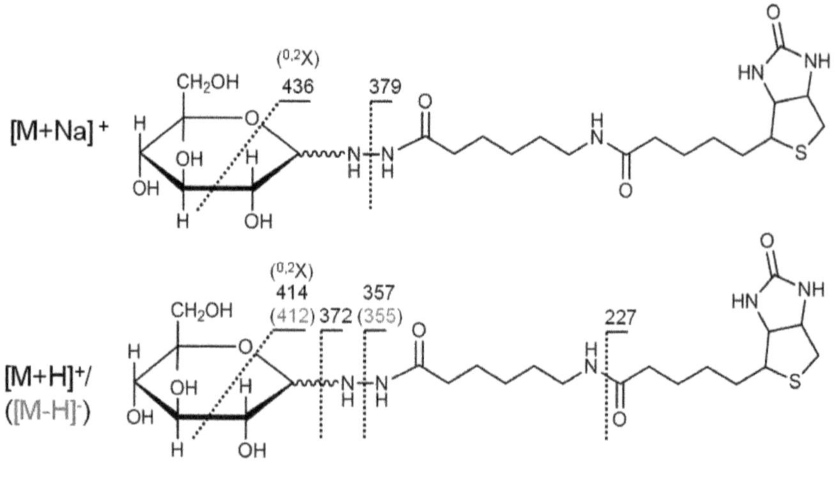

Figure 10. Scheme of fragmentation observed in MS/MS of BACH-labeled monosaccharides with different types of ions. Nominal masses are given.

The protonated ion species of the BACH-derivatized monosaccharides showed under the MS-fragmentation multiple water losses from the molecular ion (100%), $^{0,2}X$-fragment at m/z 414 (~20%) and some abundant cleavages associated with the BACH label: m/z 372 (100%) – detachment of the whole label, m/z 357 (~23%) – loss of an ammonia group and m/z 227 (~43%) which presents the acylium ion of the biotin component (Fig. 10).

The deprotonated form of the molecular ion $[M-H]^-$ of BACH monosaccharides showed similarity with the fragmentation spectrum of the $[M+Na]^+$- ion and yielded weak $^{0,2}X$-ring cleavage (m/z 412, ~7%) and the abundant cleavage of the hydrazine label (m/z 355, 100%). In the case of *N*-acetylglucosamine, on the other hand, the $^{0,2}X$-fragment was the only prominent peak in the MS/MS spectrum with 100% of rel. intensity. Hence, variation in abundances of fragment-ions of the aminosugar GlcNAc in comparison with fragments of neutral sugars was observed when performing CID of $[M+Na]^+$ and $[M-H]^-$ ions.

3.4. Hydrophilic interaction liquid chromatography of BACH-derivatized disaccharides

A common obstacle in the mass spectrometry of saccharides is the characterization of the stereochemistry of the glycosidic bond (Zhu *et al.*, 2009; Morelle & Michalski, 2005). Isomers usually have very similar MS^2- or MS^3-fragmentation and are difficult to differentiate without a previous separation. Therefore, the MS discrimination of isomers is often aided by one or more orthogonal pre-mass separation techniques.

The formation of closed-ring derivatives (glycosylhydrazides) through non-reducing derivatization has demonstrated some advantages in the chromatographic separation and in the analysis of isomers (Li & Her, 1998, 1993). Also, differentiation of anomeric configuration has been observed in some cases following derivatization (Xue *et al.*, 2004; Ashline *et al.*, 2005). Clowers et al. achieved this with reduced form of isobaric di- and trisaccharides by ion mobility mass spectrometry (Clowers *et al.*, 2005). These research groups exploring isomeric sugars with the focus on their MS fragmentation, observed differences in the negative-ion electrospray mass spectra of structurally related compounds. The presence/absence and abundance of certain fragment ions allowed for an assembly of empirical criteria for assigning either a linkage position or the anomeric configuration of a disaccharide. For example, diagnostic ions (m/z 221, 263, 281) which were characteristic for 1-4-linkage were observed by negative ion mode mass spectrometry (Mulroney *et al.*, 1995; Garozzo *et al.*, 1991). With glucose-containing disaccharides, dissociation of the m/z 221 generated product ions that allowed for the differentiation of anomeric configuration (Fang & Bendiak, 2007; Fang *et al.*, 2007).

Here we sought to examine whether the BACH-derivatized isomeric disaccharides would be separable on HILIC and would in MS/MS dissociate to yield product ions from which stereochemical information might be obtained about the glycosidic bond. For this purpose, maltose (α-D-Glc-(1-4)-D-Glc) and cellobiose (β-D-Glc-(1-4)-D-Glc), sugars which vary solely in their anomeric configuration were derivatized by non-reducing BACH-labeling according to the protocol described in the experimental part.

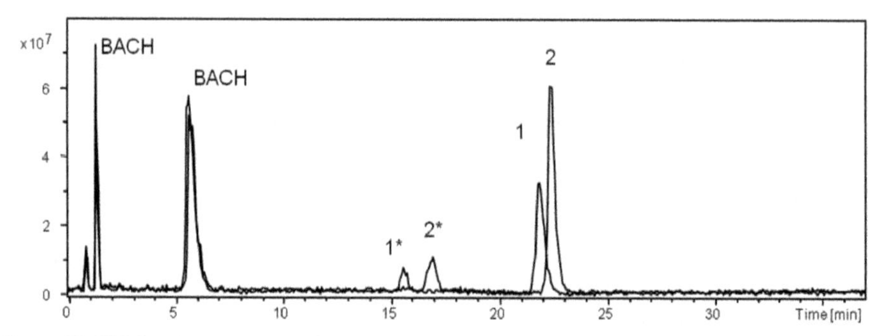

Figure 11. TIC showing separation of BACH-labeled isomeric disaccharides by means of hydrophilic interaction liquid chromatography. Peak assignements: 1 = BACH-maltose, 2 = BACH-cellobiose, * isomeric byproduct upon derivatization when BACH-label used in original unpurified form, BACH = excess of labeling reagent

With underivatized maltose and cellobiose, HPLC-separation and signal intensity were very poor (data not shown). Through BACH-derivatization, enhancement of signal intensity of two orders of magnitude (from $2x10^5$ to $6x10^7$) in comparison with native counterparts was achieved and at least a partial separation on the HILIC-phase was reached (Fig. 11). Even if they were not completely resolved, this demonstrated the feasibility of determining the presence of isomeric hits. For MS/MS investigation, the sugars were subjected separately to mass spectrometry, in order to obtain discrete fragmentation data originating only from one of the isomers. MS/MS spectra of underivatized maltose and cellobiose were without any significant difference in fragmentation pattern. Thus, no assignment could have been done with the native sugars.

In the study of BACH-labeled maltose and cellobiose, different types of molecular ions were detected: $[M+Na]^+$ (m/z 718), $[M+H]^+$ (m/z 696) and $[M-H]^-$ (m/z 694). These were subsequently subjected to tandem mass spectrometry in order to examine their fragmentation profile. Each of the precursor ions showed different fragmentation:

In positive ion mode, same fragmentation patterns for maltose and cellobiose were observed (data not shown). MS/MS of the sodium adduct contained the $^{0,2}X$-ring cleavage (m/z 436) from the reducing end of the derivatized disaccharide and the N-N-cleavage of the hydrazide group of the BACH-label (m/z 355). Fragmentation spectra of the $[M+H]^+$-ion were similar to $[M+Na]^+$. They provided week $^{0,2}X$-fragment (m/z 414) and the hydrazide group cleavage (m/z 355). Additionally to these, glycosidic cleavages (Y_0 and Y_1), which confirmed the sequence and hexose composition of the saccharide, appeared. With Y_1-ion, up to four water losses were observed. Hence, in positive ion mode, when observing either $[M+Na]^+$ or $[M+H]^+$, no difference in the fragmentation of these isomers was observed.

CID of the $[M-H]^-$ion, on the contrary (Fig. 12), yielded cleavages (m/z 412 and m/z 355) and also fragments between m/z 179 and m/z 355, having higher masses than monosaccharide and encompassing the glycosidic bond of the sugar.

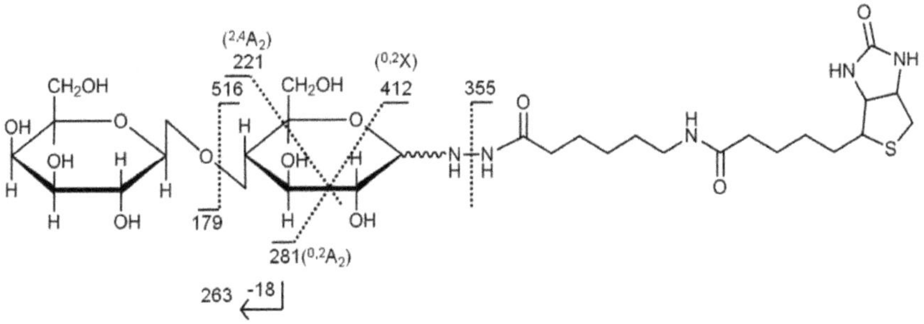

Figure 12. Fragmentation of [M-H]⁻-ion observed for BACH-labeled isomeric disaccharides maltose and cellobiose in negative ion mode

Negative MS/MS fragmentation spectra of BACH-labeled maltose and cellobiose were similar in terms of presence of diagnostic ions. However, some variation in the relative abundance of the product ions was visible. Ions observed were: m/z 179, 221, 263, 281 (Fig. 12), where m/z 179 represents the non-reducing monosaccharide containing the glycosidic oxygen; m/z 221 ($^{2,4}A_2$) comprises intact non-reducing sugar glycosidically linked to a glycoaldehyde molecule and m/z 281 corresponds to loss of one –CHOH backbone-unit and also one -CH with the BACH-label. This ion eliminated a molecule of water yielding an m/z 263 ion. This has been shown to be characteristic for 1,4-bonded underivatized glucose dimers (Dallinga & Heerma, 1991a).

In the spectrum of α-linked maltose (Fig. 13a), intensity of the m/z 221 was slightly higher than m/z 263. In the fragmentation spectra of beta-linked cellobiose, on the contrary, the ion m/z 221 appeared only in traces. With cellobiose, the intensity of m/z 263 was always higher than m/z 221. Further, the ion at m/z 281 was more abundant in the spectrum of maltose, whereas almost absent in the MS/MS of cellobiose. In order to confirm the reproducibility of the fragmentation we performed the measurements 15 times in total (5 rounds of measurements for each disaccharide on three different days) indicating the presence and intensity of respective fragments to be reproducible and constant. These observations are in accordance with the fragments reported for 4-linked disaccharides and their proportionalities correspond with dissociation studies reported for the differentiation of this type of isomeric disaccharides. In the study of Li and Her (Li & Her, 1993), where relative abundance of fragment ions of 1-4 linked disaccharides was compared, the intensity of the m/z 221 was higher than m/z 263 in alpha-linked disaccharides and smaller than m/z 263 ion in beta-linked disaccharides. As reported (Fang & Bendiak, 2007; Fang *et al.*, 2007), the cross-ring cleavage m/z 221 of the disaccharides also showed to be useful in the discrimination of anomeric configuration. Under conditions applied here, further dissociation of m/z 221 (MS^3) of the BACH- labeled maltose or cellobiose provided no additional differential information. Noneless, through the combination of BACH-labeling and tandem mass spectrometry, distinctive structural information was obtained at the MS^2-stage of fragmentation and disaccharides with the same linkage but different anomeric configurations (maltose: 1-4-α-linkage, cellobiose: 1-4-β-linkage) could have been distinguished.

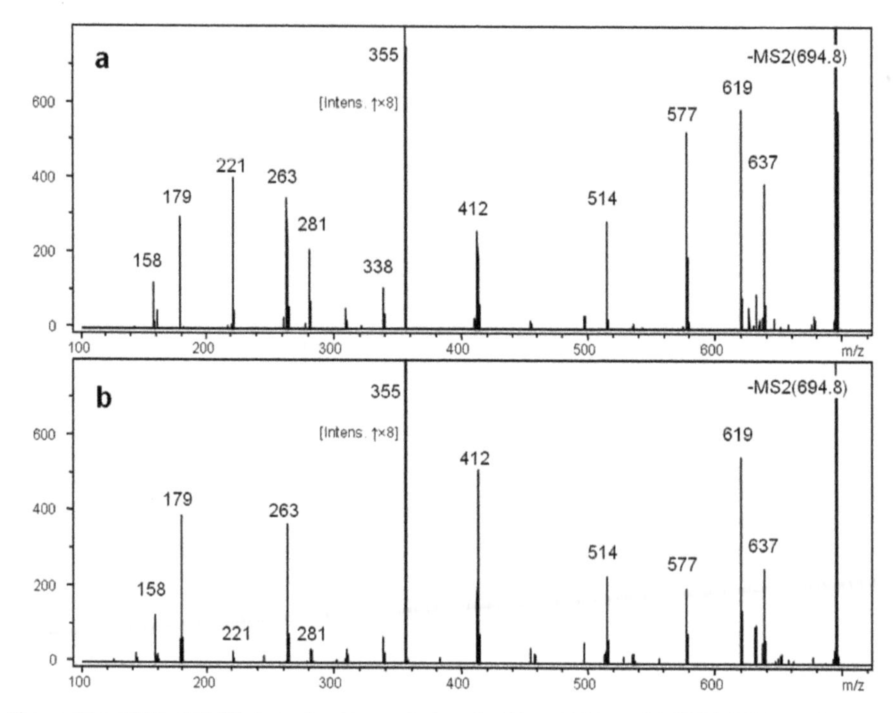

Figure 13. MS/MS of BACH-derivatized isomeric disaccharides. a) Maltose-BACH b) Cellobiose-BACH

3.5. Hydrophilic interaction liquid chromatography of isomeric non-reducing trisaccharides

In order to further examine the ability of the current analytical system to distinguish between the sugar isomers, we have analyzed two trisaccharides melezitose [α-D-Glc-(1→3)-β-D-Fru-(2→1)-α-D-Glc] and raffinose [α-D-Gal-(1→6)-α-D-Glc-(1→2)-β-D-Fru] which differ in sequence, linkage and anomeric configuration. Both of them were examined by LC-MS/MS in negative ion mode as this one showed to be the preferential mode in order to observe some distinctive fragmentation of related compounds (Section 3.4.). Figure 14a shows the total ion current of melezitose and raffinose. The chromatography of these sugars on the HILIC-phase provided a full baseline separation.

The fragmentation mass spectra of these trisaccharides were obtained from the deprotonated form of the molecular ion $[M-H]^-$ at m/z 503 (Figure 14b-e). Melezitose displayed a dominant fragment at m/z 323 which derives from cleavage of one of the two possible glycosidic bonds (Fig. 14b). This route was confirmed through the MS^3 experiment, where the dissociation of m/z 341 generated no m/z 323 but solely the ions with m/z 179 and m/z 161 (Fig. 14d). This indicates that m/z 323 occurs directly through B-fragmentation and not via formation of m/z 341 and a subsequent water loss. The peak at m/z 341 was found at

MS^2-stage of both trisaccharides and is due to the elimination of a site residue from the $[M-H]^-$-molecular ion. Interestingly, in positive-ion MS-studies on $[M+H]^+$-ions of non-reducing oligosaccharides containing fructose (including melezitose and raffinose), the formation of m/z 323 was proposed to preferably occur via the water loss from m/z 341 (Dallinga & Heerma, 1991; Perez-Victoria *et al.*, 2008).

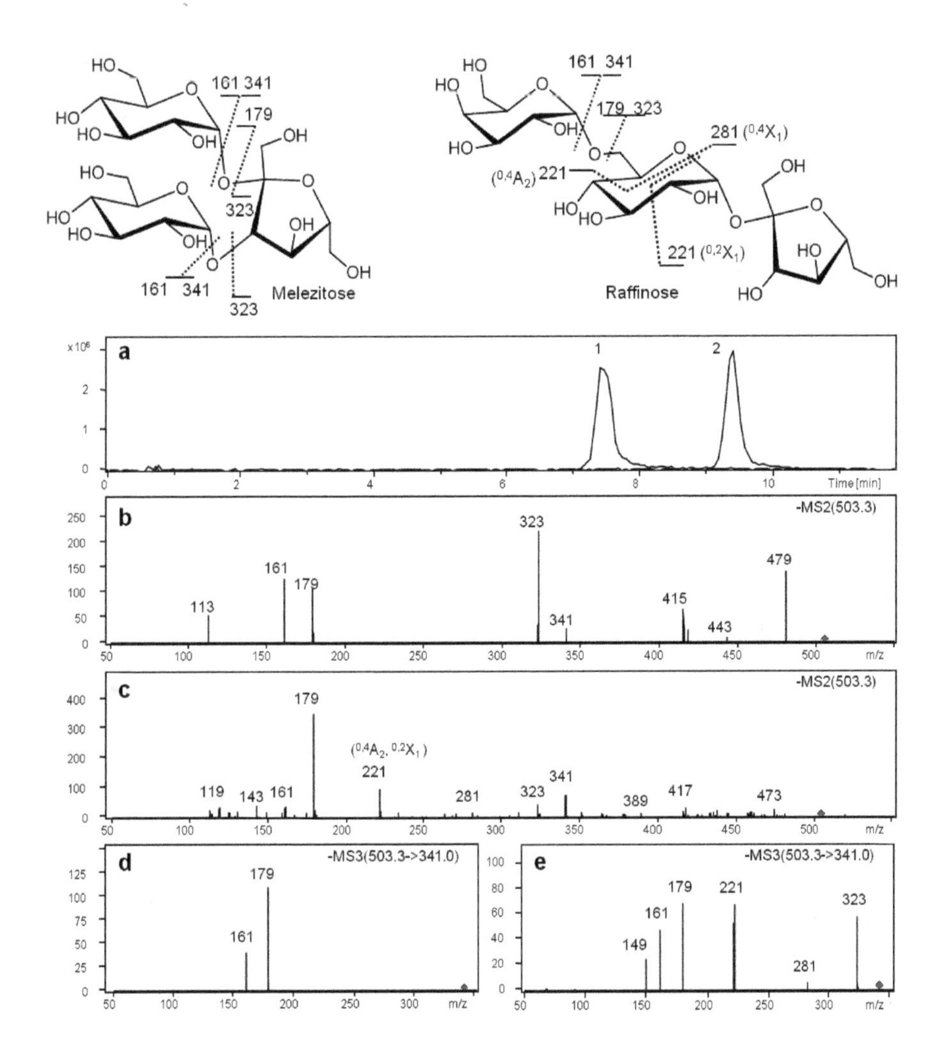

Figure 14. a) Separation of the mixture of two isomeric trisaccharides melezitose (1) and raffinose (2), demonstrating the value of coupling HILIC with ion-trap mass spectrometry performed to MS^3. b) MS^2 of melezitose c) MS^2 of raffinose d) MS^3 of melezitose e) MS^3 of raffinose

Raffinose fragmented to form preferably m/z 179, m/z 221 and glycosidic fragment at m/z 341 (Fig. 14c). Other ions (m/z 323 and 161) were present as well. At this stage of fragmentation (MS^2), the formation of cross-ring cleavage (m/z 221) involving either the 0-2 or 0-4 bonds of the center residue appeared. This ion has not been seen in the spectrum of melezitose (compare to Fig. 14b). The same fact was reported in the studies of fragmentation of trisaccharides by negative FAB (Dallinga & Heerma, 1991b) and negative ion mobility mass spectrometry (Liu & Clemmer, 1997): raffinose yielded m/z 221 in the MS^2-spectrum, whereas melezitose showed only the product ion at m/z 323 without presence of m/z 221. The presence of m/z 221 was also evident in MS^3 of raffinose (Fig. 14e). The different fragmentation routes observed may be attributed to the difference in the steric character of these isomers: Melezitose has a branched character and raffinose represents a linear molecule.

Finally, the differences in the chromatography on the HILIC-phase and in the fragmentation pattern (presence and rel. intensity of the product ions in the MS^2 and MS^3 stage) of the melezitose and raffinose indicated the occurrence of distinct isomeric structures.

This distinguished fragmentation would not have been possible to obtain without the previous pre-mass spectrometric separation on the HILIC-phase. Considering mass spectrometric techniques, such differential identification of these isomers out of their mixture was achieved up to now, only by means of ion-mobility mass spectrometry (Zhu *et al.*, 2009; Liu & Clemmer, 1997). To our best knowledge, this is the first report on separation and identification of such isomers by coupling of hydrophilic interaction chromatography and mass spectrometry.

4. Conclusion

Different types of *N*-glycans (high-mannose, complex and hybrid) have been derivatized with biotinamidocaproyl hydrazide and analyzed by chip-based LC/ESI triple-quadrupole mass spectrometry. This approach allowed for rapid mass spectrometric selection of glycans out of glycan mixtures on the base of the scanning for a common feature ion (oxonium ion). The glycans observed upon MS were primarily in the doubly charged state. For each type of glycan, product ion spectra with the characteristic B- and Y-ions were observed which allowed for elucidation of the structure and segregation of the glycans into the respective class. This methodology provided fast structural information about the nature of the glycan pool present in a glycoprotein and may be used in the profiling of *N*-linked glycans. Beyond structural mass spectrometric studies, BACH-derivatized glycans can potentially be used also in functional studies with carbohydrate binding proteins, as recently reported.

HILIC and reversed phase liquid chromatography of BACH- and 2-AB-labeled monosaccharides enabled the investigation of different analytical objectives and showed some orthogonal features in terms of the separation of monosaccharides frequently present in glycoproteins. The HILIC-system was not sufficient for separation of anomers of BACH-derivatized monosaccharides. It was, however, capable to partially resolve the

BACH-derivatized isomeric disaccharides and to fully separate non-reducing isomeric trisaccharides.

In terms of the enhancement of mass spectrometric signal of mono- and disaccharides, BACH-labeling provided higher intense ions than 2-AB-derivatization. Analysis of the BACH-derivatives in combination with the mass spectrometry can be pursued, when investigation of isomers is desired. Unlike positive ESI-MS, negative ion MS of BACH-derivatized isomeric disaccharides provided ring-cleavage ions which were helpful in the differentiation of the anomeric configuration of the glycosidic bond. Further, separation of structurally related non-derivatized trisaccharides was achieved on HILIC and their differentiation on the basis of multiple stage mass spectrometry was possible. This differentiation would not have been possible without the suitable pre-mass spectrometric separation. In this way, coupling of the HILIC-separation with MS^n-fragmentation might be a valuable approach on the way to the distinction of certain types of isomeric structures.

Author details

Petra Kapková and Stephanie Bank
University of Würzburg/Department of Pharmacy and Food Chemistry, Germany

Acknowledgement

This work was supported by the Fond der Chemischen Industrie and the Universitätsbund of University of Würzburg.

5. References

Alley, W. R., Jr., Mechref, Y. & Novotny, M. V. (2009). Use of activated graphitized carbon chips for liquid chromatography/mass spectrometric and tandem mass spectrometric analysis of tryptic glycopeptides. *Rapid Communications in Mass Spectrometry*, Vol.23, No.4, pp. 495-505, ISSN 0951-14198

Ashline, D., Singh, S., Hanneman, A. & Reinhold, V. (2005). Congruent strategies for carbohydrate sequencing. 1. Mining structural details by MSn. *Analytical Chemistry*, Vol.77, No.19, pp. 6250-6262, ISSN 0003-2700

Bigge, J. C., Patel, T. P., Bruce, J. A., Goulding, P. N., Charles, S. M. & Parekh, R. B. (1995). Nonselective and efficient fluorescent labeling of glycans using 2-amino benzamide and anthranilic acid. *Analytical Biochemistry*, Vol.230, No.2, pp. 229-238, ISSN 0003-2697

Bindila, L. & Peter-Katalinic, J. (2009). Chip-mass spectrometry for glycomic studies. *Mass Spectrometry Reviews*, Vol.28, No.2, pp. 223-253, ISSN 1098-2787

Brons, C. & Olieman, C. (1983). Study of the high-performance liquid-chromatographic separation of reducing sugars, applied to the determination of lactose in milk. *Journal of Chromatography*, Vol.259, No.1, pp. 79-86, ISSN 0021-9673

Churms, S. C. (1996). Recent progress in carbohydrate separation by high-performance liquid chromatography based on hydrophilic interaction. *Journal of Chromatography A,* Vol.720, No.1-2, pp. 75-91, ISSN 0021-9673

Clowers, B. H., Dwivedi, P., Steiner, W. E., Hill, H. H. & Bendiak, B. (2005). Separation of sodiated isobaric disaccharides and trisaccharides using electrospray ionization-atmospheric pressure ion mobility-time of flight mass spectrometry. *Journal of the American Society for Mass Spectrometry,* Vol.16, No.5, pp. 660-669, ISSN 1044-0305

Dallinga, J. W. & Heerma, W. (1991a). Positive-Ion Fast-Atom-Bombardment Mass-Spectrometry of Some Small Oligosaccharides. *Biological Mass Spectrometry,* Vol.20, No.3, pp. 99-108, ISSN 1052-9306

Dallinga, J. W. & Heerma, W. (1991b). Reaction-Mechanism and Fragment Ion Structure Determination of Deprotonated Small Oligosaccharides, Studied by Negative-Ion Fast-Atom-Bombardment (Tandem) Mass-Spectrometry. *Biological Mass Spectrometry,* Vol.20, No.4, pp. 215-231, ISSN 1052-9306

Domon, B. & Costello, C. E. (1988). A Systematic Nomenclature for Carbohydrate Fragmentations in Fab-Ms Ms Spectra of Glycoconjugates. *Glycoconjugate Journal,* Vol.5, No.4, pp. 397-409, ISSN 0282-0080

Dwivedi, P., Bendiak, B., Clowers, B. H. & Hill, H. H. (2007). Rapid resolution of carbohydrate isomers by electrospray ionization ambient pressure ion mobility spectrometry-time-of-flight mass spectrometry (ESI-APIMS-TOFMS). *Journal of the American Society for Mass Spectrometry,* Vol.18, No.7, pp. 1163-1175, ISSN 1044-0305

Fang, T. T. & Bendiak, B. (2007). The stereochemical dependence of unimolecular dissociation of monosaccharide-glycolaldehyde anions in the gas phase: A basis for assignment of the stereochemistry and anomeric configuration of monosaccharides in oligosaccharides by mass spectrometry via a key discriminatory product ion of disaccharide fragmentation, m/z 221. *Journal of the American Chemical Society,* Vol.129, No.31, pp. 9721-9736, ISSN 0002-7863

Fang, T. T., Zirrolli, J. & Bendiak, B. (2007). Differentiation of the anomeric configuration and ring form of glucosyl-glycolaldehyde anions in the gas phase by mass spectrometry: isomeric discrimination between m/z 221 anions derived from disaccharides and chemical synthesis of m/z 221 standards. *Carbohydrate Research,* Vol.342, No.2, pp. 217-235, ISSN 0008-6215

Garozzo, D., Impallomeni, G., Spina, E., Green, B. N. & Hutton, T. (1991). Linkage analysis in disaccharides by electrospray mass-spectrometry. *Carbohydrate Research,* Vol.221, pp. 253-257, ISSN 0008-6215

Geyer, H. & Geyer, R. (2006). Strategies for analysis of glycoprotein glycosylation. *Biochimica Biophysica Acta,* Vol.1764, No. 12, pp. 1853-1869, ISSN 0006-3002

Grün, C. H., van Vliet, S. J., Schiphorst, W. E., Bank, C. M., Meyer, S., van Die, I. & van Kooyk, Y. (2006). One-step biotinylation procedure for carbohydrates to study carbohydrate-protein interactions. *Analytical Biochemistry,* Vol.354, No.1, pp. 54-63, ISSN 0003-2697

Harvey, D. J. (2000). Electrospray mass spectrometry and collision-induced fragmentation of 2-aminobenzamide-labelled neutral N-linked glycans. *Analyst,* Vol.125, No.4, pp. 609-617, ISSN 0003-2654

Harvey, D. J. (2011). Derivatization of carbohydrates for analysis by chromatography; electrophoresis and mass spectrometry. *Journal of Chromatography B,* Vol.879, No.17-18, pp. 1196-1225, ISSN 1570-0232

Hermannson, G. T. (2008). *Bioconjugate Techniques,* Academic Press, New York

Hsu, J., Chang, S. J. & Franz, A. H. (2006). MALDI-TOF and ESI-MS analysis of oligosaccharides labeled with a new multifunctional oligosaccharide tag. *Journal of the American Society for Mass Spectrometry,* Vol.17, pp. 194-204, ISSN 1044-0305

Ikegami, T., Tomomatsu, K., Takubo, H., Horie, K. & Tanaka, N. (2008). Separation efficiencies in hydrophilic interaction chromatography. *Journal of Chromatography A,* Vol.1184, No.1-2, pp. 474-503, ISSN 0021-9673

Kapková, P. (2009). Mass spectrometric analysis of carbohydrates labeled with a biotinylated tag. *Rapid Communications in Mass Spectrometry,* Vol.23, No.17, pp. 2775-2784, ISSN 0951-4198

Kwon, H. & Kim, J. (1993). Determination of monosaccharides in glycoproteins by reverse-phase high-performance liquid chromatography. *Analytical Biochemistry,* Vol.215, pp. 243-252, ISSN 0003-2697

Lattová, E., Snovida, S., Perreault, H. & Krokhin, O. (2005). Influence of the labeling group on ionization and fragmentation of carbohydrates in mass spectrometry. *Journal of the American Society for Mass Spectrometry,* Vol.16, No.5, pp. 683-696, ISSN 1044-0305

Leteux, C., Childs, R. A., Chai, W., Stoll, M. S., Kogelberg, H. & Feizi, T. (1998). Biotinyl-l-3-(2-naphthyl)-alanine hydrazide derivatives of N-glycans: versatile solid-phase probes for carbohydrate-recognition studies. *Glycobiology,* Vol.8, No.3, pp. 227-236, ISSN 0959-6658

Leymarie, N. & Zaia, J. (2012). Effective use of mass spectrometry for glycan and glycopeptide structural analysis. *Analytical Chemistry,* Vol.84, pp. 3040-3048, ISSN 0003-2700

Li, D. T. & Her, G. R. (1993). Linkage Analysis of Chromophore-Labeled Disaccharides and Linear Oligosaccharides by Negative-Ion Fast-Atom-Bombardment Ionization and Collisional-Induced Dissociation with B/E Scanning. *Analytical Biochemistry,* Vol.211, No.2, pp. 250-257, ISSN 0003-2697

Li, D. T. & Her, G. R. (1998). Structural analysis of chromophore-labeled disaccharides and oligosaccharides by electrospray ionization mass spectrometry and high-performance liquid chromatography electrospray ionization mass spectrometry. *Journal of Mass Spectrometry,* Vol.33, No.7, pp. 644-652, ISSN 1076-5174

Liu, Y. S. & Clemmer, D. E. (1997). Characterizing oligosaccharides using injected-ion mobility mass spectrometry. *Analytical Chemistry,* Vol.69, No.13, pp. 2504-2509, ISSN 0003-2700

Lopes, J. F. & Gaspar, E. M. S. M. (2008). Simultaneous chromatographic separation of enantiomers, anomers and structural isomers of some biologically relevant monosaccharides. *Journal of Chromatography A,* Vol.1188, pp. 34-42, ISSN 0021-9673

Marino, K., Bones, J., Kattla, J. J. & Rudd, P. M. (2010). A systematic approach to protein glycosylation analysis: A path through the maze. *Nature Chemical Biology,* Vol.6, pp. 713-723, ISSN 1552-4450

Melmer, M., Stangler, T., Premstaller, A. & Lindner, W. (2011). Comparison of hydrophilic-interaction, reversed-phase and porous graphitic carbon chromatography for glycan analysis. *Journal of Chromatography A,* Vol.1218, pp. 118-123, ISSN 0021-9673

Morelle, W. & Michalski, J. C. (2004). Sequencing of oligosaccharides derivatized with benzlyamine using electrospray ionization-quadrupole time of flight-tandem mass spectrometry. *Electrophoresis,* Vol.25, No.14, pp. 2144-2155, ISSN 0173-0835

Morelle, W. & Michalski, J. C. (2005). Glycomics and mass spectrometry. *Current Pharmaceutical Design,* Vol.11, No.20, pp. 2615-2645, ISSN 1381-6128

Moriyasu, M., Kato, A., Okada, M. & Hashimoto, Y. (1984). HPLC separation of sugar anomers in a very low temperatur region. *Analytical Letters,* Vol.17, pp. 689-699, ISSN 0003-2719

Mulroney, B., Traeger, J. C. & Stone, B. A. (1995). Determination of Both Linkage Position and Anomeric Configuration in Underivatized Glucopyranosyl Disaccharides by Electrospray Mass-Spectrometry. *Journal of Mass Spectrometry,* Vol.30, No.9, pp. 1277-1283, ISSN 1076-5174

Nishikawa, T., Suzuki, S., Kubo, H. & Ohtani, H. (1996). On-column isomerization of sugars during high-performance liquid chromatography: analysis of the elution profile. *Journal of Chromatography A,* Vol.720, pp. 167-172, ISSN 0021-9673

North, S. J., Hitchen, P. G., Haslam, S. M. & Dell, A. (2009). Mass spectrometry in the analysis of N-linked and O-linked glycans. *Current Opinion in Structural Biology,* Vol.19, pp. 498-506, ISSN 0959-440X

Pabst, M., Kolarich, D., Poltl, G., Dalik, T., Lubec, G., Hofinger, A. & Altmann, F. (2009). Comparison of fluorescent labels for oligosaccharides and introduction of a new postlabeling purification method. *Analytical Biochemistry,* Vol.384, No.2, pp. 263-273, ISSN 0003-2697

Perez-Victoria, I., Zafra, A. & Morales, J. C. (2008). Positive-ion ESI mass spectrometry of regioisomeric nonreducing oligosaccharide fatty acid monoesters: In-source fragmentation of sodium adducts. *Journal of Mass Spectrometry,* Vol.43, No.5, pp. 633-638, ISSN 1076-5174

Raman, R., Raguram, S., Venkataraman, G., Paulson, J. C. & Sasisekharan, R. (2005). Glycomics: an integrated systems approach to structure-function relationships of glycans. *Nature Methods,* Vol.2, pp. 817-824, ISSN 1548-7091

Ridley, B. L., Spiro, M. D., Glushka, J., Albersheim, P., Darvill, A. & Mohnen, D. (1997). A method for biotin labeling of biologically active oligogalacturonides using a chemically stable hydrazide linkage. *Analytical Biochemistry*, Vol.249. No.1, pp. 10-19, ISSN 0003-2697

Rosenfeld, J. (2011). Enhancement of analysis by analytical derivatization. *Journal of Chromatography B*, Vol.879, pp. 1157-1158, ISSN 1570-0232

Rothenberg, B. E., Hayes, B. K., Toomre, D., Manzi, A. E. & Varki, A. (1993). Biotinylated diaminopyridine: an approach to tagging oligosaccharides and exploring their biology. *Proceedings of the National Academy of Sciences USA*, Vol.90, No.24, pp. 11939-11943, ISSN 1091-6490

Ruhaak, L. R., Zauner, G., Huhn, C., Bruggink, C., Deelder, A. & Wuhrer, M. (2010). Glycan labeling strategies and their use in identification and quantification. *Analytical and Bioanalytical Chemistry*, Vol.397, pp. 3457-3481, ISSN1618-2650

Schlichtherle-Cerny, H., Affolter, M. & Cerny, C. (2003). Hydrophilic interaction liquid chromatography coupled to electrospray mass spectrometry of small polar compounds in food analysis. *Analytical Chemistry*, Vol.75, pp. 2349-2354, ISSN 0003-2700

Schumacher, D. & Kroh, L. W. (1995). A rapid method for separation of anomeric saccharides using cyclodextrin bonded phase and for investigation of mutarotation. *Food Chemistry*, Vol.54, pp. 353-356, ISSN 0308-8146

Shinohara, Y., Sota, H., Gotoh, M., Hasebe, M., Tosu, M., Nakao, J., Hasegawa, Y. & Shiga, M. (1996). Bifunctional labeling reagent for oligosaccharides to incorporate both chromophore and biotin groups. *Analytical Chemistry*, Vol.68, No.15, pp. 2573-2579, ISSN 0003-2700

Strege, M. A., Stevenson, S. & Lawrence, S. M. (2000). Mixed mode anion-cation exchange/hydrophilic interaction liquid chromatography-electrospray mass spectrometry as an alternative to reversed phase for small molecule drug discovery. *Analytical Chemistry*, Vol.72, No.19, pp. 4629-4633, ISSN 0003-2700

Toomre, D. K. & Varki, A. (1994). Advances in the use of biotinylated diaminopyridine (BAP) as a versatile fluorescent tag for oligosaccharides. *Glycobiology*, Vol.4, No.5, pp. 653-663, ISSN 0959-6658

Wuhrer, M., de Boer, A. R. & Deelder, A. M. (2009). Structural glycomics using hydrophilic interaction chromatography (HILIC) with mass spectrometry. *Mass Spectrometry Reviews*, Vol.28, No.2, pp. 192-206, ISSN 1098-2787

Wuhrer, M., Koeleman, C. A. M., Hokke, C. H. & Deelder, A. M. (2004). Nano-scale liquid chromatography-mass spectrometry of 2-aminobenzamide-labeled oligosaccharides at low femtomole sensitivity. *International Journal of Mass Spectrometry*, Vol.232, No.1, pp. 51-57, ISSN 1387-3806

Wuhrer, M., van Remoortere, A., Balog, C. I. A., Deelder, A. M. & Hokke, C. H. (2010). Ligand identification of carbohydrate-binding proteins employing a biotinylated glycan binding assay and tandem mass spectrometry. *Analytical Biochemistry*, Vol. 406, No.2, pp. 132-140, ISSN 0003-2697

Xue, J., Song, L. G., Khaja, S. D., Locke, R. D., West, C. M., Laine, R. A. & Matta, K. L. (2004). Determination of linkage position and anomeric configuration in Hex-Fuc disaccharides using electrospray ionization tandem mass spectrometry. *Rapid Communications in Mass Spectrometry*, Vol.18, No.17, pp. 1947-1955, ISSN 0951-4198

Zhu, M. L., Bendiak, B., Clowers, B. & Hill, H. H. (2009). Ion mobility-mass spectrometry analysis of isomeric carbohydrate precursor ions. *Analytical and Bioanalytical Chemistry*, Vol.394, No.7, pp. 1853-1867, ISSN 1618-2650

Chapter 6

Glycosidases – A Mechanistic Overview

Natércia F. Brás , Pedro A. Fernandes,
Maria J. Ramos and Nuno M.F.S.A. Cerqueira

Additional information is available at the end of the chapter

1. Introduction

Carbohydrates are the most abundant and structurally diverse class of biological compounds in nature. However, our current understanding regarding the relationship between carbohydrate structure and its biological function is still far from what is known regarding proteins and nucleic acids. Initially, carbohydrates were only recognized as structural and energy storage molecules (e.g. cellulose, chitin and glycogen), but recent developments in the field have shown that carbohydrates are also involved in numerous biological events, such as cancer, inflammations, pathogen infections, cell-to-cell communication, etc. In addition, carbohydrate-processing enzymes have become the choice in many industrial applications due to their stereo-selectivity and efficiency[1].

Carbohydrates can be found in nature in many forms, from simple monomers to more complex oligomers, polymers or glicoconjugates. The complexity of these structures can be reasonably high since each carbohydrate monomer can accommodate multiple linkages and/or branches in its structure. Moreover, as the glycosidic linkage between each monosaccharide can have two anomeric configurations (α or β), even in small oligosaccharides, the potential number of structures that can exist is huge.

In the last decades, there has been a great effort to synthesize oligo- and poly-saccharides in the laboratory, mainly due to their key role in many biological events but also to the interest expressed by the food and technical industries. The chemical approaches to carbohydrate synthesis have been known since Arthur Michael first reported the synthesis of a natural glycoside in 1879[2]. However, the construction of complex carbohydrates and glycoconjugates in the laboratory remains a challenging endeavor. The causes for these difficulties are several but they mainly rely on the exceptional complexity and diversity that some compounds may show. Indeed, unlike the systematic processes of proteins and nucleic acids synthesis, in which the order of attachment of amino acids and nucleotides is read from a nucleic acid matrix, the synthesis of carbohydrates is a non-template-directed process that is controlled by a complex stereo- and regio-specific process. It requires a special regioselective reaction at a particular position of the sugar unit, in which the hydroxyl group that is available in such position must be distinguished from all the other hydroxyl groups in the structure that have similar properties. Additionally, the linkage between sugars must proceed through a stereoselective manner, since the linkage can produce two stereoisomers and one of them must be preferred to the other. Many carbohydrates are also found linked to protein and lipids. The synthesis of glycoconjugates has also proven to be a difficult task because it generally involves the participation of multiple transporters and enzymes. The mechanisms governing the regulation of these pathways are still being elucidated, but so far it has been found that the assembly of carbohydrates to proteins and lipids requires a specific chemistry that is far from being universal.

The production of oligosaccharides and polysaccharides has been deeply studied in the past decades and revealed to be, as expected, a challenging task[3]. In spite of the advances observed in organic chemistry, the chemical synthetic routes addressed to synthesize these compounds have proven to be inefficient in the majority of the cases. This happens because the preparation of complex oligosaccharides and polysaccharides require multiple protection/deprotection and purification steps, which often lead to a tedious and time-consuming process and normally result in poor yields. To overcome these limitations, the enzymatic synthesis rapidly gained more prominence. The attractiveness of enzymatic synthesis is that protecting groups are not required and the stereo- and regio-selectivity chemistry is always followed in the formation of the glycosidic linkages, in the majority of the cases.

Enzymatic formation and cleavage of the bond between two sugars or between a sugar and another group can occur by hydrolysis to give the free sugar (glycosidases), by transglycosylation to give a new glycoside (glycosyltransferases), by phosphorolysis to give the sugar-1-phosphate (phosphorylases) or by elimination to give unsaturated sugar products (lyases). Currently, glycosidases and glycosyltransferases are the major classes of biocatalysts that are available for the enzymatic synthesis of polysaccharides and oligosaccharides.[1] As the structure of lysozyme was first solved in 1965[4], glycosidases have long been the subject of structural biology studies in order to understand the molecular details of substrate recognition and of catalysis. As a result, about three quarters of the 113 known families of glycosidases have at least a structural representative. In contrast, progress in the structural biology of glycosyltransferases has been slower[1]. Part of the success in characterizing glycosidases is due to the high stability of these enzymes when compared with the glycosyltransferases and because they are very easy to isolate, being generally available from natural sources like seeds, micro-organisms or fungal cultures, as well as in higher organisms (typically plant seed, mollusks, etc)[5]. These facts have turned glycosidases into an attractive target for many industries involved in the food, the paper and pulp industry, as well as in organic chemistry, where glycosidases have proven to be extremely efficient catalysts, being capable of hydrolyzing the very stable glycosidic bonds in glycoconjugates, oligo- and poly-saccharides[6]. The importance of glycosidases has also attracted the attention of many pharmaceutical industries since they are involved in many biological processes such as cell-cell or cell-virus recognition, immune responses, cell growth, and viral and parasitic infections. Currently, they have been associated with many diseases, which result from the lack or dysfunction of a glycosidase and are used in the treatment of metabolic disorders, viral infections and even cancer.

Despite the current advances in the field and the exponential interest in glycosidases, many aspects of the mechanism of action of these enzymes remain hidden in the available experimental data, in particularly in the X-ray structures that figure in the protein databank. Taking this into account, we focus in this review in the current literature regarding the catalytic mechanisms of glycosidases.

2. Catalytic mechanism of glycosidases

Glycosidases (GH) are present in almost all living organisms (exceptions are some Archaeans and a few unicellular parasitic eukaryotes)[7,8] where they play diverse and different roles. Taking into account the diversity of reactions that they catalyze as well as amino acid sequence and folding, glycosidases have been classified in many different ways. According to The IUBMB (International Union of Biochemistry and Molecular Biology) glycosidases are classified based on their substrate specificity and/or their molecular mechanism.[9,10] However, this classification is far from gaining consensus. A necessary consequence of the EC classification scheme is that codes can be applied only to enzymes for which a function has been biochemically identified. Additionally, certain enzymes can catalyse reactions that fall in more than one class, which makes them bear more than one EC number. Furthermore, this classification does not reflect the structural features and evolutionary relations of enzymes. In order to overcome these limitations, a new type of classification was proposed based on the amino acid similarity within the protein. This new classification is available at the Carbohydrate-Active Enzymes database (CAZy - http://www.cazy.org/)[7,10] and provides a direct relationship between sequence and folding similarities, that can be found in 130 amino acid sequence-based families. Some families with apparently unrelated sequence similarities show some uniformity in their three-dimensional structures. In those cases, these structures have been assigned to the so-called "Clans", that have been numbered from A to N.[7] In general, GHs belonging to the families of the same clan have common ancestry, similar 3D structure and are characterized by an identical catalytic mechanism of action.[7,10-16]

The two most commonly employed mechanisms used by glycosidases to achieve glycosidic bond cleavage with overall inversion or retention of anomeric stereochemistry are shown schematically at Figure 1. These mechanisms can be generally divided in two main groups: the retaining GHs and the inverting GHs. [17,18] Generally, enzymes of the same family have the same mechanism (but not specificity) [41, 43, 63], and the only exception are the GH23 and GH97 families that combine retaining and inverting GHs.[19] Table 1 summarizes the information about all GHs discovered until now.

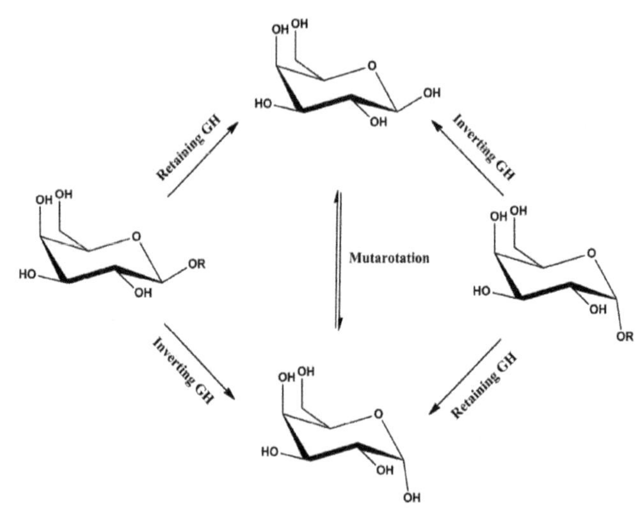

Figure 1. Mechanistic overview of retaining and inverting GHs activities.

Clan	Families (GHs)	Type of Mechanism	Three-dimensional structure
A [7,10,16,20-33]	1, 2, 5, 10, 17, 26, 30, 35, 39, 42, 50, 51, 53, 59, 72, 79, 86, 113, 128	Retaining	$(\beta/\alpha)_8$ barrel
B [34]	7, 16	Retaining	β-sandwich
C [35,36]	11, 12	Retaining	β-sandwich
D [37-43]	27, 31, 36	Retaining	$(\beta/\alpha)_8$ barrel
E [7,38]	33, 34, 83, 93	Retaining	6-bladed β-propeller
F [44,45]	43, 62	Inverting	5-bladed β-propeller
G [30]	37, 63	Inverting	$(\alpha/\alpha)_6$ barrel
H [46-52]	13, 70, 77	Retaining	$(\beta/\alpha)_8$ barrel
I [7,53-55]	24, 46, 80	Inverting	α+β lysozyme
J [26,44,45,56,57]	32, 68	Retaining	5-bladed β-propeller
K [7,25,28,30]	18, 20, 85	Retaining	$(\beta/\alpha)_8$ barrel
L [7,18,32]	15, 65, 125	Inverting	$(\alpha/\alpha)_6$ barrel
M [30]	8, 48	Inverting	$(\alpha/\alpha)_6$ barrel
N [30,32,58,59]	28, 49	Inverting	$(\beta)_3$ solenoid

Table 1. Organization of glycosidases families in clans and their correlation with the type of mechanism that they catalyze, and their protein folding.

2.1. Retaining GHs

The catalytic mechanism of retaining glycosidases was proposed about 58 years ago by Koshland *et al* [60] (Figure 2). According to this proposal, the mechanism occurs as a double displacement involving two steps: a glycosylation and a deglycosylation step. In the first step, the enzyme is glycosylated by the concerted action of the carboxylates of two residues, either Asp or Glu, or both that are found on opposite sides of the enzyme active site and are normally close to each other (around 5.5 Å). One of these residues functions as a general acid in the first step of the mechanism where the glycosidic bond starts to break. The acid residue donates a proton to the dissociated sugar. During the same step, the second deprotonated carboxylate acts as a nucleophile, attacking the anomeric carbon at the oxocarbenium ion-like transition state. This step, referred to as the glycosylation step, leads to the formation of a covalently linked glycosyl-enzyme intermediate that has an anomeric configuration opposite to that of the starting material. The second step of this reaction, the deglycosylation step, involves the hydrolytic breakdown of the glycosyl-enzyme intermediate[61]. The carboxylate that first functioned as an acid catalyst now acts as a base by abstracting a proton from the incoming nucleophile, usually a water molecule. Simultaneously, the water molecule attacks the carbohydrate-enzyme linkage in a reverse mode of the first step. At the end of the reaction, the enzymatic turnover is obtained and a hemi-acetal is formed with the same anomeric configuration as the starting material. Recent studies have shown that the transition states (TS1 and TS2) of both glycosylation and deglycosylation steps have a dissociative nature. Both reactions are favored by the distortion of the substrate during catalysis, but this effect is more evident in the first step of the reaction[62-65]. The glycosylation process is also favored by the hydrogen bond between the nucleophilic carboxylate and the hydroxyl group of position 2 in the substrate. It behaves almost as an anchor that aligns the substrate in the active site and facilitates the glycosylation process.

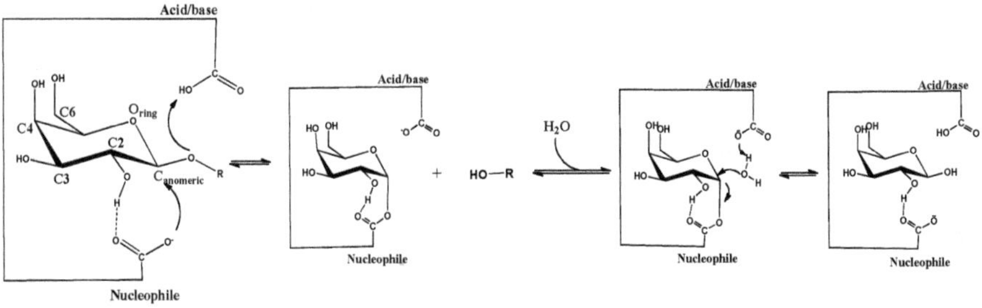

Figure 2. Catalytic mechanism of retaining GHs.

A variation of the general mechanism for retaining enzymes has been demonstrated for the N-acetyl-β-hexosaminidases, belonging to families 18 and 20[66,67]. Unlike the most retaining glycosidases, these enzymes lack a catalytic nucleophile, e.g. the water molecule. Instead, it is the acetamido substituent of the substrate that acts as an intramolecular catalytic nucleophile.[68-71] As it is shown in Figure 3, the general acid/base residue protonates the oxygen of the scissile glycosidic bond. The other charged carboxylate residue stabilizes the positive charge developed on the nitrogen of the oxazolinium ion that is formed after the intramolecular attack of the N-acetamido oxygen to the anomeric carbon.[68,72] To complete the double displacement mechanism, in the second step, an incoming water molecule attacks the anomeric carbon, resulting in a product with retention of the initial configuration.[73] In this reaction, several aromatic residues available in the active site have a key role to endorse the correct orientation of the nucleophilic carbonyl oxygen of the substrate and in this way promote and stabilize the formation of the oxazolinium ion.

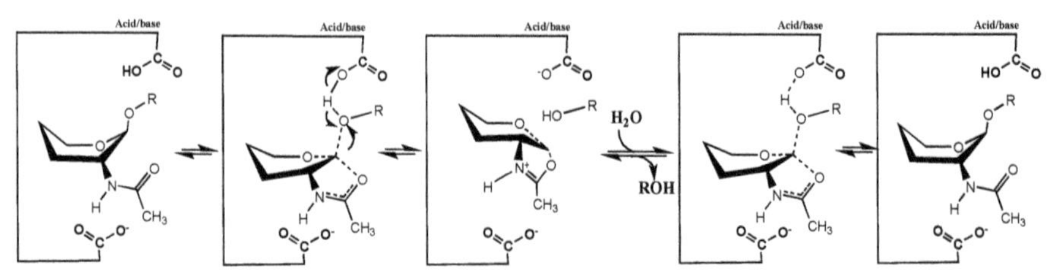

Figure 3. Catalytic mechanism of retaining β-hexosaminidase from families GH18 and GH20.

2.2. Inverting GHs

In inverting GHs, there is an inversion of the anomeric configuration of the starting material. Here, the two crucial carboxylic residues act as general acid and base catalysts and these groups are *circa* 10.5 Å apart from each other. It this specific case, this distance is larger than in retaining GHs because the substrate and the water molecule must be present simultaneously in the active site of the enzyme during the hydrolytic process.[18,66,74-78] Figure 4 shows the proposed mechanism of action for inverting GHs, which occurs via a single-displacement type of mechanism. In this case, one of the carboxylate residues protonates the scissile glycosidic oxygen atom while the other coordinates the nucleophile (i.e. the water molecule) to assist its deprotonation and in this way complete the hydrolysis reaction.[79]

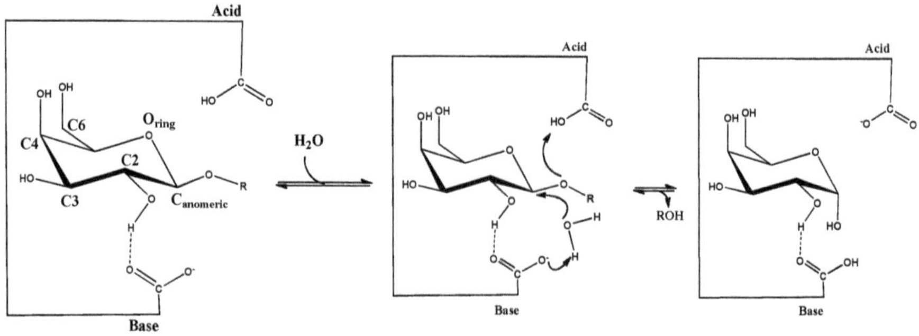

Figure 4. Catalytic mechanism of inverting GHs.

In contrast with the retaining mechanism, this reaction is completed in a single step and it is supposed that it requires the formation of a single transition state structure. Moreover, it does not involve the formation of any covalent enzyme intermediate during the course of catalysis and induces the inversion of the anomeric configuration of the starting material.

2.3. Cofactor dependent GHs

There are other GHs whose catalytic mechanism is substantially different from the mechanisms described above. One of the most interesting ones requires the presence of an NADH cofactor. The retaining 6-phospho-α-GH enzymes from family 4 are among these enzymes (Figure 5) in which the cofactor is perfectly positioned to remove the hydride from carbon C3 of the substrate[80]. Consequently, the acidity of the hydrogen atom that is attached to carbon C2 of the substrate increases and helps its abstraction by one of the tyrosine residues that is available in the active site and acts as a base. The hydroxyl group that is initially attached to carbon C3 of the substrate is subsequently oxidized to a ketone forming the 1,2-unsaturated reactive intermediate. Simultaneously, one of the carboxylate residues assists the cleavage of the sugar bound and the proton that is attached to carbon C2 of the substrate is abstracted by the base. This step is favoured by the presence of a metal ion in the active site of the enzyme that polarizes the carbonyl at carbon C3 and stabilizes the enolate species. The last step of this mechanism involves the nucleophilic attack of one water molecule to the double bond of the ketone. Simultaneously, the re-protonation of the enolate is catalysed by the close presence of the tyrosine residue, and the reduction of the ketone located at carbon C3 is accomplished by the NADH, favouring the enzymatic turnover.

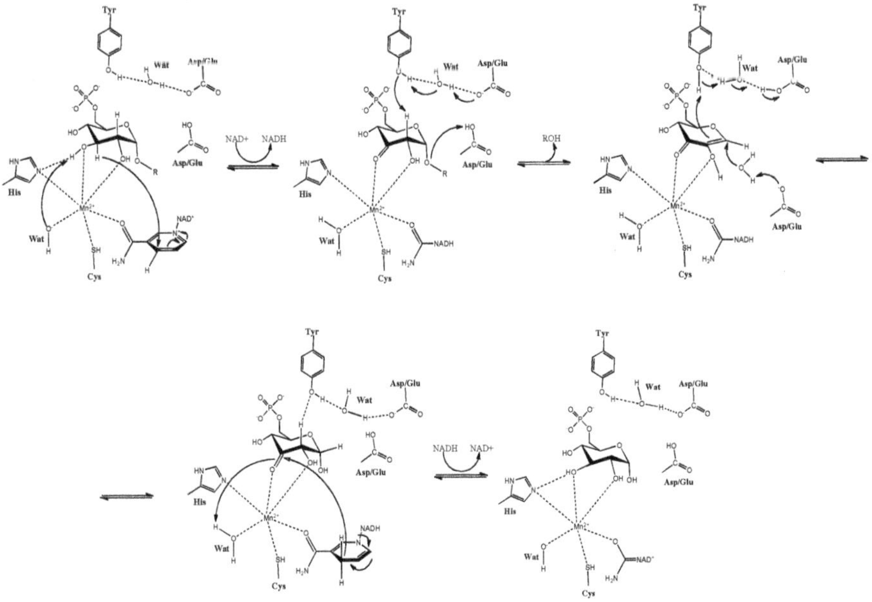

Figure 5. Catalytic mechanism for NADH dependent GHs.

2.4. Transglycosylation activity of GHs

In addition to the hydrolytic ability, GHs can also be used under appropriate conditions for the reverse reaction, thus promoting the formation of glycosidic linkages. This type of reactions are called transglycosylation[62] and generally require high concentration of substrate. The proposed catalytic mechanism is depicted in Figure 6. Similarly to the previous described mechanisms, the first step leads to the departure of the aglycon group and the formation of the covalent intermediate. The second step of the reaction involves the attack of the carbohydrate-enzyme linkage by another sugar molecule, and the proton transfer from the sugar to the active site acid/base carboxylate.

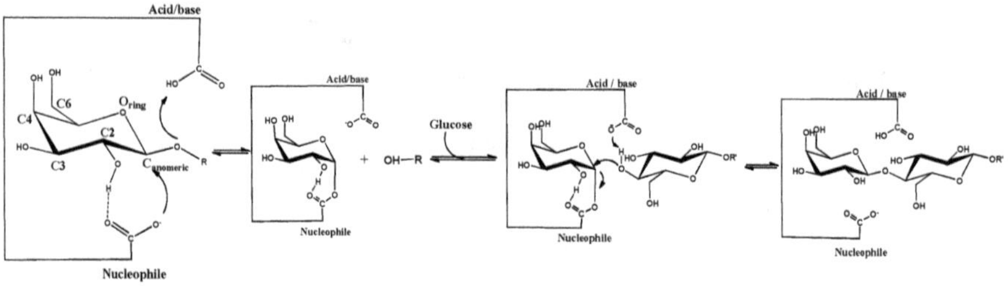

Figure 6. Transglycosylation reaction catalyzed by retaining GHs.

Usually, the synthesis of glycosidic linkages in nature is carried out natively by glycosyltransferases (that use activated glycosides as the glycosyl donors). Typical glycoside donors are expensive nucleotide sugars such as ADP glucose, UDP-glucose, and UDP-galactose. In contrast, the transglycosylation activity of GHs employs a considerably inexpensive substrate (such as simple sugars) as a glycoside donor molecule leading to large industrial interest in employing these enzymes for biotechnological synthesis. However, the yields for these transglycosylation reactions are typically low because the product itself is a substrate for the enzyme and undergoes hydrolysis. As their hydrolytic activities compete with this mechanistic pathway, it is necessary to displace the equilibrium of the glycosidic bond formation using excessive substrate concentration (thermodynamic control) or using good activated glycosyl donors, such as an aryl glycoside (kinetic control). Another disadvantage of the transglycosylation reactions catalysed by GHs is their limited efficiency for the glycosides synthesis in disaccharides or trisaccharides.[3,81] This happens because these reactions require high degrees of chemo, regio and stereo-selectivity.[82] The same is also true for oligosaccharides, but in this case the problems arise from their complex structure turning their chemical synthesis difficult to achieve, namely the production of glycosides with a mixture of various linkages (i.e., formation of 1-2, 1-3, 1-4 and 1-6 glycosidic bonds) and both anomers (α and β).[83] In this regard, the control of the stereospecificity and the regiospecificity of bond formation remains a challenging problem in the chemical synthesis of oligosaccharides [84,85]. A solution for this unsolved problem would be very important, as industrially there is only interest in the oligosaccharide target.

In order to overcome most of the limitations of the transglycosylation reaction in glycosidases, many enzymes have been mutated in the region of the active site in order to enhance the rate of this reaction. A classical example of mutated glycosidases are the glycosynthases, in which the hydrolytic activity has been inactivated through the mutation of their catalytic nucleophile residues by small non-nucleophilic residues, such as alanine, glycine or serine These enzymes possess a high activity because they are able to accept an activated glycosyl donor group (generally glycosyl fluorides or nitrophenyl glycosides) and catalyse transglycosylation reactions to an acceptor molecule.[81,85] In opposition to the native GHs, these engineered enzymes produce carbohydrates with elevated molecular weight and with higher product yields.[81] The first glycosynthase enzyme was reported in 1998 by Withers and colleagues[86] but, currently many other glycosynthases have been developed that posses specific substrate specificity. Figure 7 shows the reaction mechanisms of several types of glycosynthases. As native glycosidases, the glycosynthases can also show retaining and inverting mechanism. In the inverting α-glycosynthases, the donor group is transferred

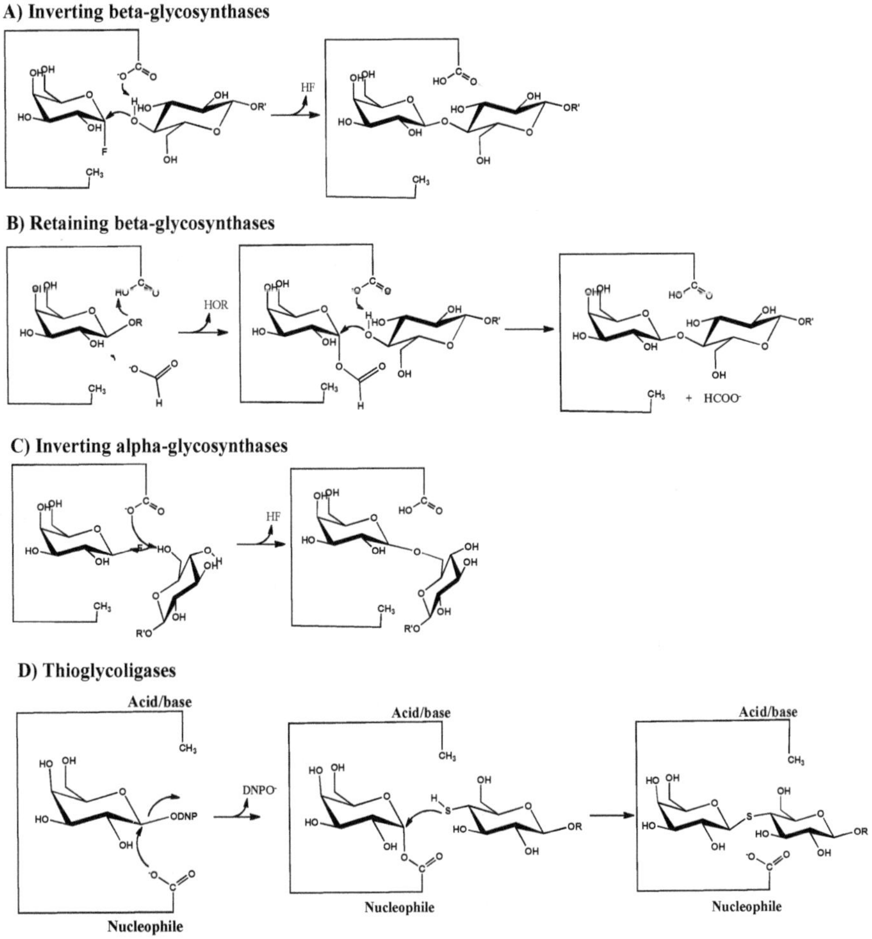

A) Inverting beta-glycosynthases

B) Retaining beta-glycosynthases

C) Inverting alpha-glycosynthases

D) Thioglycoligases

Figure 7. Catalytic mechanism of glycosynthases: a) inverting β-glycosynthase, b) retaining β-glycosynthase, c) inverting α-glycosynthase and d) thioglycoligase.

to the 4-nitrophenyl-α-glucoside acceptor group and the deglycosylation step proceeds similarly to what is observed with the retaining GHs (Figure 7a). The retaining glycosynthases act within the presence of one external nucleophile, such as sodium formate, and an activated donor group with the anomeric configuration of the native substrate (commonly 2 nitrophenyl- or 2,4-dinitrophenyl-β-glucoside) (Figure 7b). Therefore, the nucleophile mimics the catalytic active-site carboxylate of the enzyme and builds the formyl-glycoside intermediate. Subsequently, the donor carbohydrate is transferred to an acceptor sugar, promoting the polysaccharide synthesis.[85,87-90]

Some retaining glycosynthases can also have inverting mechanisms. This occurs when the donor sugar has a glucosyl fluoride in an opposite anomeric configuration relatively to the native substrate, thus mimicking the intermediate of the reaction (Figures 7a and 7c).[86]

Another type of glycosynthases are the thioglycoligase engineering enzymes, in which the mutated residue is the acid/base carboxylate instead of the nucleophile residue as in the previously described glycosynthases (Figure 7d). In these cases, a good leaving group such as dinitrophenyl, is placed in the substrate, which allows the formation of the glycosyl-enzyme intermediates that favors the catalytic process.[91]

3. Structural aspects that influence the GHs catalytic mechanism

The structural studies addressed at GHs have also provided important clues about how these enzymes enhance the catalytic process. As mentioned before, the distortion of the substrate along the full catalytic process is one of these mechanisms and this effect is found in many studies.[92-99] The available data reveals that GHs are able to selectively bind and stabilize high energy substrate conformations before the hydrolysis takes place. Such distortion of the substrate favours, in the Michaelis complex, the attack of the catalytic acid/base carboxylates to the glycosidic oxygen of the substrate. At the same time, it helps to guide the leaving group in a pseudoaxial position in relation to the substrate, facilitating the nucleophilic attack on carbon C1 and the subsequent cleavage of the glycosidic bond. It has been proposed also that the conformation changes of the ring along the catalytic process might determine the efficiency of the polysaccharides degradation. Taking into account these results and the available X-ray structures of several GHs that contain the substrates with different conformational distortions, Stoddart [100-102] proposed a diagram to classify the conformation of a α-glucopyranose molecule ring along the reaction pathway (Figure 8). In this diagram C, B, S corresponds to the chair, boat, and skew conformations, respectively. This diagram includes the most energetically stable $^{4}C_{1}$ chair, six boat-type and six skew-type conformations, as well as several transient structures (e.g. half-chair and envelope conformations) between the transition of $^{4}C_{1}$ chair to the boat/skew conformations.

The itinerary map of Stoddart gives therefore all the possible conformational pathways that a hypothetical substrate may follow as it moves from one conformation to another[94]. However, no energetically information can be extracted regarding the relative stability of different conformations, nor can it be assumed that all conformations on this map correspond to stationary points in the free energy landscape with respect to ring distortion.

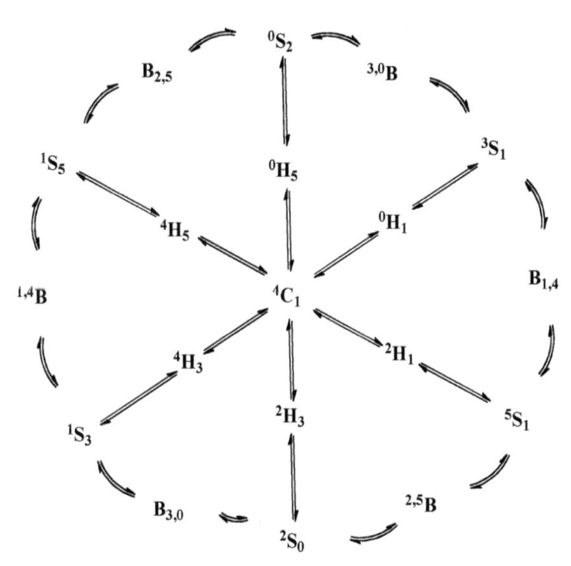

Figure 8. Stoddart's diagram.

Nevertheless, over the years, this diagram has been actively used as an "itinerary map" to design new enzyme inhibitors for therapeutic activities. In this regard, the conformational itinerary pathway of several GHs families has been studied, such as GH29 enzymes and α-xylosidases from the GH31 family that catalyze the hydrolysis using the $^{4}C_{1} \leftrightarrow {}^{3}H_{4} \leftrightarrow {}^{3}S_{1}$ glycosylation itinerary [93,103]; inverting endoglucanases from the GH8 family that use the β-$^{2}S_{0}/^{2,5}B \leftrightarrow {}^{2,5}B \leftrightarrow$ α-$^{5}S_{1}$ glycosylation itinerary of the glycon ring [104]; the glycosylation reaction of golgi α-mannosidase II from the GH38 family following an $^{0}S_{2}/B_{2,5} \leftrightarrow B_{2,5} \leftrightarrow {}^{1}S_{5}$ itinerary [105]; the catalytic itinerary of 1,3-1,4-β-glucanase from the GH16 family 16 pursue the $^{1,4}B/^{1}S_{3} \leftrightarrow {}^{4}E/^{4}H_{3} \leftrightarrow {}^{4}C_{1}$ [95,106] itinerary, and the inverting α-mannosidases from the GH47 family that follow the $^{3}S_{1} \leftrightarrow {}^{3}H_{4} \leftrightarrow {}^{4}C_{1}$ glycosylation itinerary.[107]

Many GHs also contain cations in the region of the active site. The presence of these species in the structure of GHs appears to be more common than it was initially imagined, and are believed to be very important for the stabilization of the transition states during catalysis. For instance, the Golgi α-mannosidase II from the GH38 family has a zinc ion in its 3D structure. Studies on the catalytic mechanism of this enzyme revealed that the Zn^{2+} ion is fundamental to coordinate the hydroxyl groups attached to carbons C2 and C3 of the mannosyl group, which stabilize the transition states, and thus reduces the overall activation energy required for the reaction. Furthermore, QM/MM metadynamics simulations also demonstrate that the zinc ion helps to lengthen the C2 hydroxyl bond when the substrate acquires the oxocarbenium character, facilitating the electron reduction of this species.[105] A similar role has been proposed for the calcium ion present in the structure of the endoplasmic reticulum α-mannosidase I from the GH47 family. The crystallographic structure shows that the cation coordinates with the hydroxyl groups that are attached to carbons C2 and C3 of 1-deoxymannojirimycin or kifunensin inhibitors.[108] A magnesium ion is also found close to the region of the active site of β-galactosidase from *Escherichia coli*. In these case, theoretical calculations have clearly demonstrated that the presence of the cation has a key role in lowering down the activation barrier by 14.9 kcal/mol, emphasizing its importance during the catalytic process.[63]

4. Conclusions

GHs are impressive nano-molecular machines that are present in almost all living organisms (exceptions are some Archaeans and a few unicellular parasitic eukaryotes). These enzymes catalyze the hydrolysis of the glycosidic linkage in a myriad of biological reactions and under specific conditions can also catalyse the reverse reaction promoting the formation of glycoside linkages.

The interest of GHs has started long ago but the catalytic power behind glycosidases is only now being established. Structural analyses of various enzyme complexes representing stable intermediates along the reaction coordinate together with detailed mechanistic and spectroscopic studies on wild type and mutant enzymes, have revealed that the source for their catalytic power is based on nucleophilic and general acid/base catalysis. These enzymes develop finely tuned active sites that contain two carboxylates residues (Asp and Glu) carefully aligned and positioned on opposite sides of the enzyme active site that embrace the substrate upon substrate binding. The active site also provides an extensive network of hydrogen bonds that endorse a conformational distortion of the substrate. This induces the substrate to adopt a higher energetic conformation before the hydrolysis takes place, and such configuration is maintained during the full catalytic process. This effect is very important for the stabilization of the transition-state structure and therefore to lower the activation barrier of the full process. Some GHs also possess positive ions (Zn^{2+}, Mn^{2+}, and Mg^{2+}) into their structure and these cations have been found also to be essential for the stabilization of the transition states during catalysis. The two most commonly employed mechanisms used to hydrolyze the glycoside linkage of substrates by glycosidases involve the retention or the inversion mechanisms. These mechanisms are conserved within the majority of GHs families. The only exception are the glycosidases from family 4 and 109, in which the hydrolytic process occurs via an elimination type of mechanism, and requires the presence of the NADH cofactor.

The structural and the mechanistic studies addressed to glycosidases provided and continue to provide important clues about the catalytic power of these enzymes. This knowledge is very important as it offers new ways to improve, modify or even inhibit the activity of GHs. These developments are particularly important for the biotechnology industries that have been increasing the commercial uses of glycosidases in several areas. Indeed, specific glycosidases are increasingly used for food processing, for bio-bleaching in the pulp and paper industry, as well as for biomass degradation with the potential to convert solid biomass into liquid fuels.

In the last two decades, it has also been watched an increasing interest of glycosidases for therapeutic uses. Glycosidases are involved in the biosynthesis of the oligosaccharide chains and quality control mechanisms in the endoplasmic reticulum of the N-linked glycoproteins. Inhibition of these glycosidases can have profound effects on quality control, maturation, transport, and secretion of glycoproteins and can alter cell–cell or cell–virus recognition processes. This principle is the basis for the potential use of glycosidase inhibitors in viral infection, diabetes and genetic disorders. [109,110]. Most of these drugs are glycosidase inhibitors that can bind and block the active sites of these enzymes. Some successful examples are the α-amylase inhibitor Acarbose and Miglitol that were approved by the FDA in 1990 and 1996, respectively, and are used to inhibit some of the intestinal glycosidases and pancreatic α-amylase in order to regulate the absorption of carbohydrates. These inhibitors are currently used therapeutically in the oral treatment of the non-insulin-dependent diabetes mellitus (typr II diabetes). Other glycosidases inhibitors are used as anti-viral agents[111]. Here the inhibitors were developed to inhibit the formation of glycoproteins of the viral envelopes, which are essential for virion assembly and secretion and/or infectivity. A successful example was the development of Zanamivir(Relenza) and Oseltamivir (Tamiglu), approved by FDA in 1999, that are used in the treatment and prophylaxis of influenza caused by influenza A virus and influenza B virus[112]. These compounds efficiently inhibit a glycosidases called neuraminidase. Glycosidases are also used in the therapy of human genetic disorders. The glycosphingolipid storage diseases (GSD, also glycogenosis and dextrinosis) are a rare hereditary disorders that are severe in nature and frequently fatal. These diseases result from defects in the processing of glycogen synthesis or breakdown within muscles, liver, and other cell types[113]. An example of such disorders is the Fabry disease that is caused by the deficiency of the essential enzyme α-glycosidase A, resulting in renal failure along with premature myocardial infarction and strokes. The only successful treatment is, to date, the enzyme replacement therapy. Fabrazyme was approved by FDA in 2003 and is intended to replace the missing enzyme in patients with this progressive disease.

Taking into account that almost two-thirds of all carbon that exist in the biosphere is carbohydrate, we believe that the current applications of GHs are only a small group of the many very important applications that these enzymes may find in the future. It is therefore expected that a wide variety of relevant and new applications will arise in the near future involving glycosidases. To stimulate these developments, the continuous study regarding glycosidases will be very important as they will provided crucial knowledge to turn their use more efficient and effective.

Author details

Natércia F. Brás, Pedro A. Fernandes, Maria J. Ramos and Nuno M.F.S.A. Cerqueira*
REQUIMTE, Departamento de Química e Bioquímica da Faculdade de Ciências, Universidade do Porto, Porto, Portugal

* Corresponding Author

5. References

[1] Henrissat, B.; Sulzenbacher, G.; Bourne, Y. *Current Opinion in Structural Biology* 2008, *18*, 527.

[2] Michael, A. *Am. Chem. J.* 1879, *1879*, 305.

[3] Bucke, C. *J Chem Technol Biot* 1996, *67*, 217.

[4] Blake, C. C. F.; Koenig, D. F.; Mair, G. A.; North, A. C. T.; Phillips, D. C.; Sarma, V. R. *Nature* 1965, *206*, 757.

[5] Flitsch, S. L. Current Opinion in Chemical Biology 2000, 4, 619.

[6] Bojarova, P.; Kren, V. *Trends Biotechnol* 2009, *27*, 199.

[7] *Carbohydrate Active Enzymes server,* http://www.cazy.org/ 2012.

[8] Cantarel, B. L.; Coutinho, P. M.; Rancurel, C.; Bernard, T.; Lombard, V.; Henrissat, B. *Nucleic Acids Research* 2009, *37*, D233.

[9] IUBMB: Enzyme Nomenclature. Recommendations.; Academic Press: San Diego, 1992.

[10] Henrissat, B.; Davies, G. *Current Opinion in Structural Biology* 1997, *7*, 637.

[11] Davies, G. J. Biochemical Society Transactions 1998, 26, 167.

[12] Coutinho, P. M.; Henrissat, B. In *Recent Advances in Carbohydrate Bioengineering*; Gilbert, H. J., Davies, G. J., Henrissat, B., Svensson, B., Eds. 1999, p 3.

[13] Henrissat, B. Biochemical Society Transactions 1998, 26, 153.

[14] Gebler, J.; Gilkes, N. R.; Claeyssens, M.; Wilson, D. B.; Beguin, P.; Wakarchuk, W. W.; Kilburn, D. G.; Miller, R. C.; Warren, R. A. J.; Withers, S. G. *Journal of Biological Chemistry* 1992, *267*, 12559.

[15] Henrissat, B.; Bairoch, A. *Biochemical Journal* 1996, *316*, 695.

[16] Henrissat, B.; Callebaut, I.; Fabrega, S.; Lehn, P.; Mornon, J. P.; Davies, G. *Proceedings of the National Academy of Sciences of the United States of America* 1995, *92*, 7090.

[17] Davies, G. J., Sinnott, M.L., Withers, S.G. *Glycosyl transfer. In Comprehensive Biological Catalysis. Edited by Sinnott ML.*; Academic Press: London, 1997; Vol. 1.

[18] Vasella, A.; Davies, G. J.; Bohm, M. *Current Opinion in Chemical Biology* 2002, *6*, 619.

[19] Gloster, T. M.; Turkenburg, J. P.; Potts, J. R.; Henrissat, B.; Davies, G. J. *Chemistry & Biology* 2008, *15*, 1058.

[20] Rabinovich, M. L.; Melnick, M. S.; Bolobova, A. V. *Biochemistry-Moscow* 2002, *67*, 850.

[21] Moller, P. L.; Jorgensen, F.; Hansen, O. C.; Madsen, S. M.; Stougaard, P. *Applied and Environmental Microbiology* 2001, *67*, 2276.

[22] Henrissat, B.; Teeri, T. T.; Warren, R. A. J. *Febs Letters* 1998, *425*, 352.

[23] Mian, I. S. Blood Cells Molecules and Diseases 1998, 24, 83.

[24] Juers, D. H.; Huber, R. E.; Matthews, B. W. *Protein Science* 1999, *8*, 122.

[25] Nagano, N.; Porter, C. T.; Thornton, J. M. *Protein Engineering* 2001, *14*, 845.

[26] Henrissat, B.; Bairoch, A. *Biochemical Journal* 1993, *293*, 781.

[27] Himmel, M. E.; Karplus, P. A.; Sakon, J.; Adney, W. S.; Baker, J. O.; Thomas, S. R. *Applied Biochemistry and Biotechnology* 1997, *63-5*, 315.

[28] Rigden, D. J.; Jedrzejas, M. J.; de Mello, L. V. *Febs Letters* 2003, *544*, 103.

[29] Jenkins, J.; Leggio, L. L.; Harris, G.; Pickersgill, R. *Febs Letters* 1995, *362*, 281.

[30] Pickersgill, R.; Harris, G.; Lo Leggio, L.; Mayans, O.; Jenkins, J. *Biochemical Society Transactions* 1998, *26*, 190.
[31] St John, F. J.; Gonzalez, J. M.; Pozharski, E. *Febs Letters* 2010, *584*, 4435.
[32] Stam, M. R.; Blanc, E.; Coutinho, P. M.; Henrissat, B. *Carbohydrate Research* 2005, *340*, 2728.
[33] Sakamoto, Y.; Nakade, K.; Konno, N. *Applied and Environmental Microbiology* 2011, *77*, 8350.
[34] Divne, C.; Stahlberg, J.; Reinikainen, T.; Ruohonen, L.; Pettersson, G.; Knowles, J. K. C.; Teeri, T. T.; Jones, T. A. *Science* 1994, *265*, 524.
[35] Tomme, P.; Warren, R. A. J.; Miller, R. C.; Kilburn, D. G.; Gilkes, N. R. In *Enzymatic Degradation of Insoluble Carbohydrates*; Saddler, J. N., Penner, M. H., Eds. 1995; Vol. 618, p 142.
[36] Torronen, A.; Kubicek, C. P.; Henrissat, B. *Febs Letters* 1993, *321*, 135.
[37] Naumoff, D. G. *Molecular Biology* 2004, *38*, 388.
[38] Henrissat, B. *Biochemical Journal* 1991, *280*, 309.
[39] Henrissat, B.; Romeu, A. *Biochemical Journal* 1995, *311*, 350.
[40] Naumoff, D. G.; Carreras, M. *Molecular Biology* 2009, *43*, 652.
[41] Dagnall, B. H.; Paulsen, I. T.; Saier, M. H. *Biochemical Journal* 1995, *311*, 349.
[42] MargollesClark, E.; Tenkanen, M.; Luonteri, E.; Penttila, M. *European Journal of Biochemistry* 1996, *240*, 104.
[43] Liebl, W.; Wagner, B.; Schellhase, J. *Systematic and Applied Microbiology* 1998, *21*, 1.
[44] Naumoff, D. G. Proteins-Structure Function and Bioinformatics 2001, 42, 66.
[45] Pons, T.; Naumoff, D. G.; Martinez-Fleites, C.; Hernandez, L. *Proteins-Structure Function and Bioinformatics* 2004, *54*, 424.
[46] Naumoff, D. G.; Rassb The alpha-galactosidase superfamily: Sequence based classification of alpha-galactosidases and related glycosidases, 2004.
[47] Ferretti, J. J.; Gilpin, M. L.; Russell, R. R. B. *Journal of Bacteriology* 1987, *169*, 4271.
[48] Rojas, A.; Garcia-Vallve, S.; Palau, J.; Romeu, A. *Biologia* 1999, *54*, 255.
[49] MacGregor, E. A.; Jespersen, H. M.; Svensson, B. *Febs Letters* 1996, *378*, 263.
[50] MacGregor, E. A.; Janecek, S.; Svensson, B. Biochimica Et Biophysica Acta-Protein Structure and Molecular Enzymology 2001, 1546, 1.
[51] Mooser, G.; Hefta, S. A.; Paxton, R. J.; Shively, J. E.; Lee, T. D. *Journal of Biological Chemistry* 1991, *266*, 8916.
[52] Janecek, S. *Biologia* 2005, *60*, 177.
[53] Monzingo, A. F.; Marcotte, E. M.; Hart, P. J.; Robertus, J. D. *Nature Structural Biology* 1996, *3*, 133.
[54] Wohlkonig, A.; Huet, J.; Looze, Y.; Wintjens, R. *Plos One* 2010, *5*.
[55] Tremblay, H.; Blanchard, J.; Brzezinski, R. *Canadian Journal of Microbiology* 2000, *46*, 952.
[56] Naumov, D. G.; Doroshenko, V. G. *Molecular Biology* 1998, *32*, 761.
[57] Pons, T.; Olmea, O.; Chinea, G.; Beldarrain, A.; Marquez, G.; Acosta, N.; Rodriguez, L.; Valencia, A. *Proteins-Structure Function and Genetics* 1998, *33*, 383.
[58] Jenkins, J.; Mayans, O.; Pickersgill, R. *Journal of Structural Biology* 1998, *122*, 236.
[59] Rigden, D. J.; Franco, O. L. *Febs Letters* 2002, *530*, 225.

[60] Koshland, D. E.; Stein, S. S. *Journal of Biological Chemistry* 1954, *208*, 139.
[61] Rempel, B. P.; Withers, S. G. *Glycobiology* 2008, *18*, 570.
[62] Bras, N. F.; Fernandes, P. A.; Ramos, M. J. *Theoretical Chemistry Accounts* 2009, *122*, 283.
[63] Bras, N. F.; Fernandes, P. A.; Ramos, M. J. *Journal of Chemical Theory and Computation* 2010, *6*, 421.
[64] Bras, N. F.; Moura-Tamames, S. A.; Fernandes, P. A.; Ramos, M. J. *Journal of Computational Chemistry* 2008, *29*, 2565.
[65] Bras, N. F.; Ramos, M. J.; Fernandes, P. A. *Journal of Molecular Structure-Theochem* 2010, *946*, 125.
[66] Davies, G.; Henrissat, B. *Structure* 1995, *3*, 853.
[67] Jefferson, T.; Jones, M. A.; Doshi, P.; Del Mar, C. B.; Heneghan, C. J.; Hama, R.; Thompson, M. J. *Cochrane Db Syst Rev* 2012.
[68] Passos, O.; Fernandes, P. A.; Ramos, M. J. *Journal of Physical Chemistry B* 2011, *115*, 14751.
[69] Passos, O.; Fernandes, P. A.; Ramos, M. J. *Theoretical Chemistry Accounts* 2011, *129*, 119.
[70] Bottoni, A.; Miscione, G. P.; Calvaresi, M. *Physical Chemistry Chemical Physics* 2011, *13*, 9568.
[71] He, Y.; Macauley, M. S.; Stubbs, K. A.; Vocadlo, D. J.; Davies, G. J. *Journal of the American Chemical Society* 2010, *132*, 1807.
[72] Mark, B. L.; Mahuran, D. J.; Cherney, M. M.; Zhao, D. L.; Knapp, S.; James, M. N. G. *Journal of Molecular Biology* 2003, *327*, 1093.
[73] Mark, B. L.; Vocadlo, D. J.; Knapp, S.; Triggs-Raine, B. L.; Withers, S. G.; James, M. N. G. *Journal of Biological Chemistry* 2001, *276*, 10330.
[74] Krasikov, V. V.; Karelov, D. V.; Firsov, L. M. *Biochemistry-Moscow* 2001, *66*, 267.
[75] Sinnott, M. L. *Chemical Reviews* 1990, *90*, 1171.
[76] Wang, Q. P.; Graham, R. W.; Trimbur, D.; Warren, R. A. J.; Withers, S. G. *Journal of the American Chemical Society* 1994, *116*, 11594.
[77] Rye, C. S.; Withers, S. G. *Current Opinion in Chemical Biology* 2000, *4*, 573.
[78] Zechel, D. L.; Withers, S. G. *Accounts of Chemical Research* 2000, *33*, 11.
[79] Koshland, D. E. Biological Reviews of the Cambridge Philosophical Society 1953, 28, 416.
[80] Rajan, S. S.; Yang, X. J.; Collart, F.; Yip, V. L. Y.; Withers, S. G.; Varrot, A.; Thompson, J.; Davies, G. J.; Anderson, W. F. *Structure* 2004, *12*, 1619.
[81] Faijes, M.; Planas, A. *Carbohydrate Research* 2007, *342*, 1581.
[82] Crout, D. H. G.; Vic, G. *Current Opinion in Chemical Biology* 1998, *2*, 98.
[83] Ajisaka, K.; Yamamoto, Y. Trends in Glycoscience and Glycotechnology 2002, 14, 1.
[84] Mayer, C.; Jakeman, D. L.; Mah, M.; Karjala, G.; Gal, L.; Warren, R. A. J.; Withers, S. G. *Chemistry & Biology* 2001, *8*, 437.
[85] Perugino, G.; Trincone, A.; Rossi, M.; Moracci, M. *Trends Biotechnol* 2004, *22*, 31.
[86] Mackenzie, L. F.; Wang, Q. P.; Warren, R. A. J.; Withers, S. G. *Journal of the American Chemical Society* 1998, *120*, 5583.
[87] Viladot, J. L.; de Ramon, E.; Durany, O.; Planas, A. *Biochemistry* 1998, *37*, 11332.

[88] Moracci, M.; Trincone, A.; Perugino, G.; Ciaramella, M.; Rossi, M. *Biochemistry* 1998, *37*, 17262.
[89] Trincone, A.; Perugino, G.; Rossi, M.; Moracci, M. *Bioorganic & Medicinal Chemistry Letters* 2000, *10*, 365.
[90] Perugino, G.; Trincone, A.; Giordano, A.; van der Oost, J.; Kaper, T.; Rossi, M.; Moracci, M. *Biochemistry* 2003, *42*, 8484.
[91] Witczak, Z. J. *Current Medicinal Chemistry* 1999, *6*, 165.
[92] Davies, G. J.; Planas, A.; Rovira, C. *Accounts of Chemical Research* 2012, *45*, 308.
[93] Ducros, V. M. A.; Zechel, D. L.; Murshudov, G. N.; Gilbert, H. J.; Szabo, L.; Stoll, D.; Withers, S. G.; Davies, G. J. *Angewandte Chemie-International Edition* 2002, *41*, 2824.
[94] Biarnes, X.; Ardevol, A.; Planas, A.; Rovira, C.; Laio, A.; Parrinello, M. *Journal of the American Chemical Society* 2007, *129*, 10686.
[95] Biarnes, X.; Nieto, J.; Planas, A.; Rovira, C. *Journal of Biological Chemistry* 2006, *281*, 1432.
[96] Mulakala, C.; Nerinckx, W.; Reilly, P. J. *Carbohydrate Research* 2006, *341*, 2233.
[97] Tailford, L. E.; Offen, W. A.; Smith, N. L.; Dumon, C.; Morland, C.; Gratien, J.; Heck, M. P.; Stick, R. V.; Bleriot, Y.; Vasella, A.; Gilbert, H. J.; Davies, G. J. *Nature Chemical Biology* 2008, *4*, 306.
[98] Greig, I. R.; Zahariev, F.; Withers, S. G. *Journal of the American Chemical Society* 2008, *130*, 17620.
[99] Soliman, M. E. S.; Ruggiero, G. D.; Pernia, J. J. R.; Greig, I. R.; Williams, I. H. *Organic & Biomolecular Chemistry* 2009, *7*, 460.
[100] Stoddart, J. F. *In Stereochemistry of Carbohydrates.*; Wiley-Interscience: Toronto, Canada, 1971.
[101] Davies, G. J.; Ducros, V. M. A.; Varrot, A.; Zechel, D. L. *Biochemical Society Transactions* 2003, *31*, 523.
[102] Taylor, E. J.; Goyal, A.; Guerreiro, C.; Prates, J. A. M.; Money, V. A.; Ferry, N.; Morland, C.; Planas, A.; Macdonald, J. A.; Stick, R. V.; Gilbert, H. J.; Fontes, C.; Davies, G. J. *Journal of Biological Chemistry* 2005, *280*, 32761.
[103] van Bueren, A. L.; Ardevol, A.; Fayers-Kerr, J.; Luo, B.; Zhang, Y. M.; Sollogoub, M.; Bleriot, Y.; Rovira, C.; Davies, G. J. *Journal of the American Chemical Society* 2010, *132*, 1804.
[104] Petersen, L.; Ardevol, A.; Rovira, C.; Reilly, P. J. *Journal of Physical Chemistry B* 2009, *113*, 7331.
[105] Petersen, L.; Ardevol, A.; Rovira, C.; Reilly, P. J. *Journal of the American Chemical Society* 2010, *132*, 8291.
[106] Biarnes, X.; Ardevol, A.; Iglesias-Fernandez, J.; Planas, A.; Rovira, C. *Journal of the American Chemical Society* 2011, *133*, 20301.
[107] Karaveg, K.; Siriwardena, A.; Tempel, W.; Liu, Z. J.; Glushka, J.; Wang, B. C.; Moremen, K. W. *Journal of Biological Chemistry* 2005, *280*, 16197.
[108] Vallee, F.; Karaveg, K.; Herscovics, A.; Moremen, K. W.; Howell, P. L. *Journal of Biological Chemistry* 2000, *275*, 41287.
[109] vonItzstein, M.; Colman, P. *Current Opinion in Structural Biology* 1996, *6*, 703.
[110] Taylor, G. Current Opinion in Structural Biology 1996, 6, 830.

[111] Asano, N. *Glycobiology* 2003, *13*, 93R.
[112] von Itzstein, M. *Nat Rev Drug Discov* 2007, *6*, 967.
[113] Aerts, J. M.; Hollak, C.; Boot, R.; Groener, A. *Philos T Roy Soc B* 2003, *358*, 905.

Section 2

Microbiology and Immunology

Chapter 7

IL-13, Asthma and Glycosylation in Airway Epithelial Repair

Samuel J. Wadsworth, S. Jasemine Yang and Delbert R. Dorscheid

Additional information is available at the end of the chapter

1. Introduction

1.1. Clinical assessment of asthma

Pulmonary function tests (PFTs/spirometry) are routinely used clinically to diagnose asthma. Forced expiratory volume in one second (FEV_1) is a measure of airflow and is influenced largely by the resistance in the airways. During an "asthma attack", the airways narrow owing to mucus hypersecretion and smooth muscle contraction, which results in a sharp increase in airway resistance and reduction in FEV_1. In the clinics, this can be reproduced by exposing patients with asthma to non-specific smooth muscle contractile agonists such as methacholine. A fall in FEV_1 at a relatively low dose of methacholine indicates airway hyperresponsiveness (AHR). In asthmatic patients this drop in FEV_1 can be reversed by inhaled bronchodilators such as β2-agonists (e.g. salbutamol), which relaxes airway smooth muscles. AHR is often used as a diagnostic criterion for asthma.

Asthma is increasingly considered a syndrome, with diverse overlapping pathologies and phenotypes contributing to significant heterogeneity in clinical manifestation, disease progression, and treatment response [1]. Severe uncontrolled asthmatics make up approximately 5-10% of the population of asthmatics, yet they consume around 50% of the treatment resources [2]. Spirometric tests are incapable of exposing the underlying pathologies or phenotypes which combine within an individual to produce the asthmatic disease. In contrast specific biomarkers may enable more accurate sub-phenotyping of disease by indicating the pathology within an individual, and assist clinicians to better tailor the type and/or dose of therapy. In this way biomarkers have the potential to guide more effective personalized treatment regimes in asthma.

1.2. The causes of asthma

Several studies have used lung biopsies taken from non-asthmatic patients and those who died from fatal asthma to show the airways of asthmatics possess characteristic structural differences, shown in Figure 1. These include but are not limited to; increased eosinophil numbers in asthmatics, and areas of focal bronchial epithelial denudation [3]. The contents of the bronchial lumen can also be examined using the bronchoalveolar lavage (BAL) technique, in which warm saline is instilled into a segmental bronchus of a patient and removed. BAL fluid-based studies confirmed the previous research that inflammatory cell influx and epithelial damage occur in the airways of asthmatics [3] [4]. It is a point of some debate as to whether the epithelial damage is caused by inflammation, is a cause of inflammation, or whether epithelial damage and airway inflammation are concurrent but unrelated processes. We shall discuss the potential role of the epithelium and epithelial repair in asthma in more detail later.

1.3. Inflammation in asthma

The rationale for treating severe asthmatics with steroids is centred on the assumption that the disease is driven by uncontrolled airway inflammation. The classical paradigm of the acute inflammatory response in allergic asthma is briefly; inhaled allergenic antigens are captured by professional antigen-presenting dendritic cells, macrophages, or epithelial cells. Antigens presented by these cells are recognized by T-cells which proliferate and differentiate. Primed Th_2 cells bind to B cells and release cytokines including triggering maturation of antigen-specific B cell populations into plasma cells. Plasma B cells release antigen-specific IgE that binds to IgE-receptors on mast cells in the airways, causing release of histamine-containing granules. Extracellular histamine released from mast cells binds to membrane receptors on airway smooth muscle cells triggering a rise in intracellular calcium, muscle contraction and airway narrowing. Although this paradigm has gained acceptance as an important mechanism in asthma, there is significant variation in specific cell types involved between individuals. In the setting of chronic asthma, eosinophils are thought to play a major role in maintaining airway inflammation in the long-term. Neutrophils are occasionally the predominate inflammatory cell present in the airways of chronic asthmatics, suggesting multiple underlying pathologies of disease. Neutrophilic vs. eosinophilic asthma are clinically indistinguishable by pulmonary function testing, but such pathological variation may have important implications regarding treatment as patients with "neutrophilic asthma" tend to be relatively unresponsive to steroid treatment [5]. A rapid and definitive method for determining the nature of airways inflammation in a diagnosed asthmatic is therefore needed to guide the clinician's selection of therapy.

1.4. IL-13 and asthma

Interleukin-13 (IL-13) is a pleiotropic Th_2 cytokine secreted from a variety of inflammatory cells and structural cells, including the airway epithelium. IL-13 is upregulated in the airways of patients with allergic asthma exposed to allergen; segmental allergen challenge of atopic asthmatics triggered an increase of IL-13 protein and transcript in bronchoalveolar lavage, and the source was identified as mononuclear inflammatory cells [6]. Experiments in small mammals and *in vitro* studies using human cells, have demonstrated IL-13 triggers many of the pathophysiological characteristics of asthma, independent of the IgE and eosinophil inflammatory mechanisms classically associated with allergic disease [7] [8]. Genetic studies in humans have demonstrated the gene loci for IL-13 and the IL-4 receptor portion of the IL-13Rα1 receptor have a strong genetic association to asthma susceptibility and severity [9]. IL-13 can act directly on the airway epithelium in various ways; previous studies have demonstrated IL-13 exposure of cultured human airway epithelial cells triggers an matrix-metalloproteinase-7 (MMP-7)-dependent release of the pro-inflammatory chemokine, soluble Fas-ligand, potentially leading to local inflammation and epithelial damage via removal of the protective Fas-ligand-mediated epithelial immune barrier [10]. Mucous hypersecretion is a common observation in asthmatic airways; acute treatment of differentiated cultures of primary human airway epithelial cells with IL-13 induces a hypersecretory phenotype [11], which develops into an asthma-like mucous hyperplasia with a 5-10-fold increase in goblet cell density and MUC5AC (the predominant airway mucin) detection after 14 day treatment [12]. Murine studies confirm IL-13 is required and sufficient to directly induce epithelial mucous hyperplasia *in vivo* in the absence of intermediate inflammatory cells [13]. Mucins including MUC5AC are highly glycosylated proteins, it thus follows that increased mucous production will necessitate an increase in the activity of associated transferase enzymes. In the H292 airway epithelial cell line, IL-13 treatment has been shown to increase the activity of the M and L isoforms of core 2 beta1,6 N-acetylglucosaminyltransferase (C2GnT) via activation of a JAK/STAT signaling pathway [14]. Thus, IL-13 acts directly on the airway epithelium to trigger development of an asthma-like phenotype, at least in part by increasing protein glycosylation via upregulation of glucosyltransferase activity.

1.5. Current treatments for asthma

Asthma is rarely fatal but often leads to significant patient morbidity. For over thirty years the mainstay therapy for asthma has been inhaled β2-agonists (βA), which relax the airway smooth muscles, causing bronchodilation, and inhaled glucocorticosteroids (i.e. steroids), which ameliorate airway inflammation and reduce the risk of asthma exacerbations. Steroids are anti-inflammatory drugs but also have many unwanted side-effects, particularly at high doses, including; weight-gain due to changes in body metabolism, and a reduction in bone mineral density, potentially leading to brittle bones. Therefore the decision to treat a patient with steroids should not be taken lightly. Inhaled β_2-agonists (s) are the most effective bronchodilators available and their main mechanism of action is via the relaxation of airway smooth muscle [15]. βAs interact with G-protein-coupled cell surface β_2-adrenoceptors (β_2AR) on bronchial smooth muscle cells and activate adenyl cyclase, causing a rise in intracellular cAMP, this results in relaxation of the smooth muscle and bronchodilation. Long term βA treatment can result in a loss of response which shows variation between individuals. Desensitisation due to chronic βA treatment is caused by a down-regulation of receptor transcription, secondary to a reduced activity of the CREB (cyclic AMP response element binding protein) transcription factor [15]. The β_2-adrenoceptor gene possesses at least three possible GRE binding sites for the activated GCR, and GCs have been shown to increase β_2AR transcription two-fold in human lung preparations *in vitro* and also in rat lungs *in vivo*. The downregulation of β_2AR transcription resulting from long-term βA treatment can therefore be reversed by a GC, so there is no net change in the level of β_2ARs expressed in the bronchial epithelium or airway smooth muscle of animals treated with a combination of GCs and βAs [15]. The very large β_2AR reservoir in airway smooth muscle cells may mean that this upregulation by GCs is functionally unimportant *in vivo* [15]. Many asthmatics who are unresponsive to high doses of steroids show a marked improvement in symptoms when given steroids with βA together. Thus there may be a true synergy between the two classes of drugs as β_2-adrenoceptor agonists enhance the function of GCs both *in vitro* and *in vivo*, possibly by increasing the nuclear localisation of activated GCRs. Low-dose theophylline has been prescribed as an asthma treatment for many years, however it has now been relegated to a third-line therapy according to the global asthma guidelines (reviewed in [16]). Theophylline is a weak, non-specific cyclic nucleotide phosphodiesterase (PDE) inhibitor, and recently these PDE inhibitory characteristics have re-ignited interest in this drug as a possible asthma treatment.

Monoclonal antibodies have gained recent credence as a potential means to target therapy against specific receptors and signalling pathways. Omalizumab, a biologic agent that binds to IgE, has shown some promise as a means of controlling some difficult to treat asthma patients who are poorly controlled with standard inhaled steroid and β-agonist combination therapy (reviewed in [17]). Interleukin-5 (IL-5) is an important cytokine for eosinophil differentiation, maturation, migration into the circulation and survival. Various studies have implicated the eosinophil as being the primary cell responsible for airway hyperresponsiveness in asthma, thus anti-IL-5 antibodies have been developed (such as mepolizumab) with the intention of reducing airways hyperresponsiveness by preventing eosinophil recruitment and survival in the airway. Clinical trials have demonstrated such inhaled anti-IL-5 therapy does indeed reduce airway eosinophilia, but does not have a clinically-beneficial effect on lung function [18, 19]. Considering the highly important role IL-13 is thought to play in asthma, it is unsurprising that therapies are being developed to specifically target IL-13. Lebrikizumab, an anti-IL-13 monoclonal antibody, has just completed phase II clinical trials in which patients with poorly-controlled asthma were given the drug subcutaneously at 4 weekly intervals for 6 months [20]. This study found anti-IL-13 therapy did significantly improve lung function, but only in a sub-group of asthma patients who exhibited high serum periostin levels, a surrogate marker of high airway IL-13 [20]. These studies demonstrate highly-targeted therapies may be useful in asthma, but each individual drug will probably only prove effective in a specific sub-group of asthmatic patients. The potential for such personalised medicine highlights the importance for accurate phenotyping of asthmatic individuals, discussed in the biomarkers section of this chapter (Section 5).

1.6. Carbohydrates and asthma

Carbohydrate decoration of proteins has been associated with asthmatic disease. Two case-control studies have investigated the role of the histoblood group antigens; H(O), A, B, or AB in asthma susceptibility. They found in human subjects the O-secretor mucin glycan (H-antigen) phenotype is associated with an increased susceptibility to recurrent asthma exacerbations [21].

2. Normal airway structure

From the level of the bronchioles upwards, the stratified epithelium lining the airway lumen rests on a basal lamina or 'true' basement membrane, a specialised 'mat' of extracellular matrix proteins. Below the basal lamina is a layer of collagenous matrix, termed the reticular basement membrane, in which fibroblasts are embedded in a sporadic arrangement. Farther below the fibroblastic layer are arranged bundles of smooth muscle myocytes, blood vessels and afferent nerve endings.

2.1. Structural changes to the airway in asthma

Although asthma is considered an inflammatory disease, there are many structural changes in the airways. Figure 1. shows cross-sections through the large airways of two patients, one normal and one severe asthmatic. The Movat's pentachrome stain clearly highlights the various architectural remodeling events occurring in the asthmatic airway. Obstruction of the airways from excessive mucus production is a common finding in severe asthmatics, blue staining in the epithelium and the lumen demonstrates mucous cell hyperplasia, with excessive mucus deposition into the airway. Under the asthmatic epithelium a thicker basement membrane is present which contains several different extracellular matrix (ECM) factors (including tenascin-C) compared to normals [22]. Deeper into the airway, red-stained muscle mass is increased in the asthmatic owing to a combination of smooth muscle hypertrophy and hyperplasia (reviewed in [23]). The airways of severe asthmatics also demonstrate a fibrotic response with increased connective tissue deposition and fibroblast and myofibroblast proliferation. Opinion is divided as to whether inflammation precedes airway remodeling, or whether the two occur in parallel (reviewed in [15, 24, 25]). Evidence tends to favour the latter because; firstly, remodeling occurs very early on in the disease, and in some cases in the absence of inflammation [26], secondly, there is only a weak link between airway inflammation and symptoms [27], and thirdly, epidemiological data demonstrate steroids do not work in all asthmatics [5]. In reality it is likely effective therapies will need to target both airway inflammation and remodeling.

2.2. Airway epithelial structure and function

The bronchial epithelium lines the inner wall of the respiratory tract in a continuous layer; it forms the interface between inspired air and the internal milieu as well as being the primary target for inhaled respiratory drugs. In the larger airways and down the respiratory tree to the level of the bronchioles the epithelium is at least two cells thick (reviewed. in [28]). Although there are at least eight morphologically and functionally distinct epithelial cell types present in the respiratory tract, they can be classified into three main groups; basal, ciliated columnar, and secretory columnar (reviewed in [29]).

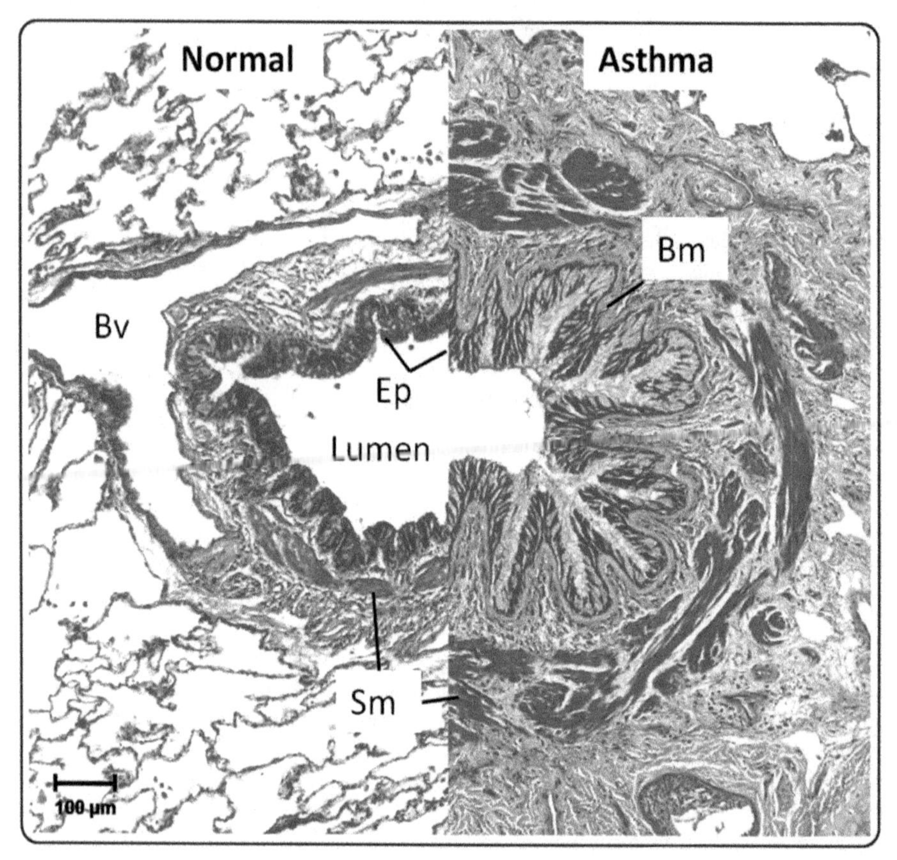

Figure 1. The airways in asthma undergo significant structural remodeling. Medium-sized airways from a normal and severe asthmatic patient were sectioned and stained using Movat's pentachrome stain. The epithelium (Ep) in asthma shows mucous hyperplasia and hyper secretion (blue), and significant basement membrane (Bm) thickening. Smooth muscle (Sm) volume is also increased in asthma. Bv = blood vessel. Scale bar = 100μm.

Basal cells

Basal epithelial cells are pyramidal-shaped cells with a low cytoplasm to nucleus ratio, although found throughout the airway, their contribution to epithelial volume decreases with airway size (reviewed in [29]). Basal cells are anchored to the underlying basement membrane and to other cells via specialised adhesion structures discussed below (Figure 2.)

[4, 30]. Historically the basal cell has been considered to be the stem cell of the bronchial epithelium (reviewed in [31]), however animal studies suggest there may be alternative stem cells located in the airway epithelium, including secretory cells, and cells located at specialised niches including the broncho-alveolar junction and the neuroepithelial bodies.

Columnar ciliated cells

Columnar epithelial cells lie above the basal cell layer and line the airway lumen, in direct contact with the inspired air. Ciliated cells are terminally differentiated columnar cells and are the most common cell-type in the bronchial epithelium, accounting for around 50% of all epithelial cells and 80% of terminally-differentiated apical cells in the human trachea [32], reviewed in [29]. Their main function is the removal of particulate matter by means of the mucociliary pathway. Ciliated cells arise from either a basal or secretory cell pre-cursor and possess around 300 cilia per cell, directly beneath the cilia are observed numerous mitochondria and a dense microtubule system, reflecting the high metabolic demands of particle clearance [29] [33].

Secretory cells

Secretory cells are also present at the apical surface, comprising 15 to 20% of the normal tracheobronchial epithelium [32]. In the larger airways the mucus-secreting goblet cell represents the predominant secretory cell and is the main source of airway mucus. These cells are characterised by membrane-bound electron-lucent acidic-mucin granules which are released into the airway lumen to trap inhaled particles and pathogens, prior to removal from the respiratory tract by coordinated cilia beating. Serous cells and Clara cells are relatively rare secretory cells in the larger airways present in large airways, but Clara cells are the main secretory epithelial cell type of the bronchioles, particularly the most distal bronchioles (reviewed in [34]). Several studies in rodents indicate the progenitor cell of the bronchial epithelium is not the basal cell but it is in fact the non-ciliated, secretory columnar cell. Both goblet and clara cells have been observed to undergo de-differentiation and proliferation during *in vivo* wound repair [4] [35-38].

2.3. Airway epithelial carbohydrate expression

The airway epithelial surface is covered by a layer of airway surface liquid (ASL), mainly of epithelial origin. A layer of mucins form a mucous layer that overlies the thin watery periciliary layer (PCL). The predominant mucins in the airways are MUC5AC and MUC5B secreted by goblet cells and submucosal glands respectively. Mucins are highly glycosylated proteins; 70-80% of their molecular weight are carbohydrate, and the structural characteristics of the mucous layer depends on interaction between the carbohydrate side-chains of mucin proteins. Studies of airway mucins in disease, demonstrate mucins are both oversulphated and hyper-sialylated in patients with cystic fibrosis and in chronic bronchitis patients with significant infections [39]. Pseudomonas aeruginosa, the pathogen responsible for the majority of morbidity and mortality in cystic fibrosis patients, uses the sialylated and sulphated Lewis-x determinants as attachment sites. Thus mucin glycosylation patterns may significantly affect infection in the airway. Specific lectin-binding assays have been used to characterise carbohydrate expression patterns in differentiated human airway epithelium [40]. Out of 38 lectin probes tested, seven bound specifically to basal cells, seven to columnar cells, and three specifically labelled secretory cells. The 1HAEo- and 16HBE14o- airway epithelial cell lines were also probed with the same lectin panel, revealing identical carbohydrate expression to the basal epithelial cells in the human tissue [40]. Thus the different cell types in the pseudostratified epithelium of the airways, express specific patterns of carbohydrates, possibly reflecting their different functions.

2.4. Airway epithelial functions

The primary function of the bronchial epithelium is to serve as a continuous physical barrier and as such it functions as part of the non-specific immune system, preventing pollutants, bacteria, viruses, allergens, and other potentially noxious substances transferring from inspired air into the underlying mesenchyme. This defence is mediated by several adhesive mechanisms which have been elucidated through transmission electron microscopy and immunohistochemical studies of bronchial epithelial biopsies [4] [30] (Figure 2.). Belt-like tight junctions (zonula occludens) seal the lateral apices of columnar cells, regulating para-cellular transport, whilst columnar cells adhere to each other via classical E-cadherin mediated adherens junctions [30] Desmosomal cadherins mediate strong cell-cell adhesion and desmosomes are present between basal cells and particularly at the junction of basal and columnar cell layers [30]. Basal cells are in turn anchored to the basement membrane via specialised integrin-mediated cell-ECM junctions. Integrin heterodimers function by dynamically linking the contractile machinery of the cell's cytoskeletal network to the external matrix at specialised sites termed focal adhesions or focal contacts. The airway epithelium expresses several integrin heterodimers, including the $\alpha6\beta4$ integrin which binds to laminin-5 in the basement membrane forming a hemidesmosome anchor with the ECM.

The bronchial epithelium is more than a passive barrier; it performs a variety of roles allowing it to function as a dynamic regulator of the innate immune system [39]. The mucociliary pathway represents the principle mechanism of particle clearance in the airway. Inhaled particles and pathogens (bacteria, viruses) are trapped in mucus prior to removal from the respiratory tract by coordinated cilia beating. In addition to producing mucus, secretory cells are the source of a variety of mediators capable of having direct effects on inhaled noxious agents; including anti-oxidants and a variety of anti-bacterial agents; lactoferrin, lysozyme, β-defensins and also opsonins, components of the complement system that coat bacteria to facilitate phagocytosis by macrophages (reviewed in [39]). Various immunoglobin isotypes are secreted onto mucosal surfaces where they act in a protective manner by opsonising bacteria and other pathogens, rendering them relatively harmless and activating phagocytosis and digestion by tissue macrophages. Allergen-specific IgA is secreted by activated B lymphocytes (plasma cells) in the sub-mucosa, dimeric IgA (dIgA) selectively binds to the polymeric Ig receptor (pIgR) expressed on the basal surface of basal airway epithelial cells. The pIgR-IgA complex is subsequently internalised and is transported to the apical surface via the endosomal pathway ready for secretion into the airway lumen. Dimeric secretory IgA (sIgA) is released into the airway lumen as a 1:1 combination of dIgA and a portion of the pIgR known as secretory component (SC), which protects the IgA from digestion by proteases on the mucosal surface [40, 41]. SC is a highly glycosylated protein containing 15% N-linked carbohydrate. Glycosylated SC binds IL-8, inhibiting IL-8-mediated neutrophil chemotaxis and transendothelial migration. This interaction is dependent on SC glycosylation state as de-glycosylation with peptide N-glycosidase F abolishes the SC-IL-8 complex [42].

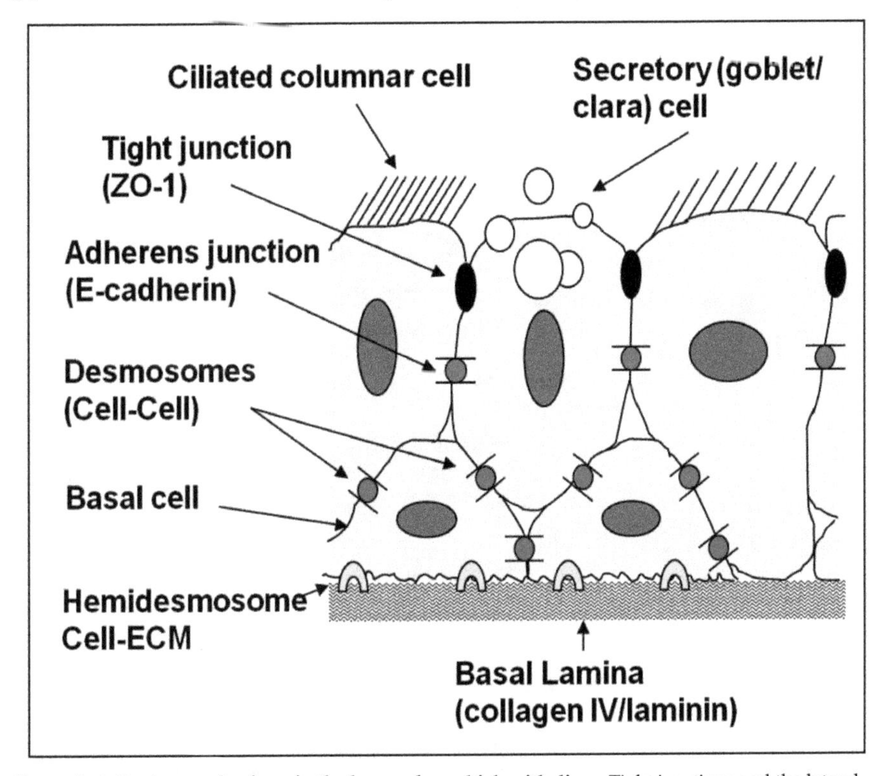

Figure 2. Adhesion mechanisms in the human bronchial epithelium. Tight junctions seal the lateral apices of columnar cells which adhere to each other via adherens junctions. Desmosomes mediate basal-basal and columnar-basal cell attachments, whilst hemidesmosomes anchor basal cells to the basal lamina. (Adapted from Roche et al. [30]).

3. Epithelial injury and repair

All epithelial tissues are regularly damaged as a consequence of exposure to environmental insults. Rapid repair following injury is crucial for restoring adequate barrier function, with subsequent cellular differentiation required to regenerate normal epithelial structure and function. The repair processes in all epithelial types have common elements producing an orderly progression of events including; cell spreading at the margins of the wound, cell migration into the wound, cell proliferation and finally re-differentiation (Figure 3). *In vivo* investigations of wound repair have demonstrated that migration and proliferation are both critical for the rapid restitution of the bronchial epithelium after injury [35-38]. Hamster tracheal epithelia were mechanically denuded to leave behind a bare (and sometimes damaged) basal lamina and wound repair was monitored until a morphologically normal epithelium was restored. Following epithelial removal, plasma promptly exudes into the injured site from the underlying vasculature to cover the denuded basement membrane, facilitating the binding of serum proteins to cellular receptors that stimulate repair of the damaged epithelium. At 12h post-wounding, viable secretory and basal cells at the wound margins de-differentiate, flatten into a squamous morphology, and migrate at about 0.5μm/min to cover the wound site.

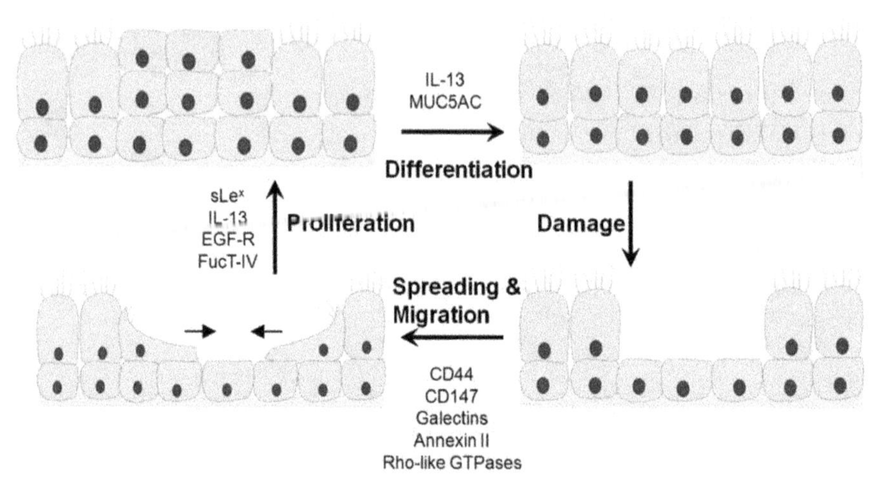

Figure 3. Airway epithelial repair involves a series of events including, cell spreading, migration, proliferation and differentiation. Following epithelial injury, neighboring cells flatten, spread and migrate to cover the site of injury. Subsequent proliferation and differentiation restores the epithelial structure and barrier function. The different stages of repair are mediated and facilitated by a diverse array of growth factors, cytokines and cell surface receptors, adhesion molecules, and intracellular enzymes, many of which are glycosylated. These glycans can modulate the function of proteins and lipids thereby regulating the repair process.

Proliferation is also an intrinsic component of epithelial repair. In the aforementioned hamster tracheal injury studies, mitosis rates at 12h post-injury are low (0.4%) but by 24h a wave of proliferation occurs primarily in secretory cells, resulting in a multi-layered epidermoid metaplasia [35]. At 48h post-wounding proliferation rates decrease (although mitosis is still significantly above basal levels), from this time point onwards the upper-most cells of the metaplastic epidermis start to slough off, resulting in a gradual loss of the metaplastic phenotype, with subsequent regeneration of a functional mucociliated phenotype achieved by apical cell re-differentiation. In small wounds normal mucociliary structure is restored by 120h but some persistent metaplasia persists in larger wounds through 168h post-wounding [38]. Although both secretory and basal cells are involved in epithelial repair, secretory cells play a dominant role in these experiments, consistently demonstrating higher mitosis rates than any other epithelial cell type [35-38]. *In vitro* studies using muco-ciliated human bronchial epithelial cells grown in air-liquid interface (ALI) culture, demonstrate that scrape wound repair is achieved by rapid spreading and migration of cells at the wound edge, followed by proliferation of basal cells outside of the wound area [43]. Acute wound repair speeds were inhibited slightly by the steroid dexamethasone, but the same drug potentiated the ability of long-term ALI cultures to repair repeated wounds [43].

In vivo studies demonstrate epithelial repair is accompanied by a corresponding proliferation of the underlying mesenchyme. Hamster tracheal epithelial damage is followed at 24h by capillary endothelial cell division, and at 36h many fibroblasts were observed to be in mitosis [35-37]. Mesenchymal proliferation persisted for a longer period than the epithelial response and supports several *in vitro* studies demonstrating epithelial damage can influence the phenotype and proliferation of underlying fibroblasts [44, 45].

3.1. Airway epithelial wound repair is compromised in asthmatics

The patho-physiological changes observed in the asthmatic airway may be due to reactivation of the epithelial-mesenchymal trophic unit (EMTU), triggered by an abnormal epithelium that is held in a repair phenotype [24]. *In vitro* studies support the hypothesis that epithelial-derived factors are capable of altering the phenotype of adjacent mesenchymal cells [43, 44]. The mechanisms involved in initiating and maintaining the epithelial repair response in asthma are largely unknown, however recent studies have demonstrated the airway epithelium in asthmatics is intrinsically altered compared to normals, and that these differences persist for extended periods in culture [45, 46]. There is considerable evidence to suggest the airway epithelium is capable of orchestrating the inflammatory immune response, as following injury, epithelial cells release a variety of pro-inflammatory cytokines, chemokines and growth factors. Acute inflammation is a necessary aspect of wound repair, however if epithelial damage persists, or the epithelial repair response is not shut-off, then potentially deleterious chronic inflammation may result (Figure 4.).

3.2. Mechanisms of cellular migration

Many disorders characterized by impaired re-epithelialization, are not a result of inadequate proliferation, but are due to impaired cell migration over the denuded site [47-49]. Cell migration requires a coordinated, highly complex series of events including; changes to the cell cytoskeleton, modification of the surrounding ECM and modulation of adhesions to ECM substrate and to other cells. The majority of these processes are regulated by the Rho-like family of small GTPases including, RhoA, B, C, E, Rac1, 2, 3 and Cdc42 (Figure 3.) (reviewed in [50, 51]) that function by acting as "molecular switches", cycling between an

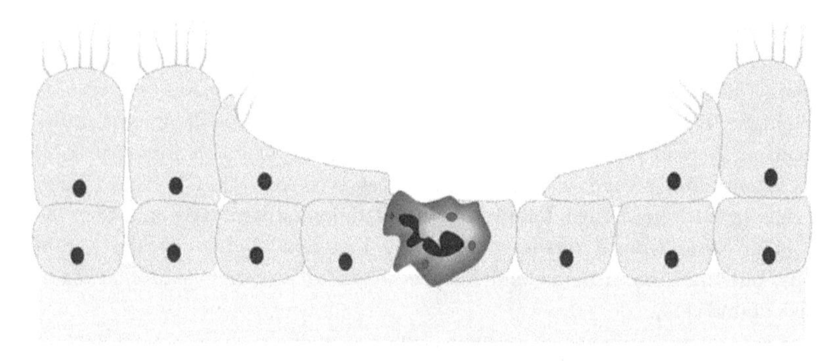

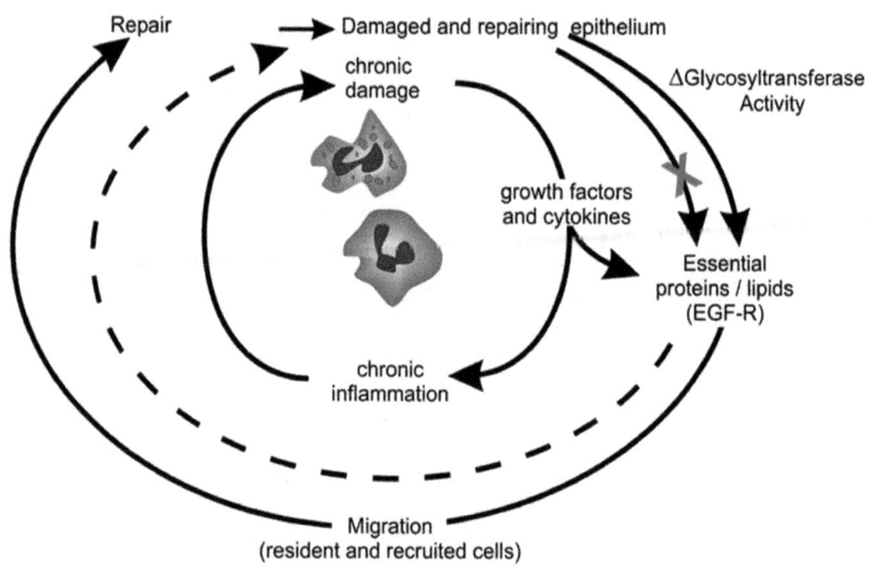

Figure 4. Injury and repair cycle of the airway epithelium in asthma. Complete repair requires glycosylation as a means of regulation of essential elements. Aberrant glycosylation would result in a defect in the mechanisms of repair, the accumulation of epithelial damage and persistent airway inflammation. Modified from Davies [46].

inactive GDP-bound state and an active GTP-bound form at the plasma membrane [52]. GTPases are able to regulate the transmission of signals from cell surface receptors, such as integrin-mediated cell-ECM adhesion, or growth factor ligation, to downstream intracellular signalling pathways, influencing the cell cytoskeleton, gene transcription, and the cell cycle (reviewed in [53]). Rho induces formation of stress fibres, integrin-mediated focal adhesions, and actomyosin-mediated cell body contraction [54], Rac-dependent actin polymerisation leads to lamellipodium extension and membrane ruffles at the leading edge of the cell [55], and Cdc42 induces formation of microspikes/filopodia [56].

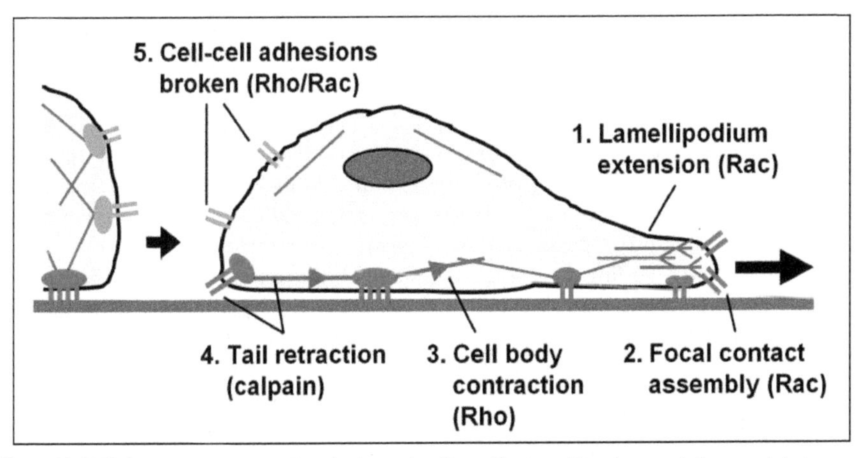

Figure 5. Cellular processes occurring during migration. The transition from a stationary state to a migratory phenotype involves several distinct mechanisms, Rho-like GTPases regulate the majority of these processes and the individual GTPases most directly linked to each are: (1.) Rac-dependent actin (blue) polymerisation leads to lamellipodium extension at the leading edge of the cell, (2.) new ECM adhesions are formed at the lamellipodium via focal contacts (red), also regulated by Rac, (3.) cell body contraction is dependent on actomyosin contraction (blue arrows) and is dependent on Rho activity, (4.) tail detachment is possibly affected by GTPase activity but the protease calpain is important for degradation of focal adhesions at the rear of migrating cells, (5.) cadherin-mediated cell-cell adhesions (green) are dependent on Rho and Rac activity and are often downregulated during migration. MMP-mediated ECM degradation is necessary for migration, even in cells migrating over a basement membrane.

Several *in vitro* studies have demonstrated activity of the RhoGTPase family is essential for airway epithelial wound repair; activation of protein kinase A (PKA) facilitates scrape wound closure in cultured bovine airway epithelial cells via inhibition of Rho activity and decreased focal contact attachments [57]. The 16HBE human airway epithelial cell line has been used to demonstrate coordination of RhoA and Rac1 activities are vital for efficient wound closure [58]. Detachment of the trailing edge is often a limiting factor of cellular migration speed, detachment is possibly affected by GTPase activity but the protease calpain is important for degradation of focal adhesions at the rear of migrating cells. Rho-GTPases are also involved in cell proliferation during airway epithelial repair; the scaffold IQ domain GTPase-activating protein (IQGAP1), an effector of Rho GTPases, is upregulated during 16HBE scrape wound repair and facilitates closure by increasing via a ß-catenin-mediated transcription of genes involved in cell cycle regulation such as CyclinD1 [59, 60].

3.3. The effect of intracellular protein glycosylation on epithelial repair

The activity of the Rho-GTPase family is highly regulated by glycosylation. *Clostridium difficile* toxins A and B (TcdA and TcdB) are major virulence factors for the induction of antibiotic-associated diarrhea and pseudomembranous colitis. These toxins possess potent cytotoxicity against cultured cell lines by inducing disaggregation of the cytoskeleton. TcdA and TcdB cellular toxicity is dependent on their intrinsic glucosyltransferase activity, they modify Rho-like GTPases by using UDP-glucose as a co-substrate [61]. RhoA undergoes mono-O-glucosylation at threonine-37, and Rac1 and Cdc42 are glucosylated at the homologous amino acid threonine-35. The threonine residue at position 37 of Rho (position 35 of Rac or Cdc42) is highly conserved and is located in the so-called effector region, it participates in co-ordination of the bound magnesium ion and thereby plays a pivotal role in nucleotide binding [62]. Glucose binding to residues within the effector region inhibits Rho-GTPase activity leading to drastic alteration of the cell cytoskeleton via several potential mechanisms including; inhibition of effector coupling, inhibition of nucleotide exchange (thereby directly blocking GTPase activity), by altering the sub-cellular localization of the GTPase, and finally by altering the stability of the enzyme within the cell (Reviewed in [52], [63]). TcdA has also been shown to induce apoptosis of intestinal epithelial cells via activation of caspases-3, -8 and -9 [64]. To date there is no direct experimental evidence linking Rho-GTPase glucosylation to the inhibition of wound repair in airway epithelial cells. However as this family of enzymes play such a fundamental role in repair of the airway epithelium, and the structure and function of the Rho-family of GTPases is very highly conserved, it is logical to presume that glycosylation would also play a pivotal role in regulating GTPase activity and thus wound repair.

3.4. Role of cell surface glycoconjugates in epithelial repair

Many proteins essential for normal cell physiology including adhesion molecules and cell surface receptors are glycosylated [65,66]. Alteration in the glycosylation pattern of many glycoproteins leads to changes in their function. It has been shown that impaired glycosylation of receptors often leads to abnormal intracellular trafficking, ligand binding and downstream signal transduction ability [67-70]. Tsuda *et al.* demonstrated that removal of sialic acid from erythropoietin leads to loss of its *in vivo* activity [71]. Complex carbohydrate structures attached to cell surface proteins and lipids have functional roles in cell motility [72], adhesion [73], proliferation [74], and growth potential [75] in several cell types. It also has been demonstrated that certain apical cell-surface carbohydrates are altered during cellular differentiation [76]. Thus many cellular functions are regulated and dependent upon glycoproteins. Because of the role of glycosylated structures in cell-cell and cell-matrix interaction, it is not surprising that there is a growing interest to explore the role for and regulation of these cell-surface carbohydrates in epithelial repair.

Lectins are naturally occurring proteins that can be isolated from a variety of plants and animals. Each lectin binds to a specific sugar moiety. As such, lectins are exquisitely selective tools to identify or block specific glycoconjugate motifs and have been extensively employed to study the role of cell-surface sugars and complex carbohydrates in cellular function [77, 78]. A variety of approaches have been employed to unravel the role(s) of carbohydrates in the multiple steps of the repair process. Some studies have investigated the expression pattern of different carbohydrates after injury using lectins as probes. Using three methods for localizing or quantifying lectin-binding sites Gipson *et al.* compared cell-surface of normal and migrating corneal epithelium of the rat. In their study they found that cell-surface of migrating epithelia express different sugar moieties relative to cell membranes of stratified stationary epithelia. There was a dramatic increase in concanavalin A (ConA) and wheat germ agglutinin (WGA) binding on migrating cells relative to stationary cells. The authors also showed that migrating cells have an increase in glycoprotein production determined by an increase in the incorporation of radiolabelled leucine and glucosamine. In addition they found that N-glycosylation of epithelial cells is necessary for epithelial cell migration [79, 80]. Sweatt *et al.* demonstrated an increase in N-acetylgalactosamine in cell-surface glycoconjugates at the site of epithelial injury in pig cornea [81]. Dorscheid *et al.* has previously characterized cell-surface glycosylation in non-secretory cells of central human airway epithelium and airway epithelial cell lines utilizing lectin-binding patterns [82]. In this study it was shown that galactose- or galactosamine-specific lectins labeled basal epithelial cells and cell lines derived from basal cells. Lectins specific for several different carbohydrate structures bound columnar epithelial cells, and certain fucose-specific lectins labeled subsets of the airway epithelial cells. The cellular specificity of these differences suggests they may be relevant in various cellular functions. It has also been demonstrated that following mechanical injury of guinea pig tracheal epithelium, glycosylation profiles in the repairing epithelium change over time [83]. These changes may represent either the expression of one or more new glycoproteins, or changes in glycosylation of constitutive proteins, required for activation or a change in cell function needed for repair to proceed. Studies have shown that injury of the respiratory epithelium enhances *P. aeruginosa* adhesion and it has been speculated that changes of cell surface glycoconjugates related to wound repair, cell migration and/or spreading may favor *P. aeruginosa* adhesion [84].

In a previous study, Dorscheid *et al.* examined the functional role of cell surface carbohydrates in an *in vitro* model of wound repair after mechanical injury of human airway epithelial cells. The results demonstrated that N-glycosylated glycoproteins, particularly those with a terminal fucose residue, are essential in the adhesion and migration of airway epithelial cells and facilitate closure of epithelial wounds in monolayer culture [85]. Recently, Allahverdian *et al.* studied the role of a fucose containing tetrasaccharide, sialyl-Lewis X (sLex) [NeuAcα2-3Galβ1-4(Fucα1-3)GlcNAc], in airway epithelial repair. Increased presentation of sLex was observed in areas of epithelial damage relative to areas of intact epithelium. In an *in vitro* model of bronchial epithelial repair, cell surface expression of sLex was shown to be significantly increased following mechanical injury of airway epithelial cell monolayers and inhibition of sLex completely prevented repair [86]. Further studies demonstrated that sLex decoration of epidermal growth factor receptor (EGFR) plays an important role in mediating airway epithelial wound repair [87]. Allahverdian *et al.* demonstrated that sLex co-localizes with EGFR and blocking of sLex with a neutralizing antibody resulted in reduced phosphorylation of EGFR and prevented repair. The final step in the synthesis of sLex is catalysed by a specific α-1,3-fucosyltransferase, fucosyltransferase-IV (FucT-IV). Reduction in EGFR phosphorylation and repair were similarly observed when FucT-IV gene expression was knocked down using small interfering RNA (siRNA). These studies demonstrate the importance of cell surface carbohydrates in mediating airway epithelial repair.

It has been shown previously by Donaldson *et al.* that the plant lectin concanavalin A (ConA), which binds glucose and mannose moieties, can inhibit migration of newt epidermal cells [88]. Gipson *et al.* cultured rat corneas with 3 mm central epithelial abrasions in the presence of four plant lectins. In this study the authors demonstrated that blocking glucose, mannose, and glucosamine sites on corneal epithelial cell surfaces and/ or the epithelial basement membrane reversibly slows or inhibits epithelial migration [78]. Using the same culture model of bronchial epithelial repair, Patchell *et al.* has demonstrated that following mechanical wounding of intact monolayers, the lectins *Allomyrina dichotoma* (AlloA) and chick pea agglutinin (CPA) differentially stain human airway epithelial cells in damaged areas relative to the staining of intact epithelial monolayers [89]. While AlloA positive staining cells are those that appear to be migrating from areas distant to the wound and accumulating in the wound, CPA positive staining cells are restricted to the leading edge of the wound. Moreover, the addition of the above lectins following mechanical wounding inhibited the repair. These results suggest that AlloA and CPA bind specific carbohydrate structures involved in normal epithelial repair. Further work is being carried out to determine the identity of the relevant proteins associated with these carbohydrate ligands.

In a study by Trinkaus-Randall *et al.*, the effect of specific carbohydrate moieties of the basal lamina on the attachment and spreading of rabbit corneal epithelial cells was studied. Corneal epithelial basal cells were plated onto freshly denuded basal lamina and three lectins WGA, ConA and RCA were used to block specific sugar moieties in the basal lamina. This study showed that lectin binding of glucose, mannose, and galactose moieties on the basal lamina significantly altered the extent of cellular spreading, while the binding of glucosamine inhibited attachment. This study demonstrated that alteration of specific sugar moieties on the native basal lamina dramatically affects the ability of basal cells of corneal epithelium to attach or spread [90]. Using an *in vitro* model of airway epithelial repair Adam *et al.* demonstrate that lectin WGA which binds to N-acetyl glucosamine residues inhibits the repair of epithelial damage without altering cell viability, while other N-acetyl glucosamine binding lectins do not affect the repair process [81].

These studies clearly demonstrate the critical role of carbohydrates in the process of epithelial repair. However, carbohydrates must either modify a protein or lipid to regulate its function or participate in binding to a specific receptor to effect the desired action. As such, the biological role of glycans can be broadly divided into two groups. One group depends on the structural and modulatory properties of glycans and the other relies on specific recognition of glycan structures (generally receptor proteins or lectins).

3.5. Galectins, annexins and epithelial injury – repair

Galectins and annexins are two of possibly many families of proteins with carbohydrate binding capabilities. Although their role in epithelial repair remains unclear, recent studies highlight their importance in these processes. Galectins are a specific family of lectins with an affinity for β-galactose-containing oligosaccharides that have no enzymatic activity. To date, as many as 14 galectins have been identified, and they have been found in non-mammalian species such as birds, amphibians, fish, worms, sponges and fungi. The binding affinities of galectins are often greater for the oligosaccharides than the monosaccharide galactose. The expression of galectins is conserved however; their expression is often tissue specific and developmentally regulated. Different cells express a unique complement of galectins. Of the 14 galectins, eight have been identified in the nucleus and cytoplasm and participate in specific cellular functions. Nine galectins have been found extracellularly. Their presentation on the cell surface remains a mystery since they lack a signal sequence for secretion via the classical secretory pathway for membrane proteins.

Studies have shown that galectin-1 and galectin-3 have the potential to mediate cell-matrix interactions [92-95]. Their ubiquitous expression makes it difficult to understand the mechanism of their involvement in this process. In corneal epithelial cells, galectin-3 and galectin-7 have been characterized as mediators of epithelial cell migration, an essential component of epithelial wound repair [96, 97]. Galectin immunostaining in healing wounds is more intense relative to normal epithelium, primarily at the leading edge and at areas of cell-matrix interactions [97]. Furthermore, wound repair was impaired in gal-$3^{-/-}$ mice in the wound model systems of either alkali burn wounds that leave the corneal stroma intact or laser ablation wounds that result in damage to the underlying basement membrane. This impaired repair was not seen in gal-$1^{-/-}$ mice. The authors propose that this effect is mediated by the lectin domain binding complementary glycoconjugates in the ECM and cell surface molecules resulting in enhanced cell-cell and cell-matrix interactions [97]. Similar work has demonstrated that galectin-7 also plays a role in corneal epithelial repair [96]. Although obvious differences exist between tissues, galectins may play a similar role in other epithelial organs and epithelial repair such as the lung and intestinal tract.

Annexins are a family of proteins that bind acidic phospholipids in a calcium dependant manner. Similar to galectins, annexins are largely cytoplasmic, however; several annexins have been detected on the cell surface of a variety of cells. Like galectins, annexins also lack a signal sequence for the cell surface presentation [98]. Annexins are proteins that have been associated with many cellular functions however their role remains poorly understood. Recent work has suggested that annexins possess carbohydrate binding abilities. This was first observed with annexin A4 when it was shown that it binds a variety of glycans [99]. Recent studies have suggested that annexin II (AII) is also capable of binding carbohydrates however this work is still in preliminary stages. Previously, the binding of tissue-type plasminogen activator (t-PA) was shown to bind HepG2 cells, however, following enzymatic removal of α-fucose residues on t-PA, this binding was dramatically decreased [100]. Subsequently, AII has been shown to be a cell surface receptor for t-PA [101]. The importance of α-fucose on t-PA may be a result of the carbohydrate binding of AII. Patchell *et al.* has shown that AII is presented on the cell surface and acts as a mediator of epithelial repair. AII was shown to be actively involved in wound repair in cells at the leading edge of the wound. Interestingly, the initial isolation purification of AII was achieved using a lectin, chick pea agglutinin, suggesting that itself, or an associated protein is glycosylated. Their tight association with the membrane and potential carbohydrate binding may influence the several functions associated with these relatively poorly understood proteins. On the surface of cells, AII has previously been shown to be involved in the migration of prostate cancer and lung carcinoma cells [102, 103]. In the case of metastatic cells, the regulation of cell migration and cell death is lost, however, under the appropriate levels of control, similar mechanisms of migration may be involved in the normal processes of wound repair. Following mechanical injury to rat cornea, at the leading edge of the wound AII translocates to the cell surface and interacts with the matrix protein tenascin-C [104]. The interaction of AII with tenascin-C results in the loss of focal adhesions and cell migration in endothelial cells and has been shown to initiate downstream signaling pathways. These findings combined with previous results suggest that AII is involved in cell migration and repair. However, the mechanism of AII involvement in these cellular events is unknown but likely involves a fucose binding lectin activity to coordinate the needed response.

As a cell surface molecule, it has been suggested that AII is the human cytomegalovirus receptor [105]. Following the initial interaction with heparan sulfate proteoglycan, the virus particle is primed for membrane fusion and infection [106, 107]. As a membrane fusion protein, AII could potentially act as a receptor to other enveloped viruses such as respiratory syncytial virus (RSV). Recently, AII has been shown to bind RSV and was characterized as a potential cell surface protein that can promote RSV infection [108]. These findings along with data that has identified AII on the surface of airway epithelial cells provide strong evidence AII may be a cell fusion receptor allowing RSV infection of the respiratory tract. Viruses are a common source of lung injury, specifically the epithelium. Epithelial cells, as a protective mechanism, become highly apoptotic following virus infection to prevent further infection and viral persistence [109]. Viral infectivity of several enveloped viruses is dependent upon carbohydrate interactions between the viral proteins and their receptors. The end result is injury to the epithelium that requires repair. Altered glycosylation then may either increase the resulting injury from RSV infection or impair the AII coordinated repair.

3.6. MMPs in asthma, glycosylation and repair

Matrix-metalloproteinases (MMPs) are a family of zinc-dependant endopeptidases that digest all components of the extracellular matrix (ECM) and many non-ECM substrates including growth factors, cytokines, and their receptors [110, 111]. Several MMPs are increased in airway tissue and broncho-alveolar lavage (BAL) fluid of asthmatics including MMP-1, -2, -3, -7, -8 and -9 [112, 113], suggesting a link between increased airway proteolytic activity and the asthma phenotype. *In vivo* studies have shown MMPs play a direct role in the development of allergic airway inflammation [114] and hyperresponsiveness [115-117]. Inflammatory cells, particularly neutrophils and alveolar macrophages, were thought to be the major source of increased proteinases in asthma and other inflammatory lung diseases [118, 119], however several groups have demonstrated structural cells of the lung including the epithelium, also synthesize MMPs and their endogenous tissue inhibitors (TIMPs) [10, 120-125]. MMPs perform many biological functions, they are integral to connective tissue homeostasis and their activities are strictly coordinated during epithelial wound repair [124-

126]. Many growth factors such as TGF-β and VEGF bind to sites in the ECM forming "reservoirs", and are released during ECM digestion [112]. Recent studies have provided strong evidence that dysregulated airway MMP activity in the lung epithelium may contribute to airway inflammation and the progression of tissue remodeling in asthma. MMP-7 expression is altered in the epithelium of asthmatics which may contribute to airway inflammation via cleavage and release of cytokines and chemokines including IL-25 and soluble FasL [10, 127]. It is possible that aberrant MMP expression in the airway epithelium contributes to inflammation and lung remodeling in asthma via proteolytic alterations in cytokine and chemokine signaling.

3.7. IgCAMs in asthma, repair and the effects of IL-13

Immunoglobulin-cell adhesion molecules (Ig-CAMs) are a diverse family of cell adhesion molecules characterised by the presence of one or more copies of a structure known as the Ig fold [128]. These receptors are expressed on a variety of cell types and are vital for many different biological processes, including neural development and immune system function [129]. In the adult, Ig-CAMs regulate the recruitment and subsequent activation of circulating lymphocytes, studies show they are also expressed in the epithelium and vasculature of the bronchial mucosa and levels of two Ig-CAMs, ICAM-1 and CD44 (hyaluronic acid receptor), are increased in the epithelium of asthmatics [130, 131]. Recent data also suggests a third Ig-CAM, CD147 (EMMPRIN – Extracellular matrix Metalloprotease Inducer) is increased in the epithelium of patients with chronic obstructive pulmonary disease (COPD) [132]. CD147 is over-expressed by metastatic cells of the skin, bladder, breast, and lung [92]. In cancer CD147 stimulates synthesis of various MMPs via homophilic interaction of CD147 oligomers on adjacent cells [134]. CD147 is potentially involved in various signaling pathways relevant to remodeling in airways disease, for an in depth review see [135]. Studies have shown CD147 is endogenously expressed at low levels in the basal cells of various stratified epithelia including the airway. Airway epithelial CD147 expression is increased in murine models of lung injury [136] and in several inflammatory airway diseases including pulmonary fibrosis [137] and interstitial pneumonias, where increased CD147 is associated with MMP-2, -7 (matrilysin) and -9 levels in the lungs of patients [138]. Recent studies have demonstrated CD147 is increased in the airway epithelium and BAL of patients with COPD, and signaling via CD147 regulates MMP-9 expression in cultured bronchial epithelial cells [132]. CD147 is potentially important in asthma; in a mouse model of allergic asthma, inhalation of a CD147-neutralizing monoclonal antibody reduced ovalbumin (OVA)-induced airway mucin production and airway hyperreactivity, whilst partially inhibiting the recruitment of eosinophils and effector $CD4^+$ T cells into lung tissue [139]. The authors attributed the reduced airways inflammation to an inhibition of the interaction between CD147 on inflammatory cells with cyclophilin chemokines. The reduction in OVA-induced mucous production suggests CD147 inhibition directly affects the epithelium.

MMP activity is dependent upon the close spatial juxtaposition of enzyme and substrate, therefore targeted MMP localization is a further mechanism of regulation. Specific MMP-substrate interactions can be facilitated and regulated at the cell surface via association of MMPs with chaperone molecules. CD44 is strongly upregulated in injured asthmatic airways and at the leading edge of repairing bronchial epithelial cells [43, 140], yet its function during airway epithelial repair is unknown. Studies suggest CD44 may co-ordinate MMP-substrate interactions at the plasma membrane by functioning as a docking molecule. In repairing reproductive epithelium, the v3 isoform of CD44 (CD44v3) forms a signaling complex with MMP-7 which is presented at the cell surface in close proximity to a potential substrate, pro-Heparin-Binding Epidermal-like Growth Factor (pro-HB-EGF). MMP-7 cleaves pro-HB-EGF into an active soluble form which triggers recruitment and activation of the Erb-B4 receptor (a member of the EGFR family) [141]. The EGF pathway is essential for efficient repair of the bronchial epithelium [43], and a similar mechanism of targeted MMP-7 presentation may occur during lung epithelial repair. Carbohydrate modification of proteins is essential for the creation of such micro-environments at the cell surface; in this case the interaction of MMP-7 (an non-glycosylated protease) with CD44 is dependent on the presence of branched glycosylated side-chains on the extracellular portion of the CD44 molecule. It is logical to extrapolate that a deficiency in the post-translational modification of CD44 could lead to a lack of MMP-7 binding, a reduction in HB-EGF activation, and a consequent inhibition of epithelial repair.

3.8. How carbohydrates mediate cell-cell interaction and migration

The glycans attached to matrix molecules such as collagens and proteoglycans are important for the maintenance of tissue structure, porosity, and integrity. Such molecules also contain binding sites for specific types of glycans, which in turn help with the overall organization of the matrix. Glycans are also involved in the proper folding of newly synthesized polypeptides in the endoplasmic recticulum and /or in the subsequent maintenance of protein solubility and conformation [66]. In this manner, altered glycosylation of matrix proteins may change adhesion properties and the potential of cell to migrate. This is due to either an altered conformation of the extracellular matrix of a change in the carbohydrate ligands available to bind.

Glycosylation of growth factor receptors has tremendous effects on the receptor function. Several studies have focused on the role of EGF and its family of receptors on epithelial repair [24, 142]. EGF, a well known mitogen for epithelial cells, has been used in to stimulate epithelial wound healing in guinea pigs as well as human airway cell monolayers and differentiated cells *in vitro* [43, 143]. EGFR is known to be up-regulated upon the creation of a wound on airway epithelial monolayers in culture and correlated to the damaged areas of epithelium [24]. Increased tyrosine phosphorylation of the EGFR has also been observed after mechanical injury even in the absence of exogenous ligand [24]. While a critical role of signaling mediated by EGFR in repairing damaged epithelium has been well demonstrated in many epithelial systems, an important role for core glycosylation of extracellular domain of EGFR in ligand binding and tyrosine kinase activity has also been documented [105-107]. The extracellular domain of EGFR has 10 to 11 potential site for N-glycosylation. It has been shown that glycosylation is necessary for ligand binding and tyrosine kinase activity of

EGFR [144-146]. Carbohydrate moieties of EGFR are critical for the direct interaction of the receptor with other structures [147]. Moreover, glycosylation defines localization of EGFR to specific domain of plasma membrane which could facilitate association of the receptor with other molecules and its transactivation [148]. Modification of receptor N-glycans can also regulate receptor trafficking and duration of cell surface residency [149]. A direct association between changes in glycosylation of EGFR and cellular events involved in epithelial repair such as migration and proliferation remains to be identified. It could be inferred that role of carbohydrates in epithelial repair is partly through their remarkable role in modulation of EGFR and other growth factor receptors.

The specificity of carbohydrate interactions allows for the high degree of selectivity found within the cell. Whether it is in regulating receptor activation, cell-cell or cell-matrix interactions or cell migration, defects in these processes may result in abnormalities and altered phenotypes. Their understanding will provide new avenues for research and therapies.

4. IL-13 structure

Interleukin-13 (IL-13) is a type I cytokine, comprised of four short α-helical hydrophobic bundles joined by two disulfide bonds [111]. The cDNA of IL-13 has a single open reading frame that encodes 132 amino acids, including a 20-amino acid signal sequence that is cleaved from the mature secreted protein [112]. IL-13 has four predicted N-linked glycosylation sites and can either exist as a 17kDa glycoprotein or 10-12kDa unglycosylated protein [113]. IL-13 has been demonstrated to be secreted predominantly as the 10-12kDa unglycosylated protein [113]. In a study conducted by Liang *et al.*, the distribution and expression levels of unglycosylated and glycosylated IL-13 was demonstrated to differ between human bronchial smooth muscle cells (BSMC) and lymphocytes [114]. Intracellularly, BSMC were shown to express higher levels of 12kDa unglycosylated IL-13, while lymphocytes expressed more of the 17kDa glycosylated form. However, lymphocytes secreted significantly higher levels of 12kDa unglycosylated IL-13 compared to BSMC. These findings suggest that cellular distribution of unglycosylated and glycosylated forms of IL-13 is tissue and cell type-dependent and may reflect the role of IL-13 in each cell type. The study also demonstrated that IL-13 secretion is regulated by human sulfatase modifying factor 2 (SUMF2), a protein predominantly found in the endoplasmic reticulum and highly expressed in BSMC [114]. SUMF2 was shown to physically interact with IL-13 and regulate its secretion irrespective of the glycosylation status of IL-13 in BSMC. Lymphocytes were shown to express lower levels of SUMF2 and to secrete higher levels of IL-13 compared to BSMC, further supporting the role of SUMF2 in regulating IL-13 secretion.

The glycosylation state of IL-13 can also affect its activity and potency to induce certain effector functions such as the up-regulation of low affinity IgE receptor, CD23 on the surface of monocytes. A functional study conducted by Vladich *et al.* [115] demonstrated that eukaryotic COS-7 cell-derived IL-13 was significantly more active in inducing the surface expression of CD23 on peripheral blood mononuclear cells (PBMC) compared to two different commercially available prokaryotic *E. Coli*-derived IL-13. The difference in bio-activity between the prokaryotic and eukaryotic-derived IL-13 molecules can be attributed to the fact that prokaryotic systems often lack post-translational modification steps such as glycosylation that take place in eukaryotic cells. These findings suggest the importance of glycosylation for IL-13 bio activity.

4.1. IL-13 function

IL-13 is a pleiotropic cytokine that exerts its effect on a wide variety of cell types. IL-13 has a variety of functions which are critical to immune homeostasis and repair, but when uncontrolled contribute to the asthma phenotype. IL-13 is predominantly produced by T helper type 2 (Th_2) cells [116] and to a lesser extent by mast cells, basophils, and eosinophils [117, 118]. IL-13 is also produced by a variety of non-hematopoietic cells including airway epithelial cells [119] and bronchial smooth muscle cells [114]. As a Th_2 cytokine, IL-13 shares numerous overlapping characteristics with IL-4, where both cytokines promote B cell proliferation, IgE class switching and synthesis in B cells, and induce surface expression of antigens such as CD23 and major histocompatibility complex (MHC) class II [120]. These similarities between IL-13 and IL-4 function can be explained by the fact that the cytokines share a common receptor subunit, IL-4 receptor α subunit (IL-4Rα). However unlike IL-4, IL-13 does not exert direct effects on T cells and cannot elicit Th_2 differentiation in naïve T cells; IL-13 is instead involved in mediating downstream Th_2-effector functions [121]. In monocytes and macrophages, IL-13 can inhibit the production of various pro-inflammatory mediators such as prostaglandins, reactive oxygen species, IL-1, and TNF-α [113, 122]. IL-13 can also enhance the expression of several members of the integrin family that play important roles in adhesion, including CD11b, CD11c, CD18, and CD29 as well as induce surface expression of CD23 and MHC class II in monocytes and macrophages [121]. Furthermore, IL-13 has been reported to have direct effects on eosinophils, including promoting eosinophil survival, activation, and recruitment [123, 124].

IL-13 has important functions on non-hematopoietic cells, including endothelial cells, smooth muscle cells, fibroblasts, and epithelial cells. In endothelial cells IL-13 is a potent inducer of vascular cell adhesion molecule 1 (VCAM-1), which plays a role in eosinophil recruitment [125]. IL-13 has been shown to enhance proliferation and cholinergic-induced contractions of smooth muscle cells *in vitro* [126, 127]. In epithelial cells, IL-13 is a potent inducer of growth factors [128, 129] and chemokine expression [130]. It also induces epithelial cell proliferation [128], alters mucociliary differentiation [131], resulting in mucin production and goblet cell metaplasia [7, 132, 133].

4.2. IL-13Rα1 structure and function

IL-13 signals via a receptor system that consists of IL-13 receptor α1 (IL-13Rα1)/IL-4 receptor α subunit (IL-4Rα) and IL-13 receptor α2 (IL-13Rα2). Both IL-13Rα1 and IL-13Rα2 belong to the type I cytokine receptor family, which possess several definitive features, including a W-S-X-W-S motif, four conserved cysteine residues, fibronectin type II modules in the extracellular domain and proline-rich box regions in the intracellular domain [134]. The two receptors share 33% homology and 21% identity at the amino acid level [135] and their respective genes have both been mapped to the X chromosome [136]. Kinetic analysis have revealed that IL-13Rα1 and IL-13Rα2 both bind specifically only to IL-13 and not IL-4, while IL-13 alone has no measurable affinity for IL-4Rα [137].

IL-13Rα1 is widely expressed on both hematopoietic and nonhematopoietic cells except human and mouse T cells and mouse B cells [112]. It is a 65-70kDa glycosylated trans-membrane receptor protein with 10 predicted N-glycosylation sites in the extracellular domain [138]. The cDNA for human IL-13Rα1 encodes a 427-amino acid sequence, including a 60-amino acid intracellular domain and 21-amino acid signal sequence [139]. A mutational analysis demonstrated that Leu319 and Tyr321 in the cytokine receptor homology module (CRH) of human IL-13Rα1 are critical residues for binding to IL-13 [135]. Interestingly, these residues are in close proximity to the predicted glycosylation sites of IL-13Rα1, suggesting that the glycosylation state of IL-13Rα1 could potentially influence receptor-ligand interactions.

IL-13Rα1 binds IL-13 with low affinity by itself but binds IL-13 with high affinity when heterodimerized with IL-4Rα to form a functional signalling receptor [138]. This signalling complex is known as the type II IL-4/IL-13 receptor and it also serves as an alternate receptor for IL-4. IL-13 is known to primarily signal through IL-13Rα1/IL-4Rα by first binding to IL-13Rα1 chain with low affinity and then heterodimerizing with IL-4Rα to become a stable, high affinity signalling complex [137]. The formation of IL-13Rα1/IL-4Rα results in the activation of the Jak kinases, Jak1 and Tyk2, followed by the recruitment of the transcription factor, signal transducer and activator of transcription 6 (STAT6) to the receptor. STAT6 is then phosphorylated and forms functional dimers that translocate to the nucleus to bind specific canonic DNA elements and initiate transcription of downstream genes [140]. Some examples of downstream targets of IL-13Rα1 activation include eotaxin [141], MUC5AC [142] , and, arginase I which is an enzyme important in the development of airway hyperreactivity [143].

4.3. IL-13Rα2 structure and function

IL-13Rα2 is a 56kDa glycoprotein that is expressed by a variety of cell types including monocytes, airway epithelial cells, fibroblasts, and keratinocytes [144, 145]. The cDNA of human IL-13Rα2 encodes a 380-amino acid protein with a 26-amino acid signal sequence and a 17-amino acid intracellular domain [146]. IL-13Rα2 contains four predicted glycosylation sites [147] and the glycosylation state of IL-13Rα2 has been demonstrated to be important in its interactions with IL-13 [148]. IL-13Rα2 exists in three cellular compartments: on the cell surface as a single trans-membrane receptor, in the cytosol within large intracellular pools, and in the extracellular space as a soluble form [149]. In fibroblasts and airway epithelial cells, IL-13Rα2 has been demonstrated to be predominantly localized to intracellular pools that can be rapidly mobilized to the membrane upon stimulation by IL-13 and IL-4 [150].

IL-13Rα2 has been considered for a long time as a decoy receptor which does not directly contribute to IL-13 signaling, but serves as a negative regulator to terminate IL-13 responses by directly binding to IL-13. The notion that this receptor had no signaling function arose from the fact that it has a short cytoplasmic tail that is missing two known signalling motifs [151] and does not bind JAKs or STATs [152]. IL-13Rα2 has also been shown to be internalized quickly upon IL-13 binding [152]. Unlike IL-13Rα1, IL-13Rα2 alone has very high affinity for IL-13 [146] and this is further increased when IL-13Rα2 exists as a soluble form [153]. N-linked glycosylation in the extracellular domain of IL-13Rα2 has shown to be necessary for optimal IL-13 binding, demonstrating the importance of glycosylation in IL-13 signalling [148].

Recent investigations have suggested that IL-13Rα2 acts as a signaling receptor not merely as a decoy receptor. Dienger *et al.* showed that IL-13Rα2 knockout mice have attenuated rather than enhanced allergic airway responses, suggesting that under some circumstances, IL-13Rα2 may contribute to IL-13 signaling [118] and reduce inflammation. Further evidence of anti-inflammatory properties comes from models of allergic asthma and chronic helminth infection where IL-10 and IL-13Rα2 coordinately suppress Th_2-mediated inflammation and pathology [154]. Fichtner-Feigl *et al.* has demonstrated that IL-13 signals through the IL-13Rα2 to mediate TGF-β production via the transcription factor, AP-1 in macrophages. [129]. These studies provide evidence that IL-13Rα2 can serve as a functional signalling receptor but the role of glycosylation in regulating its signalling function has yet to be elucidated.

4.4. IL-13 genetic and clinical linkage to asthma

The gene encoding IL-13 consists of four exons and three introns and is located on chromosome 5q31 [155]. The chromosomal region 5q31 also contains genes for other molecules associated with asthma such as IL-4, IL-3, IL-9 and GM-CSF [156]. Previous studies have demonstrated strong associations between asthma and several single nucleotide polymorphisms (SNP) in the IL-13 gene, including +2043G>A and -1111C>T [156, 157]. +2043G>A is within the IL-13 coding sequence and causes a positively charged arginine (R) to be substituted by a neutral glutamine (Q) at position 130, forming the IL-13 R130Q variant [158]. Position 130 is in the α-D segment of the IL-13 molecule, where it has been demonstrated to interact with IL-13Rα1 [159]. Compared to wildtype IL-13, the R130Q variant has previously been reported to have lower affinity for soluble IL-13Rα2 due to slower association rates with the receptor [176]. IL-13R130Q is also more active compared to wildtype IL-13 in inducing STAT-6 activation and CD23 expression in monocytes [115]. The enhanced activity of IL-13R130Q may be explained by its decreased affinity to IL-13Rα2.

4.5. Roles of IL-13 in airway epithelial repair

IL-13 has been demonstrated to be critical in mediating normal airway epithelial wound repair. In an *in vitro* model, Allahverdian *et al.* demonstrated that normal bronchial epithelial cells produce and secrete IL-13 in response to mechanical injury. IL-13 then mediates repair by inducing the production and autocrine/ paracrine release of HB-EGF, which subsequently activates EGFR and downstream signalling pathways required for repair [119]. In another *in vitro* study, thymic stromal lymphopoietin (TSLP) was shown to promote bronchial epithelial proliferation and repair via the upregulation of IL-13 production, providing further evidence for the critical role of IL-13 in repair [160]. IL-13 also plays a role in promoting normal airway epithelial health. Treatment of normal bronchial epithelial cells with IL-13 and IL-9 alone and in combination was shown to significantly reduce spontaneous cell apoptosis and to be protective against dexamethasone-induced cell apoptosis [161].

5. The use of biomarkers in asthma diagnosis and therapeutic intervention

Asthma is increasingly recognised as a heterogeneous syndrome with multiple patient phenotypes reflecting the varied underlying pathologies present in each individual. Pulmonary function testing has severe limitations; spirometry can identify a broad spectrum of asthmatics, but it is incapable of discerning sub-types of disease and therefore which individuals will respond to specific treatment regimes. Tissue biopsies and inflammatory cell counting in induced sputum, are accepted measures of determining airway inflammation, however both techniques are invasive, expensive and difficult to standardize, making them unsuitable for routine clinical use. Biomarkers hold the promise of being able to rapidly and specifically diagnose and monitor various sub-types of asthma in a non-invasive manner. The advantages and disadvantages of current and potential future biomarkers of airway inflammation for the diagnosis and monitoring of asthma in the clinic are reviewed in [162].

5.1. "Invasive" airway biomarkers

Tissue biopsies: To date the most accurate method to assess lung inflammation (and remodeling) is by histological examination of lung tissue. Sub-types of inflammatory cells can be identified in tissue sections using specific stains. Unfortunately the process of taking bronchial biopsies via bronchoscopy is invasive and requires experienced, skilled pathologists for tissue examination. In addition, clinical studies demonstrate a disconnect between the numbers of inflammatory cells counted in airway biopsies, and lung function in asthmatics [27].

Induced sputum: One alternative method of directly assessing airway inflammation is by sputum induction. A patient inhales nebulized hypertonic saline to trigger sputum production in the airways. The sputum is then coughed out along with any inflammatory cells present in the airway lumen. The main assumption is that the inflammatory infiltrate in the airway lumen reflects that in the tissue. Cytospins of the resulting sputum samples are then stained using similar techniques to tissue biopsies to examine the specific type of cellular infiltrate in the sputum. The principal output is the differential inflammatory cell count, expressed as a percentage, based on the manual counting of cells (eosinophils, neutrophils, lymphocytes, macrophages, and epithelial cells). The protein content of induced sputum may also have some value in diagnosing asthma. High mobility group box-1 (HMGB-1), a ligand of the receptor for advanced glycation end products (RAGE), is a mediator in many inflammatory disorders. HMGB-1 is increased in the sputum of asthmatic patients, along with endogenous secretory RAGE, a soluble receptor that inhibits RAGE signalling [163]. Sputum induction is considerably less invasive than a tissue biopsy, but this technique is nonetheless uncomfortable for the patient. As many children are unwilling to undergo sputum collection in follow-up visits this technique tends to be limited to patients aged 8 years and above (although sputum induction has been performed successfully on younger children)[164]. Although the techniques of sputum induction and processing are well validated, they are time-consuming, require skilled people, and results are often difficult to reproduce and vary across centers.

5.2. Exhaled biomarkers

Less invasive means of obtaining airway biomarkers are desirable, particularly as diagnostic or disease monitoring tools in pediatric patients. The composition of exhaled breath is correlated to various disease states. This has been exploited by cancer researchers who have pioneered the use of gas sensor arrays, or "electronic noses", to aid the diagnosis of lung cancer [165]. As asthma is also a disease of the airways, it follows that the composition of exhaled breath will also be altered in the disease. The advent of so-called "breathomics" promises to revolutionize the way clinicians diagnose and treat asthmatics.

Fractional exhaled nitric oxide (FeNO): To date FeNO is the most widely used exhaled biomarker of airway inflammation in asthma. Levels of NO in exhaled breath can be measured relatively quickly in the clinic although the gas analyzers required are expensive [164]. FeNO is often increased in steroid naïve asthmatics and severe asthmatics, and is correlated with airway eosinophilia [166, 167]. FeNO is derived from the action of inducible NO synthase (iNOS) expressed by the airway epithelium [168] however the precise mechanism of how eosinophilia triggers iNOS activity in epithelial cells is undefined. It is unlikely that FeNO measurements per se would give any indication of the carbohydrate modifications of proteins in a patient, therefore this method will not be discussed further herein. For a more in-depth discussion of FeNO as a biomarker in asthma readers are referred to a recent review by Wadsworth *et al.* [162].

Exhaled Breath Condensate: The collection of exhaled breath condensate (EBC) and subsequent measure of inflammatory biomarkers is a relatively recent development. Exhaled breath condenses when it comes into contact with a cooled collector, allowing the collection of respiratory particles, droplets and water vapour. The pH of the EBC has been shown to relate to airway inflammation, low EBC pH indicates poorly controlled eosinophilic asthma in a similar manner to high FeNO [169]. It is unlikely that a single biomarker will be able to reflect the various pathologies which are present in a heterogeneous disease such as asthma. Therefore additional markers of airway inflammation will be needed to provide information complementary to that gained from FeNO or pH measurements. Many proteins are present in EBC and this method has been postulated as a means to allow the objective proteomic analysis of exhaled breath. As with pH, several of these proteins are markers of oxidative stress including, cysteinyl leukotrienes, leukotriene B4, 8-isoprostane and hydrogen peroxide, although inflammatory proteins such as IL-6, IL-8, TNF-α may also be useful markers [170] (reviewed in [164]). Other proteins which do not obviously fit into any inflammatory pathway including; actin, cytokeratins, albumin, and hemoglobin have also been shown to be increased in the EBC of asthmatic patients [171].

5.3. Non-exhaled biomarkers

Serum proteins: As described above, biomarker studies in asthma have tended to concentrate on changes to the composition of exhaled breath. Although primarily a disease of the airways, there is mounting evidence to suggest there is also a systemic component to asthma [172]. If this is the case then circulating metabolites may be able to act as biomarkers of disease. Blood collection, serum isolation and analysis are highly standardized techniques of a minimally invasive nature and are therefore an ideal source of reproducible data. Serum proteins are already gaining credence as biomarkers in other inflammatory lung diseases. In COPD for example, not all smokers develop the disease, so being able to identify those who are at risk would be useful. Studies have shown circulating levels of pulmonary and activation-regulated chemokine (PARC)/CCL-18 are increased in COPD patients [173] and COPD patients who exhibit a rapid fall in FEV_1 tend to have high circulating levels of fibrinogen [174]. Even accepted markers of cardiac dysfunction, cardiac troponin-T and N-terminal pro-brain natriuretic peptide (NT-proBNP) have been associated with increased mortality in patients with COPD, suggesting a cardiac component to the disease [175]. Several serum biomarkers have been demonstrated to be associated with asthma, including eosinophil cationic protein (ECP). ECP levels increase in response to allergen challenge, and decrease after allergen avoidance or inhaled corticosteroids (ICS) therapy, albeit in a less responsive manner than sputum eosinophils or FeNO (reviewed in [164]). In the clinic however, serum ECP levels do not reflect treatment-induced functional changes in chronic asthmatics, and serum ECP is unable to predict steroid responsiveness (reviewed in [164]). One randomized trial demonstrated that patients whose asthma management was based on serum ECP levels, experienced no improvement in symptoms compared to those treated using traditional monitoring techniques [176]. In addition, raised serum ECP levels may not be a specific marker of asthma; studies in pediatric patients have demonstrated serum ECP is also raised in cystic fibrosis and viral bronchiolitis [177]. Studies using purified ECP demonstrate it undergoes N-linked glycosylation and its cytotoxic activity against a cultured human small cell lung cancer cell line is dependent on both gene polymorphisms and glycosylation state [178]. Thus, rather than measuring total ECP, it is likely that analysing the proportion of total ECP which is glycosylated would prove to be a more specific biomarker for monitoring asthma progression. Other novel biomarkers are also altered in the circulation of patients with asthma. CCL-17 is a chemokine released from dendritic cells and epithelial cells after allergen contact, and is involved in the recruitment of Th_2 cells into the lungs. Studies have shown sputum levels of CCL-17 is increased in the lungs of asthmatic adults,[179] whilst serum levels of CCL-17 is increased in children with asthma and is lowered in steroid-treated children with asthma [180]. In asthma, structural cells of the airway including the epithelium, release various pro-inflammatory chemokines such as IL-6 and IL-8 which trigger subsequent infiltration by immune cells. The epithelium also releases proteinases which are capable and necessary for the cleavage and activation of chemokinetic activity of molecules such as Fas-ligand and IL-25 [10] [127]. It is possible circulating levels of these chemokines and proteinases may provide an early warning for an imminent asthma exacerbation, or even indicate the particular sub-type of inflammation occurring in the patient.

Urinary metabolites: Data mentioned above suggests that the metabolism of patients with asthma is altered compared to normals. Urine is possibly the least invasive biofluid for biomarker measurements and is therefore highly suitable for the study and assessment of asthma in young children. Clinical studies have shown urinary biomarkers are potentially useful in asthma; levels of the downstream histamine metabolite, N-methylhistamine, is increased in the urine of patients with asthma, is increased after allergen challenge or exacerbation, and is reduced in asthmatic children taking anti-allergy medication [183, 184]. The human histamine receptor (hH1R) is a G-protein-coupled receptor, GPCRs are a large family of transmembrane receptors that are dependent on glycosylation for correct expression, trafficking and signaling. Studies using hH1R-expressing insect cells have shown that the receptor is N-glycosylated at a specific asparagene residue, Asn_5 [185], whether glycosylation affects receptor activity, and how hH1R glycosylation could be measured in patients is unknown.

One recent study used nuclear magnetic resonance spectroscopy (^{1}H-NMR) to measure levels of 70 metabolites in the urine of children with and without asthma. NMR is an attractive method for urinary biomarker examination as it is able to provide qualitative and quantitative data on multiple compounds in a complex biofluid, without requiring significant pre-treatment of the sample. Urine was collected from control children without asthma (C), with stable asthma in the outpatient department (AO), and in children with unstable disease hospitalized for an asthma exacerbation (AED). NMR examination of urinary metabolites showed a 94% success rate in identifying AO children versus C, and a similar success rate was seen when diagnosing AED versus AO asthma [186]. This study suggests the measurement of urinary metabolites is a potentially valuable technique to help clinicians diagnose and monitor asthma in children.

6. Closing remarks

In addition to forming a barrier against inhaled toxins, pathogens and allergens, the airway epithelium performs numerous innate immune and transport functions necessary for airway health and homeostasis. Damage to the epithelial layer is a common occurrence, thus complex repair mechanisms have evolved to rapidly restore the epithelial barrier, with subsequent regeneration of the fully functional differentiated tissue. In asthmatic airways, these epithelial repair mechanisms are compromised.

The post-translational modification of proteins by glycosylation is capable of drastically affecting function via changes to; sub-cellular localisation and secretion, enzyme substrate binding, receptor-ligand binding, protein stability and degradation amongst others. Many of the proteins involved in the complex repair mechanisms defined in the airway epithelium are dependent on correct glycosylation in order to function correctly. There is accumulating evidence to suggest glycosylation is altered in airway diseases such as asthma. In addition, an important cytokine in asthma, IL-13, has a significant effect on the glycosylation state of the epithelial lining of the airway, and IL-13 signalling itself is affected by glycosylation of its receptors. Thus altered glycosylation of the airway epithelium, will have significant effects on the repair and regeneration of a functional airway epithelial barrier.

Asthma is a complex syndrome, individuals with asthma can be classified into a number of sub-phenotypes who will respond to different types of therapies. In order to facilitate the accurate diagnosis of asthmatics the development of specific bio-markers is required. Considerable research is being conducted into the use of non-invasive exhaled, urinary, or blood-borne markers of asthmatic disease to aid diagnosis in adults and children. To date the majority of biomarker research has concentrated on the identification of proteins linked to disease, rather than identifying differences in post-translational modifications. As glycosylation plays a fundamental role in regulating protein function, it is highly likely that future biomarkers for asthma will need focus on identifying alterations to the carbohydrate structures of proteins, particularly those involved in airway epithelial repair.

Author details

S. Jasemine Yang and Delbert R. Dorscheid and Samuel J. Wadsworth*
James Hogg Research Centre, Institute for Heart + Lung Health, St. Paul's Hospital, University of British Columbia, Vancouver, Canada

7. References

[1] Wenzel, S.E., *Asthma: defining of the persistent adult phenotypes.* Lancet, 2006. 368(9537): p. 804-13.

[2] Adcock, I.M. and K. Ito, *Steroid resistance in asthma: a major problem requiring novel solutions or a non-issue?* Curr Opin Pharmacol, 2004. 4(3): p. 257-62.

[3] Djukanovic, R., et al., *Quantitation of mast cells and eosinophils in the bronchial mucosa of symptomatic atopic asthmatics and healthy control subjects using immunohistochemistry.* Am Rev Respir Dis, 1990. 142(4): p. 863-71.

[4] Montefort, S., et al., *The site of disruption of the bronchial epithelium in asthmatic and non-asthmatic subjects.* Thorax, 1992. 47(7): p. 499-503.

[5] Green, R.H., et al., *Analysis of induced sputum in adults with asthma: identification of subgroup with isolated sputum neutrophilia and poor response to inhaled corticosteroids.* Thorax, 2002. 57(10): p. 875-9.

* Correspong Author

[6] Huang, S.K., et al., *IL-13 expression at the sites of allergen challenge in patients with asthma.* J Immunol, 1995. 155(5): p. 2688-94.
[7] Wills-Karp, M., et al., *Interleukin-13: central mediator of allergic asthma.* Science, 1998. 282(5397): p. 2258-61.
[8] Cohn, L., et al., *Th2-induced airway mucus production is dependent on IL-4Ralpha, but not on eosinophils.* J Immunol, 1999. 162(10): p. 6178-83.
[9] Beghe, B., et al., *Polymorphisms in IL13 pathway genes in asthma and chronic obstructive pulmonary disease.* Allergy, 2010. 65(4): p. 474-81.
[10] Wadsworth, S.J., et al., *IL-13 and TH2 cytokine exposure triggers matrix metalloproteinase 7-mediated Fas ligand cleavage from bronchial epithelial cells.* J Allergy Clin Immunol, 2010. 126(2): p. 366-74, 374 e1-8.
[11] Danahay, H., et al., *Interleukin-13 induces a hypersecretory ion transport phenotype in human bronchial epithelial cells.* Am J Physiol Lung Cell Mol Physiol, 2002. 282(2): p. L226-36.
[12] Atherton, H.C., G. Jones, and H. Danahay, *IL-13-induced changes in the goblet cell density of human bronchial epithelial cell cultures: MAP kinase and phosphatidylinositol 3-kinase regulation.* Am J Physiol Lung Cell Mol Physiol, 2003. 285(3): p. L730-9.
[13] Whittaker, L., et al., *Interleukin-13 mediates a fundamental pathway for airway epithelial mucus induced by CD4 T cells and interleukin-9.* Am J Respir Cell Mol Biol, 2002. 27(5): p. 593-602.
[14] Beum, P.V., et al., *Mucin biosynthesis: upregulation of core 2 beta 1,6 N-acetylglucosaminyltransferase by retinoic acid and Th2 cytokines in a human airway epithelial cell line.* Am J Physiol Lung Cell Mol Physiol, 2005. 288(1): p. L116-24.
[15] Barnes, P.J., *The role of inflammation and anti-inflammatory medication in asthma.* Respir Med, 2002. 96 Suppl A: p. S9-15.
[16] Barnes, P.J., *Update on asthma.* Isr Med Assoc J, 2003. 5(1): p. 68-72.
[17] Kraft, M., *Asthma phenotypes and interleukin-13--moving closer to personalized medicine.* N Engl J Med, 2011. 365(12): p. 1141-4.
[18] Flood-Page, P., et al., *A study to evaluate safety and efficacy of mepolizumab in patients with moderate persistent asthma.* Am J Respir Crit Care Med, 2007. 176(11): p. 1062-71.
[19] O'Byrne, P.M., *The demise of anti IL-5 for asthma, or not.* Am J Respir Crit Care Med, 2007. 176(11): p. 1059-60.
[20] Corren, J., et al., *Lebrikizumab treatment in adults with asthma.* N Engl J Med, 2011. 365(12): p. 1088-98.
[21] Innes, A.L., et al., *The H antigen at epithelial surfaces is associated with susceptibility to asthma exacerbation.* Am J Respir Crit Care Med, 2011. 183(2): p. 189-94.
[22] Laitinen, A., et al., *Tenascin is increased in airway basement membrane of asthmatics and decreased by an inhaled steroid.* Am J Respir Crit Care Med, 1997. 156(3 Pt 1): p. 951-8.
[23] Hirst, S.J., *Airway smooth muscle as a target in asthma.* Clin Exp Allergy, 2000. 30 Suppl 1: p. 54-9.
[24] Holgate, S.T., et al., *Epithelial-mesenchymal interactions in the pathogenesis of asthma.* J Allergy Clin Immunol, 2000. 105(2 Pt 1): p. 193-204.
[25] Hackett, T.L. and D.A. Knight, *The role of epithelial injury and repair in the origins of asthma.* Curr Opin Allergy Clin Immunol, 2007. 7(1): p. 63-8.

[26] Malmstrom, K., et al., *Lung function, airway remodelling and inflammation in symptomatic infants: outcome at 3 years.* Thorax, 2011. 66(2): p. 157-62.

[27] Sont, J.K., et al., *Relationship between the inflammatory infiltrate in bronchial biopsy specimens and clinical severity of asthma in patients treated with inhaled steroids.* Thorax, 1996. 51(5): p. 496-502.

[28] Jeffery, P.K., *Remodeling in asthma and chronic obstructive lung disease.* Am J Respir Crit Care Med, 2001. 164(10 Pt 2): p. S28-38.

[29] Knight, D.A. and S.T. Holgate, *The airway epithelium: structural and functional properties in health and disease.* Respirology, 2003. 8(4): p. 432-46.

[30] Roche, W.R., et al., *Cell adhesion molecules and the bronchial epithelium.* Am Rev Respir Dis, 1993. 148(6 Pt 2): p. S79-82.

[31] Evans, M.J., et al., *Cellular and molecular characteristics of basal cells in airway epithelium.* Exp Lung Res, 2001. 27(5): p. 401-15.

[32] Rhodin, J.A., *The ciliated cell. Ultrastructure and function of the human tracheal mucosa.* Am Rev Respir Dis, 1966. 93(3): p. Suppl:1-15.

[33] Philippou, S., et al., *The morphological substrate of autonomic regulation of the bronchial epithelium.* Virchows Arch A Pathol Anat Histopathol, 1993. 423(6): p. 469-76.

[34] Nettesheim, P., J.S. Koo, and T. Gray, *Regulation of differentiation of the tracheobronchial epithelium.* J Aerosol Med, 2000. 13(3): p. 207-18.

[35] Keenan, K.P., J.W. Combs, and E.M. McDowell, *Regeneration of hamster tracheal epithelium after mechanical injury. I. Focal lesions: quantitative morphologic study of cell proliferation.* Virchows Arch B Cell Pathol Incl Mol Pathol, 1982. 41(3): p. 193-214.

[36] Keenan, K.P., J.W. Combs, and E.M. McDowell, *Regeneration of hamster tracheal epithelium after mechanical injury. III. Large and small lesions: comparative stathmokinetic and single pulse and continuous thymidine labeling autoradiographic studies.* Virchows Arch B Cell Pathol Incl Mol Pathol, 1982. 41(3): p. 231-52.

[37] Keenan, K.P., J.W. Combs, and E.M. McDowell, *Regeneration of hamster tracheal epithelium after mechanical injury. II. Multifocal lesions: stathmokinetic and autoradiographic studies of cell proliferation.* Virchows Arch B Cell Pathol Incl Mol Pathol, 1982. 41(3): p. 215-29.

[38] Keenan, K.P., T.S. Wilson, and E.M. McDowell, *Regeneration of hamster tracheal epithelium after mechanical injury. IV. Histochemical, immunocytochemical and ultrastructural studies.* Virchows Arch B Cell Pathol Incl Mol Pathol, 1983. 43(3): p. 213-40.

[39] Tam, A., et al., *The airway epithelium: more than just a structural barrier.* Ther Adv Respir Dis, 2011. 5(4): p. 255-73.

[40] Johansen, F.E. and C.S. Kaetzel, *Regulation of the polymeric immunoglobulin receptor and IgA transport: new advances in environmental factors that stimulate pIgR expression and its role in mucosal immunity.* Mucosal Immunol, 2011. 4(6): p. 598-602.

[41] Woof, J.M. and M.W. Russell, *Structure and function relationships in IgA.* Mucosal Immunol, 2011. 4(6): p. 590-7.

[42] Martin, N. and I.D. Pavord, *Bronchial thermoplasty for the treatment of asthma.* Curr Allergy Asthma Rep, 2009. 9(1): p. 88-95.

[43] Wadsworth, S.J., H.S. Nijmeh, and I.P. Hall, *Glucocorticoids increase repair potential in a novel in vitro human airway epithelial wounding model.* J Clin Immunol, 2006. 26(4): p. 376-87.

[44] Zhang, S., et al., *Growth factors secreted by bronchial epithelial cells control myofibroblast proliferation: an in vitro co-culture model of airway remodeling in asthma.* Lab Invest, 1999. 79(4): p. 395-405.

[45] Morishima, Y., et al., *Triggering the induction of myofibroblast and fibrogenesis by airway epithelial shedding.* Am J Respir Cell Mol Biol, 2001. 24(1): p. 1-11.

[46] Davies, D.E., *The bronchial epithelium: translating gene and environment interactions in asthma.* Curr Opin Allergy Clin Immunol, 2001. 1(1): p. 67-71.

[47] Woodley, D., *The Molecular and Cellular Biology of Wound Repair.* 2nd ed, ed. R.A.F. Clark1996, New York: Plenum Publishing Corp. 339-354.

[48] Seiler, W.O., et al., *Impaired migration of epidermal cells from decubitus ulcers in cell cultures. A cause of protracted wound healing?* Am J Clin Pathol, 1989. 92(4): p. 430-4.

[49] Hanna, C., *Proliferation and migration of epithelial cells during corneal wound repair in the rabbit and the rat.* Am J Ophthalmol, 1966. 61(1): p. 55-63.

[50] Ridley, A.J., *Rho family proteins: coordinating cell responses.* Trends Cell Biol, 2001. 11(12): p. 471-7.

[51] Ridley, A.J., *Rho GTPases and cell migration.* J Cell Sci, 2001. 114(Pt 15): p. 2713-22.

[52] Schirmer, J. and K. Aktories, *Large clostridial cytotoxins: cellular biology of Rho/Ras-glucosylating toxins.* Biochim Biophys Acta, 2004. 1673(1-2): p. 66-74.

[53] Sander, E.E. and J.G. Collard, *Rho-like GTPases: their role in epithelial cell-cell adhesion and invasion.* Eur J Cancer, 1999. 35(9): p. 1302-8.

[54] Ridley, A.J. and A. Hall, *The small GTP-binding protein rho regulates the assembly of focal adhesions and actin stress fibers in response to growth factors.* Cell, 1992. 70(3): p. 389-99.

[55] Ridley, A.J., et al., *The small GTP-binding protein rac regulates growth factor-induced membrane ruffling.* Cell, 1992. 70(3): p. 401-10.

[56] Kozma, R., et al., *The Ras-related protein Cdc42Hs and bradykinin promote formation of peripheral actin microspikes and filopodia in Swiss 3T3 fibroblasts.* Mol Cell Biol, 1995. 15(4): p. 1942-52.

[57] Spurzem, J.R., et al., *Activation of protein kinase A accelerates bovine bronchial epithelial cell migration.* Am J Physiol Lung Cell Mol Physiol, 2002. 282(5): p. L1108-16.

[58] Desai, L.P., et al., *RhoA and Rac1 are both required for efficient wound closure of airway epithelial cells.* Am J Physiol Lung Cell Mol Physiol, 2004. 287(6): p. L1134-44.

[59] Wang, Y.P., et al., *IQ domain GTPase-activating protein 1 mediates the process of injury and repair in bronchial epithelial cells.* Sheng Li Xue Bao, 2008. 60(3): p. 409-18.

[60] Wang, Y., et al., *IQGAP1 promotes cell proliferation and is involved in a phosphorylation-dependent manner in wound closure of bronchial epithelial cells.* Int J Mol Med, 2008. 22(1): p. 79-87.

[61] Just, I., et al., *Glucosylation of Rho proteins by Clostridium difficile toxin B.* Nature, 1995. 375(6531): p. 500-3.

[62] Herrmann, C., et al., *Functional consequences of monoglucosylation of Ha-Ras at effector domain amino acid threonine 35.* J Biol Chem, 1998. 273(26): p. 16134-9.

[63] Sehr, P., et al., *Glucosylation and ADP ribosylation of rho proteins: effects on nucleotide binding, GTPase activity, and effector coupling.* Biochemistry, 1998. 37(15): p. 5296-304.

[64] Gerhard, R., et al., *Glucosylation of Rho GTPases by Clostridium difficile toxin A triggers apoptosis in intestinal epithelial cells.* J Med Microbiol, 2008. 57(Pt 6): p. 765-70.

[65] Zanetta, J.P., et al., *Glycoproteins and lectins in cell adhesion and cell recognition processes.* Histochem J, 1992. 24(11): p. 791-804.

[66] Lis, H. and N. Sharon, *Protein glycosylation. Structural and functional aspects.* Eur J Biochem, 1993. 218(1): p. 1-27.

[67] Hoe, M.H. and R.C. Hunt, *Loss of one asparagine-linked oligosaccharide from human transferrin receptors results in specific cleavage and association with the endoplasmic reticulum.* J Biol Chem, 1992. 267(7): p. 4916-23.

[68] Rands, E., et al., *Mutational analysis of beta-adrenergic receptor glycosylation.* J Biol Chem, 1990. 265(18): p. 10759-64.

[69] Feige, J.J. and A. Baird, *Glycosylation of the basic fibroblast growth factor receptor. The contribution of carbohydrate to receptor function.* J Biol Chem, 1988. 263(28): p. 14023-9.

[70] Leconte, I., et al., *N-linked oligosaccharide chains of the insulin receptor beta subunit are essential for transmembrane signaling.* J Biol Chem, 1992. 267(24): p. 17415-23.

[71] Tsuda, E., et al., *The role of carbohydrate in recombinant human erythropoietin.* Eur J Biochem, 1990. 188(2): p. 405-11.

[72] Schnaar, R.L., *Glycosphingolipids in cell surface recognition.* Glycobiology, 1991. 1(5): p. 477-85.

[73] Phillips, M.L., et al., *ELAM-1 mediates cell adhesion by recognition of a carbohydrate ligand, sialyl-Lex.* Science, 1990. 250(4984): p. 1130-2.

[74] Chammas, R., et al., *Functional hypotheses for aberrant glycosylation in tumor cells.* Braz J Med Biol Res, 1994. 27(2): p. 505-7.

[75] Lowe, J.B., et al., *ELAM-1--dependent cell adhesion to vascular endothelium determined by a transfected human fucosyltransferase cDNA.* Cell, 1990. 63(3): p. 475-84.

[76] Mann, P.L., I. Lopez-Colberg, and R.O. Kelley, *Cell surface oligosaccharide modulation during differentiation. I. Modulation of lectin binding.* Mech Ageing Dev, 1987. 38(3): p. 207-17.

[77] Lis, H. and N. Sharon, *The biochemistry of plant lectins (phytohemagglutinins).* Annu Rev Biochem, 1973. 42(0): p. 541-74.

[78] Gipson, I.K. and R.A. Anderson, *Effect of lectins on migration of the corneal epithelium.* Invest Ophthalmol Vis Sci, 1980. 19(4): p. 341-9.

[79] Gipson, I.K. and T.C. Kiorpes, *Epithelial sheet movement: protein and glycoprotein synthesis.* Dev Biol, 1982. 92(1): p. 259-62.

[80] Gipson, I.K., et al., *Lectin binding to cell surfaces: comparisons between normal and migrating corneal epithelium.* Dev Biol, 1983. 96(2): p. 337-45.

[81] Sweatt, A.J., R.M. Degi, and R.M. Davis, *Corneal wound-associated glycoconjugates analyzed by lectin histochemistry.* Curr Eye Res, 1999. 19(3): p. 212-8.

[82] Dorscheid, D.R., et al., *Characterization of cell surface lectin-binding patterns of human airway epithelium.* Histochem J, 1999. 31(3): p. 145-51.

[83] Xiantang, L., et al., *Glycosylation profiles of airway epithelium after repair of mechanical injury in guinea pigs.* Histochem J, 2000. 32(4): p. 207-16.

[84] Plotkowski, M.C., et al., *Differential adhesion of Pseudomonas aeruginosa to human respiratory epithelial cells in primary culture.* J Clin Invest, 1991. 87(6): p. 2018-28.

[85] Dorscheid, D.R., et al., *Role of cell surface glycosylation in mediating repair of human airway epithelial cell monolayers.* Am J Physiol Lung Cell Mol Physiol, 2001. 281(4): p. L982-92.

[86] Allahverdian, S., B.J. Patchell, and D.R. Dorscheid, *Carbohydrates and epithelial repair - more than just post-translational modification.* Current Drug Targets, 2006. 7(5): p. 597-606.

[87] Allahverdian, S., et al., *Sialyl Lewis X modification of the epidermal growth factor receptor regulates receptor function during airway epithelial wound repair.* Clin Exp Allergy, 2010. 40(4): p. 607-18.

[88] Donaldson, D.J. and J.M. Mason, *Inhibition of epidermal cell migration by concanavalin A in skin wounds of the adult newt.* J Exp Zool, 1977. 200(1): p. 55-64.

[89] Patchell, B.J. and D.R. Dorscheid, *Repair of the injury to respiratory epithelial cells characteristic of asthma is stimulated by Allomyrina dichotoma agglutinin specific serum glycoproteins.* Clinical and experimental allergy : journal of the British Society for Allergy and Clinical Immunology, 2006. 36(5): p. 585-593.

[90] Trinkaus-Randall, V., et al., *Carbohydrate moieties of the basal lamina: their role in attachment and spreading of basal corneal epithelial cells.* Cell Tissue Res, 1988. 251(2): p. 315-23.

[91] Adam, E.C., et al., *Role of carbohydrates in repair of human respiratory epithelium using an in vitro model.* Clin Exp Allergy, 2003. 33(10): p. 1398-404.

[92] Sato, S. and R.C. Hughes, *Binding specificity of a baby hamster kidney lectin for H type I and II chains, polylactosamine glycans, and appropriately glycosylated forms of laminin and fibronectin.* J Biol Chem, 1992. 267(10): p. 6983-90.

[93] Woo, H.J., et al., *The major non-integrin laminin binding protein of macrophages is identical to carbohydrate binding protein 35 (Mac-2).* J Biol Chem, 1990. 265(13): p. 7097-9.

[94] Warfield, P.R., et al., *Adhesion of human breast carcinoma to extracellular matrix proteins is modulated by galectin-3.* Invasion Metastasis, 1997. 17(2): p. 101-12.

[95] Matarrese, P., et al., *Galectin-3 overexpression protects from apoptosis by improving cell adhesion properties.* Int J Cancer, 2000. 85(4): p. 545-54.

[96] Cao, Z., et al., *Galectin-7 as a potential mediator of corneal epithelial cell migration.* Arch Ophthalmol, 2003. 121(1): p. 82-6.

[97] Cao, Z., et al., *Galectins-3 and -7, but not galectin-1, play a role in re-epithelialization of wounds.* J Biol Chem, 2002. 277(44): p. 42299-305.

[98] Hughes, R.C., *Secretion of the galectin family of mammalian carbohydrate-binding proteins.* Biochim Biophys Acta, 1999. 1473(1): p. 172-85.

[99] Kojima, K., et al., *Affinity purification and affinity characterization of carbohydrate-binding proteins in bovine kidney.* J Chromatogr, 1992. 597(1-2): p. 323-30.

[100] Hajjar, K.A. and C.M. Reynolds, *alpha-Fucose-mediated binding and degradation of tissue-type plasminogen activator by HepG2 cells.* J Clin Invest, 1994. 93(2): p. 703-10.

[101] Kim, J. and K.A. Hajjar, *Annexin II: a plasminogen-plasminogen activator co-receptor.* Front Biosci, 2002. 7: p. d341-8.

[102] Liu, J.W., et al., *Annexin II expression is reduced or lost in prostate cancer cells and its re-expression inhibits prostate cancer cell migration*. Oncogene, 2003. 22(10): p. 1475-85.
[103] Balch, C. and J.R. Dedman, *Annexins II and V inhibit cell migration*. Exp Cell Res, 1997. 237(2): p. 259-63.
[104] Matsuda, A., et al., *Identification and immunohistochemical localization of annexin II in rat cornea*. Curr Eye Res, 1999. 19(4): p. 368-75.
[105] Raynor, C.M., et al., *Annexin II enhances cytomegalovirus binding and fusion to phospholipid membranes*. Biochemistry, 1999. 38(16): p. 5089-95.
[106] Compton, T., R.R. Nepomuceno, and D.M. Nowlin, *Human cytomegalovirus penetrates host cells by pH-independent fusion at the cell surface*. Virology, 1992. 191(1): p. 387-95.
[107] Compton, T., D.M. Nowlin, and N.R. Cooper, *Initiation of human cytomegalovirus infection requires initial interaction with cell surface heparan sulfate*. Virology, 1993. 193(2): p. 834-41.
[108] Malhotra, R., et al., *Isolation and characterisation of potential respiratory syncytial virus receptor(s) on epithelial cells*. Microbes Infect, 2003. 5(2): p. 123-33.
[109] Kotelkin, A., et al., *Respiratory syncytial virus infection sensitizes cells to apoptosis mediated by tumor necrosis factor-related apoptosis-inducing ligand*. J Virol, 2003. 77(17): p. 9156-72.
[110] Chakraborti, S., et al., *Regulation of matrix metalloproteinases: an overview*. Mol Cell Biochem, 2003. 253(1-2): p. 269-85.
[111] Van Lint, P. and C. Libert, *Chemokine and cytokine processing by matrix metalloproteinases and its effect on leukocyte migration and inflammation*. J Leukoc Biol, 2007. 82(6): p. 1375-81.
[112] Gueders, M.M., et al., *Matrix metalloproteinases (MMPs) and tissue inhibitors of MMPs in the respiratory tract: potential implications in asthma and other lung diseases*. Eur J Pharmacol, 2006. 533(1-3): p. 133-44.
[113] Greenlee, K.J., Z. Werb, and F. Kheradmand, *Matrix metalloproteinases in lung: multiple, multifarious, and multifaceted*. Physiol Rev, 2007. 87(1): p. 69-98.
[114] Trifilieff, A., et al., *Pharmacological profile of PKF242-484 and PKF241-466, novel dual inhibitors of TNF-alpha converting enzyme and matrix metalloproteinases, in models of airway inflammation*. Br J Pharmacol, 2002. 135(7): p. 1655-64.
[115] Lee, K.S., et al., *Doxycycline reduces airway inflammation and hyperresponsiveness in a murine model of toluene diisocyanate-induced asthma*. J Allergy Clin Immunol, 2004. 113(5): p. 902-9.
[116] Kumagai, K., et al., *Inhibition of matrix metalloproteinases prevents allergen-induced airway inflammation in a murine model of asthma*. J Immunol, 1999. 162(7): p. 4212-9.
[117] Demedts, I.K., et al., *Matrix metalloproteinases in asthma and COPD*. Curr Opin Pharmacol, 2005. 5(3): p. 257-63.
[118] Russell, R.E., et al., *Alveolar macrophage-mediated elastolysis: roles of matrix metalloproteinases, cysteine, and serine proteases*. Am J Physiol Lung Cell Mol Physiol, 2002. 283(4): p. L867-73.
[119] Bracke, K., et al., *Matrix metalloproteinase-12 and cathepsin D expression in pulmonary macrophages and dendritic cells of cigarette smoke-exposed mice*. Int Arch Allergy Immunol, 2005. 138(2): p. 169-79.

[120] Miller, T.L., et al., *Expression of matrix metalloproteinases 2, 7 and 9, and their tissue inhibitors 1 and 2, in developing rabbit tracheae.* Biol Neonate, 2006. 89(4): p. 236-43.

[121] Prikk, K., et al., *In vivo collagenase-2 (MMP-8) expression by human bronchial epithelial cells and monocytes/macrophages in bronchiectasis.* J Pathol, 2001. 194(2): p. 232-8.

[122] Lopez-Boado, Y.S., C.L. Wilson, and W.C. Parks, *Regulation of matrilysin expression in airway epithelial cells by Pseudomonas aeruginosa flagellin.* J Biol Chem, 2001. 276(44): p. 41417-23.

[123] Puchelle, E., et al., *[Regeneration of injured airway epithelium].* Ann Pharm Fr, 2006. 64(2): p. 107-13.

[124] Parks, W.C., Y.S. Lopez-Boado, and C.L. Wilson, *Matrilysin in epithelial repair and defense.* Chest, 2001. 120(1 Suppl): p. 36S-41S.

[125] Coraux, C., et al., *Differential expression of matrix metalloproteinases and interleukin-8 during regeneration of human airway epithelium in vivo.* J Pathol, 2005. 206(2): p. 160-9.

[126] Adiseshaiah, P., et al., *A Fra-1-dependent, matrix metalloproteinase driven EGFR activation promotes human lung epithelial cell motility and invasion.* J Cell Physiol, 2008. 216(2): p. 405-12.

[127] Goswami, S., et al., *Divergent functions for airway epithelial matrix metalloproteinase 7 and retinoic acid in experimental asthma.* Nat Immunol, 2009. 10(5): p. 496-503.

[128] Nabeshima, K., et al., *Emmprin (basigin/CD147): matrix metalloproteinase modulator and multifunctional cell recognition molecule that plays a critical role in cancer progression.* Pathol Int, 2006. 56(7): p. 359-67.

[129] Kadomatsu, K. and T. Muramatsu, *[Role of basigin, a glycoprotein belonging to the immunoglobulin superfamily, in the nervous system].* Tanpakushitsu Kakusan Koso, 2004. 49(15 Suppl): p. 2417-24.

[130] Vignola, A.M., et al., *HLA-DR and ICAM-1 expression on bronchial epithelial cells in asthma and chronic bronchitis.* Am Rev Respir Dis, 1993. 148(3): p. 689-94.

[131] Lackie, P.M., et al., *Expression of CD44 isoforms is increased in the airway epithelium of asthmatic subjects.* Am J Respir Cell Mol Biol, 1997. 16(1): p. 14-22.

[132] Jouneau, S., et al., *EMMPRIN (CD147) regulation of MMP-9 in bronchial epithelial cells in COPD.* Respirology, 2011. 16(4): p. 705-12.

[133] Caudroy, S., et al., *Expression of the extracellular matrix metalloproteinase inducer (EMMPRIN) and the matrix metalloproteinase-2 in bronchopulmonary and breast lesions.* J Histochem Cytochem, 1999. 47(12): p. 1575-80.

[134] Caudroy, S., et al., *EMMPRIN-mediated MMP regulation in tumor and endothelial cells.* Clin Exp Metastasis, 2002. 19(8): p. 697-702.

[135] Huet, E., et al., *Role of emmprin/CD147 in tissue remodeling.* Connect Tissue Res, 2008. 49(3): p. 175-9.

[136] Betsuyaku, T., et al., *Increased basigin in bleomycin-induced lung injury.* Am J Respir Cell Mol Biol, 2003. 28(5): p. 600-6.

[137] Guillot, S., et al., *Increased extracellular matrix metalloproteinase inducer (EMMPRIN) expression in pulmonary fibrosis.* Exp Lung Res, 2006. 32(3-4): p. 81-97.

[138] Odajima, N., et al., *Extracellular matrix metalloproteinase inducer in interstitial pneumonias.* Hum Pathol, 2006. 37(8): p. 1058-65.

[139] Gwinn, W.M., et al., *Novel approach to inhibit asthma-mediated lung inflammation using anti-CD147 intervention.* J Immunol, 2006. 177(7): p. 4870-9.

[140] Leir, S.H., et al., *Increased CD44 expression in human bronchial epithelial repair after damage or plating at low cell densities.* Am J Physiol Lung Cell Mol Physiol, 2000. 278(6): p. L1129-37.

[141] Yu, W.H., et al., *CD44 anchors the assembly of matrilysin/MMP-7 with heparin-binding epidermal growth factor precursor and ErbB4 and regulates female reproductive organ remodeling.* Genes Dev, 2002. 16(3): p. 307-23.

[142] Polosa, R., et al., *Expression of c-erbB receptors and ligands in the bronchial epithelium of asthmatic subjects.* J Allergy Clin Immunol, 2002. 109(1): p. 75-81.

[143] White, S.R., et al., *Role of very late adhesion integrins in mediating repair of human airway epithelial cell monolayers after mechanical injury.* Am J Respir Cell Mol Biol, 1999. 20(4): p. 787-96.

[144] Bishayee, S., *Role of conformational alteration in the epidermal growth factor receptor (EGFR) function.* Biochem Pharmacol, 2000. 60(8): p. 1217-23.

[145] Stroop, C.J., et al., *Characterization of the carbohydrate chains of the secreted form of the human epidermal growth factor receptor.* Glycobiology, 2000. 10(9): p. 901-17.

[146] Soderquist, A.M., G. Todderud, and G. Carpenter, *The role of carbohydrate as a post-translational modification of the receptor for epidermal growth factor.* Adv Exp Med Biol, 1988. 231: p. 569-82.

[147] Wang, X.Q., et al., *Epidermal growth factor receptor glycosylation is required for ganglioside GM3 binding and GM3-mediated suppression [correction of suppresion] of activation.* Glycobiology, 2001. 11(7): p. 515-22.

[148] Konishi, A. and B.C. Berk, *Epidermal growth factor receptor transactivation is regulated by glucose in vascular smooth muscle cells.* J Biol Chem, 2003. 278(37): p. 35049-56. Epub 2003 Jun 26.

[149] Partridge, E.A., et al., *Regulation of cytokine receptors by Golgi N-glycan processing and endocytosis.* Science, 2004. 306(5693): p. 120-4.

[150] Eisenmesser, E.Z., et al., *Solution structure of interleukin-13 and insights into receptor engagement.* Journal of Molecular Biology, 2001. 310(1): p. 231-241.

[151] Hershey, G.K.K., *IL-13 receptors and signaling pathways: An evolving web.* Journal of Allergy and Clinical Immunology, 2003. 111(4): p. 677-690.

[152] Minty, A., et al., *Interleukin-13 is a new human lymphokine regulating inflammatory and immune responses.* Nature, 1993. 362(6417): p. 248-250.

[153] Liang, H., et al., *SUMF2 interacts with interleukin-13 and inhibits interleukin-13 secretion in bronchial smooth muscle cells.* Journal of cellular biochemistry, 2009. 108(5): p. 1076-1083.

[154] Vladich, F.D., et al., *IL-13 R130Q, a common variant associated with allergy and asthma, enhances effector mechanisms essential for human allergic inflammation.* The Journal of clinical investigation, 2005. 115(3): p. 747-754.

[155] McKenzie, A.N., et al., *Interleukin 13, a T-cell-derived cytokine that regulates human monocyte and B-cell function.* Proceedings of the National Academy of Sciences of the United States of America, 1993. 90(8): p. 3735-3739.

[156] Schmid-Grendelmeier, P., et al., *Eosinophils express functional IL-13 in eosinophilic inflammatory diseases.* Journal of immunology (Baltimore, Md.: 1950), 2002. 169(2): p. 1021-1027.

[157] Wills-Karp, M., *Interleukin-13 in asthma pathogenesis.* Immunological reviews, 2004. 202(Journal Article): p. 175-190.

[158] Allahverdian, S., et al., *Secretion of IL-13 by airway epithelial cells enhances epithelial repair via HB-EGF.* Am J Respir Cell Mol Biol, 2008. 38(2): p. 153-60.

[159] Zurawski, G. and J.E. de Vries, *Interleukin 13 elicits a subset of the activities of its close relative interleukin 4.* Stem cells (Dayton, Ohio), 1994. 12(2): p. 169-174.

[160] Zurawski, G. and J.E. de Vries, *Interleukin 13, an interleukin 4-like cytokine that acts on monocytes and B cells, but not on T cells.* Immunology today, 1994. 15(1): p. 19-26.

[161] Hart, P.H., et al., *Potential antiinflammatory effects of interleukin 4: suppression of human monocyte tumor necrosis factor alpha, interleukin 1, and prostaglandin E2.* Proceedings of the National Academy of Sciences of the United States of America, 1989. 86(10): p. 3803-3807.

[162] Pope, S.M., et al., *IL-13 induces eosinophil recruitment into the lung by an IL-5- and eotaxin-dependent mechanism.* The Journal of allergy and clinical immunology, 2001. 108(4): p. 594-601.

[163] Luttmann, W., et al., *Activation of human eosinophils by IL-13. Induction of CD69 surface antigen, its relationship to messenger RNA expression, and promotion of cellular viability.* Journal of immunology (Baltimore, Md.: 1950), 1996. 157(4): p. 1678-1683.

[164] Bochner, B.S., et al., *IL-13 selectively induces vascular cell adhesion molecule-1 expression in human endothelial cells.* Journal of immunology (Baltimore, Md.: 1950), 1995. 154(2): p. 799-803.

[165] Grunstein, M.M., et al., *IL-13-dependent autocrine signaling mediates altered responsiveness of IgE-sensitized airway smooth muscle.* American journal of physiology.Lung cellular and molecular physiology, 2002. 282(3): p. L520-8.

[166] Akiho, H., et al., *Role of IL-4, IL-13, and STAT6 in inflammation-induced hypercontractility of murine smooth muscle cells.* American journal of physiology.Gastrointestinal and liver physiology, 2002. 282(2): p. G226-32.

[167] Booth, B.W., et al., *Interleukin-13 induces proliferation of human airway epithelial cells in vitro via a mechanism mediated by transforming growth factor-alpha.* American journal of respiratory cell and molecular biology, 2001. 25(6): p. 739-743.

[168] Fichtner-Feigl, S., et al., *IL-13 signaling through the IL-13alpha2 receptor is involved in induction of TGF-beta1 production and fibrosis.* Nature medicine, 2006. 12(1): p. 99-106.

[169] Li, L., et al., *Effects of Th2 cytokines on chemokine expression in the lung: IL-13 potently induces eotaxin expression by airway epithelial cells.* Journal of immunology (Baltimore, Md.: 1950), 1999. 162(5): p. 2477-2487.

[170] Laoukili, J., et al., *IL-13 alters mucociliary differentiation and ciliary beating of human respiratory epithelial cells.* The Journal of clinical investigation, 2001. 108(12): p. 1817-1824.

[171] Zhu, Z., et al., *Pulmonary expression of interleukin-13 causes inflammation, mucus hypersecretion, subepithelial fibrosis, physiologic abnormalities, and eotaxin production.* The Journal of clinical investigation, 1999. 103(6): p. 779-788.

[172] Shim, J.J., et al., *IL-13 induces mucin production by stimulating epidermal growth factor receptors and by activating neutrophils.* American journal of physiology.Lung cellular and molecular physiology, 2001. 280(1): p. L134-40.

[173] Leonard, W.J. and J.X. Lin, *Cytokine receptor signaling pathways.* The Journal of allergy and clinical immunology, 2000. 105(5): p. 877-888.

[174] Arima, K., et al., *Characterization of the interaction between interleukin-13 and interleukin-13 receptors.* The Journal of biological chemistry, 2005. 280(26): p. 24915-24922.

[175] Guo, J., et al., *Chromosome mapping and expression of the human interleukin-13 receptor.* Genomics, 1997. 42(1): p. 141-145.

[176] Andrews, A.L., et al., *Kinetic analysis of the interleukin-13 receptor complex.* The Journal of biological chemistry, 2002. 277(48): p. 46073-46078.

[177] Miloux, B., et al., *Cloning of the human IL-13Rα1 chain and reconstitution with the IL-4Rα of a functional IL-4/IL-13 receptor complex.* FEBS letters, 1997. 401(2-3): p. 163-166.

[178] Aman, M.J., et al., *cDNA cloning and characterization of the human interleukin 13 receptor alpha chain.* The Journal of biological chemistry, 1996. 271(46): p. 29265-29270.

[179] Wang, I.M., et al., *STAT-1 is activated by IL-4 and IL-13 in multiple cell types.* Molecular immunology, 2004. 41(9): p. 873-884.

[180] Hirst, S.J., et al., *Selective induction of eotaxin release by interleukin-13 or interleukin-4 in human airway smooth muscle cells is synergistic with interleukin-1beta and is mediated by the interleukin-4 receptor alpha-chain.* American journal of respiratory and critical care medicine, 2002. 165(8): p. 1161-1171.

[181] Yasuo, M., et al., *Relationship between calcium-activated chloride channel 1 and MUC5AC in goblet cell hyperplasia induced by interleukin-13 in human bronchial epithelial cells.* Respiration; international review of thoracic diseases, 2006. 73(3): p. 347-359.

[182] Wei, L.H., et al., *IL-4 and IL-13 upregulate arginase I expression by cAMP and JAK/STAT6 pathways in vascular smooth muscle cells.* American journal of physiology.Cell physiology, 2000. 279(1): p. C248-56.

[183] Daines, M.O. and G.K. Hershey, *A novel mechanism by which interferon-gamma can regulate interleukin (IL)-13 responses. Evidence for intracellular stores of IL-13 receptor alpha -2 and their rapid mobilization by interferon-gamma.* The Journal of biological chemistry, 2002. 277(12): p. 10387-10393.

[184] David, M., et al., *Induction of the IL-13 receptor alpha2-chain by IL-4 and IL-13 in human keratinocytes: involvement of STAT6, ERK and p38 MAPK pathways.* Oncogene, 2001. 20(46): p. 6660-6668.

[185] Caput, D., et al., *Cloning and characterization of a specific interleukin (IL)-13 binding protein structurally related to the IL-5 receptor alpha chain.* The Journal of biological chemistry, 1996. 271(28): p. 16921-16926.

[186] Lupardus, P.J., M.E. Birnbaum, and K.C. Garcia, *Molecular basis for shared cytokine recognition revealed in the structure of an unusually high affinity complex between IL-13 and IL-13Ralpha2.* Structure (London, England : 1993). 18(3): p. 332-342.

[187] Kioi, M., S. Seetharam, and R.K. Puri, *N-linked glycosylation of IL-13R alpha2 is essential for optimal IL-13 inhibitory activity.* FASEB journal : official publication of the Federation of American Societies for Experimental Biology, 2006. 20(13): p. 2378-2380.

[188] Konstantinidis, A.K., et al., *Cellular localization of interleukin 13 receptor alpha2 in human primary bronchial epithelial cells and fibroblasts.* Journal of investigational allergology & clinical immunology : official organ of the International Association of Asthmology (INTERASMA) and Sociedad Latinoamericana de Alergia e Inmunologia, 2008. 18(3): p. 174-180.

[189] Andrews, A.L., et al., *IL-13 receptor alpha 2: a regulator of IL-13 and IL-4 signal transduction in primary human fibroblasts.* The Journal of allergy and clinical immunology, 2006. 118(4): p. 858-865.

[190] Donaldson, D.D., et al., *The Murine IL-13 Receptor {alpha}2: Molecular Cloning, Characterization, and Comparison with Murine IL-13 Receptor {alpha}1.* The Journal of Immunology, 1998. 161(5): p. 2317-2324.

[191] Kawakami, K., et al., *The interleukin-13 receptor alpha2 chain: an essential component for binding and internalization but not for interleukin-13-induced signal transduction through the STAT6 pathway.* Blood, 2001. 97(9): p. 2673-2679.

[192] Zhang, J.G., et al., *Identification, purification, and characterization of a soluble interleukin (IL)-13-binding protein. Evidence that it is distinct from the cloned Il-13 receptor and Il-4 receptor alpha-chains.* The Journal of biological chemistry, 1997. 272(14): p. 9474-9480.

[193] Wilson, M.S., et al., *IL-13Ralpha2 and IL-10 coordinately suppress airway inflammation, airway-hyperreactivity, and fibrosis in mice.* The Journal of clinical investigation, 2007. 117(10): p. 2941-2951.

[194] Smirnov, D.V., et al., *Tandem arrangement of human genes for interleukin-4 and interleukin-13: resemblance in their organization.* Gene, 1995. 155(2): p. 277-281.

[195] Heinzmann, A., et al., *Genetic variants of IL-13 signalling and human asthma and atopy.* Human molecular genetics, 2000. 9(4): p. 549-559.

[196] Howard, T.D., et al., *Gene-gene interaction in asthma: IL4RA and IL13 in a Dutch population with asthma.* American Journal of Human Genetics, 2002. 70(1): p. 230-236.

[197] Tarazona-Santos, E. and S.A. Tishkoff, *Divergent patterns of linkage disequilibrium and haplotype structure across global populations at the interleukin-13 (IL13) locus.* Genes and immunity, 2005. 6(1): p. 53-65.

[198] Madhankumar, A.B., A. Mintz, and W. Debinski, *Alanine-scanning mutagenesis of alpha-helix D segment of interleukin-13 reveals new functionally important residues of the cytokine.* The Journal of biological chemistry, 2002. 277(45): p. 43194-43205.

[199] Semlali, A., et al., *Thymic stromal lymphopoietin-induced human asthmatic airway epithelial cell proliferation through an IL-13-dependent pathway.* The Journal of allergy and clinical immunology, 2010. 125(4): p. 844-850.

[200] Singhera, G.K., R. MacRedmond, and D.R. Dorscheid, *Interleukin-9 and -13 inhibit spontaneous and corticosteroid induced apoptosis of normal airway epithelial cells.* Experimental lung research, 2008. 34(9): p. 579-598.

[201] Wadsworth, S., D. Sin, and D. Dorscheid, *Clinical update on the use of biomarkers of airway inflammation in the management of asthma.* J Asthma Allergy, 2011. 4: p. 77-86.

[202] Watanabe, T., et al., *Increased levels of HMGB-1 and endogenous secretory RAGE in induced sputum from asthmatic patients.* Respir Med, 2011. 105(4): p. 519-25.

[203] Reddel, H.K., et al., *An official American Thoracic Society/European Respiratory Society statement: asthma control and exacerbations: standardizing endpoints for clinical asthma trials and clinical practice.* Am J Respir Crit Care Med, 2009. 180(1): p. 59-99.

[204] D'Amico, A., et al., *An investigation on electronic nose diagnosis of lung cancer.* Lung Cancer, 2010. 68(2): p. 170-6.

[205] Payne, D.N., et al., *Relationship between exhaled nitric oxide and mucosal eosinophilic inflammation in children with difficult asthma, after treatment with oral prednisolone.* Am J Respir Crit Care Med, 2001. 164(8 Pt 1): p. 1376-81.

[206] Schleich, F.N., et al., *Exhaled nitric oxide thresholds associated with a sputum eosinophil count >/=3% in a cohort of unselected patients with asthma.* Thorax, 2010. 65(12): p. 1039-44.

[207] Jiang, J., et al., *Nitric oxide gas phase release in human small airway epithelial cells.* Respir Res, 2009. 10: p. 3.

[208] Kostikas, K., et al., *Exhaled NO and exhaled breath condensate pH in the evaluation of asthma control.* Respir Med, 2011. 105(4): p. 526-32.

[209] Gessner, C., et al., *Angiogenic markers in breath condensate identify non-small cell lung cancer.* Lung Cancer, 2010. 68(2): p. 177-84.

[210] Bloemen, K., et al., *A new approach to study exhaled proteins as potential biomarkers for asthma.* Clin Exp Allergy, 2011. 41(3): p. 346-56.

[211] Lessard, A., et al., *Obesity and asthma: a specific phenotype?* Chest, 2008. 134(2): p. 317-23.

[212] Sin, D.D., et al., *Serum PARC/CCL-18 Concentrations and Health Outcomes in Chronic Obstructive Pulmonary Disease.* Am J Respir Crit Care Med, 2011. 183(9): p. 1187-1192.

[213] Wedzicha, J.A., et al., *Acute exacerbations of chronic obstructive pulmonary disease are accompanied by elevations of plasma fibrinogen and serum IL-6 levels.* Thromb Haemost, 2000. 84(2): p. 210-5.

[214] Chang, C.L., et al., *Biochemical markers of cardiac dysfunction predict mortality in acute exacerbations of COPD.* Thorax, 2011.

[215] Lowhagen, O., et al., *The inflammatory marker serum eosinophil cationic protein (ECP) compared with PEF as a tool to decide inhaled corticosteroid dose in asthmatic patients.* Respir Med, 2002. 96(2): p. 95-101.

[216] Dosanjh, A., et al., *Elevated Serum Eosinophil Cationic Protein Levels in Cystic Fibrosis, Pediatric Asthma, and Bronchiolitis.* Pediatric Asthma, Allergy & Immunology, 1996. 10(4): p. 169-173.

[217] Trulson, A., et al., *The functional heterogeneity of eosinophil cationic protein is determined by a gene polymorphism and post-translational modifications.* Clin Exp Allergy, 2007. 37(2): p. 208-18.

[218] Sekiya, T., et al., *Increased levels of a TH2-type CC chemokine thymus and activation-regulated chemokine (TARC) in serum and induced sputum of asthmatics.* Allergy, 2002. 57(2): p. 173-7.

[219] Leung, T.F., et al., *Plasma concentration of thymus and activation-regulated chemokine is elevated in childhood asthma.* J Allergy Clin Immunol, 2002. 110(3): p. 404-9.

[220] Stephan, V., et al., *Determination of N-methylhistamine in urine as an indicator of histamine release in immediate allergic reactions.* J Allergy Clin Immunol, 1990. 86(6 Pt 1): p. 862-8.

[221] Takei, S., et al., *Urinary N-methylhistamine in asthmatic children receiving azelastine hydrochloride.* Ann Allergy Asthma Immunol, 1997. 78(5): p. 492-6.

[222] Sansuk, K., et al., *GPCR proteomics: mass spectrometric and functional analysis of histamine H1 receptor after baculovirus-driven and in vitro cell free expression.* J Proteome Res, 2008. 7(2): p. 621-9.

[223] Saude, E.J., et al., *Metabolomic profiling of asthma: diagnostic utility of urine nuclear magnetic resonance spectroscopy.* J Allergy Clin Immunol, 2011. 127(3): p. 757-64 e1-6.

Chapter 8

Metabolism of Carbohydrates in the Cell of Green Photosynthesis Sulfur Bacteria

M. B. Gorishniy and S. P. Gudz

Additional information is available at the end of the chapter

1. Introduction

Green bacteria - are phylogenetic isolated group photosyntetic microorganisms. The peculiarity of the structure of their cells is the presence of special vesicles - so-called chlorosom containing bacteriochlorofils and carotenoids. These microorganisms can not use water as a donor of electrons to form molecular oxygen during photosynthesis. Electrons required for reduction of assimilation CO_2, green bacteria are recovered from the sulfur compounds with low redox potential.

Ecological niche of green bacteria is low. Well known types of green bacteria - a common aquatic organisms that occur in anoxic, was lit areas of lakes or coastal sediments. In some ecosystems, these organisms play a key role in the transformation of sulfur compounds and carbon. They are adapted to low light intensity. Compared with other phototrophic bacteria, green bacteria can lives in the lowest layers of water in oxygen-anoxic ecosystems.

Representatives of various genera and species of green bacteria differ in morphology of cells, method of movement, ability to form gas vacuoles and pigment structure of the complexes. For most other signs, including metabolism, structure photosyntetic apparats and phylogeny, these families differ significantly. Each of the two most studied families of green bacteria (*Chlorobium* families and *Chloroflexus*) has a unique way of assimilation of carbon dioxide reduction. For species of the genus *Chlorobium* typical revers tricarboxylic acids cycle, and for members of the genus *Chloroflexus* - recently described 3-hidrocsypropionatn cycle. Metabolism of organic compounds, including carbohydrates in the cells of representatives of genera *Chlorobium* and *Chloroflexus* remains poorly understood. Anabolism and catabolism of monomeric and polymeric forms of carbohydrates in the cells of green bacteria is discussed.

Representatives of the green sulfur bacteria family *Chlorobiaceae* and green nonsulfur bacteria family *Chloroflexaceae* grow with CO_2 as the sole carbon source. In addition, the growth process, they can use some organic carbon sources, particularly carbohydrates. Species of the genus *Chlorobium* able to assimilate organic compounds is limited only in the presence of CO_2 and inorganic electron donors. Instead, representatives of the genus *Chloroflexus* grow on different carbon sources anaerobically and aerobically under illumination in the dark. Significant differences between species and genera *Chlorobium, Chloroflexus* due to the nature of their photosyntetic apparat. Despite the fact that members of both families contain the same types of chlorosoms, species of the genus *Chlorobium* reaction centers have photosystemy I (PS I), while both species of the genus *Chloroflexus* contain reaction centers photosystemy II (PS II). Because of the low redox potential (~ -0,9 V) of the primary electron acceptor in green sulfur bacteria, reaction center capable reduce ferredocsyn. In green bacteria nonsulfur bacteria redox potential of the primary acceptor is less negative (~ -0,5 V), resulting in these organisms synthesize reduction equivalents by reverse electron transport, like the purple bacteria. Thus, differences in the structure of the apparatus of green sulfur and green nonsulfur bacteria are reflected in their exchange carbohidrats compounds, and there fore also in their evolution and ecology.

In the evolution of autotrophic organisms formed several ways to assimilate CO_2, each of which is characterized by biochemical reactions that require the appropriate enzymes and reduction equivalents [9, 11]. The most common mechanism for CO_2 assimilation is Calvin cycle, which was found in most plants, algae and most famous groups of autotrophic prokaryotes. In green bacteria described two alternative ways of assimilation of CO_2. Revers cycle of tricarboxylic acids (RTAC) in green sulfur bacteria, first proposed by Evans in 1966. In 1989, Holo described 3-hidroksypropionat way that is characteristic of green non sulfur bacteria.

Larsen, using washed cells of *C. thiosulfatophilum*, for the first time found that they can absorb light in only small amounts of CO_2. They found that most carbon dioxide *C. thiosulfatophilum* records in an atmosphere containing H_2S, H_2 and CO_2. Most data on how carbon dioxide conversion and other compounds related to green sulfur bacteria genus *Chlorobium*, including *C. thiosulfatophilum*, *C. phaeobacteroides* and *C. limicola*.

Green sulfur bacteria can use some organic compounds (sugars, amino acids and organic acids). However, adding these compounds to the environment leads only to a slight stimulation of growth of culture in the presence of CO_2 and is to ensure that they are used only as additional sources of carbon [13]. In any case they are electron donors or major source of carbon. The use of these substances only if there among CO_2 and H_2S.

In the cells of *C. thiosulfatophilum* not identified Calvin cycle enzyme activity. The main role in the transformation of CO_2 is open in this group of bacteria (RTAC). Here in green sulfur bacteria is reduction of CO_2 assimilation. The cycle was proposed by the opening of phototrophic bacteria and other anaerobs two new ferredocsyn - dependent carboxylation reactions [13]:

$$\text{Acetyl} - \text{CoA} + CO_2 \rightarrow \text{pyruvate} + \text{CoA}$$
$$\text{Succinyl} - \text{CoA} + \text{CO2} \rightarrow \alpha - \text{ketoglutarate} + \text{CoA}$$

They make possible (RTAC), in which two molecules of CO_2 formed a molecule of acetyl - CoA (Fig. 1).

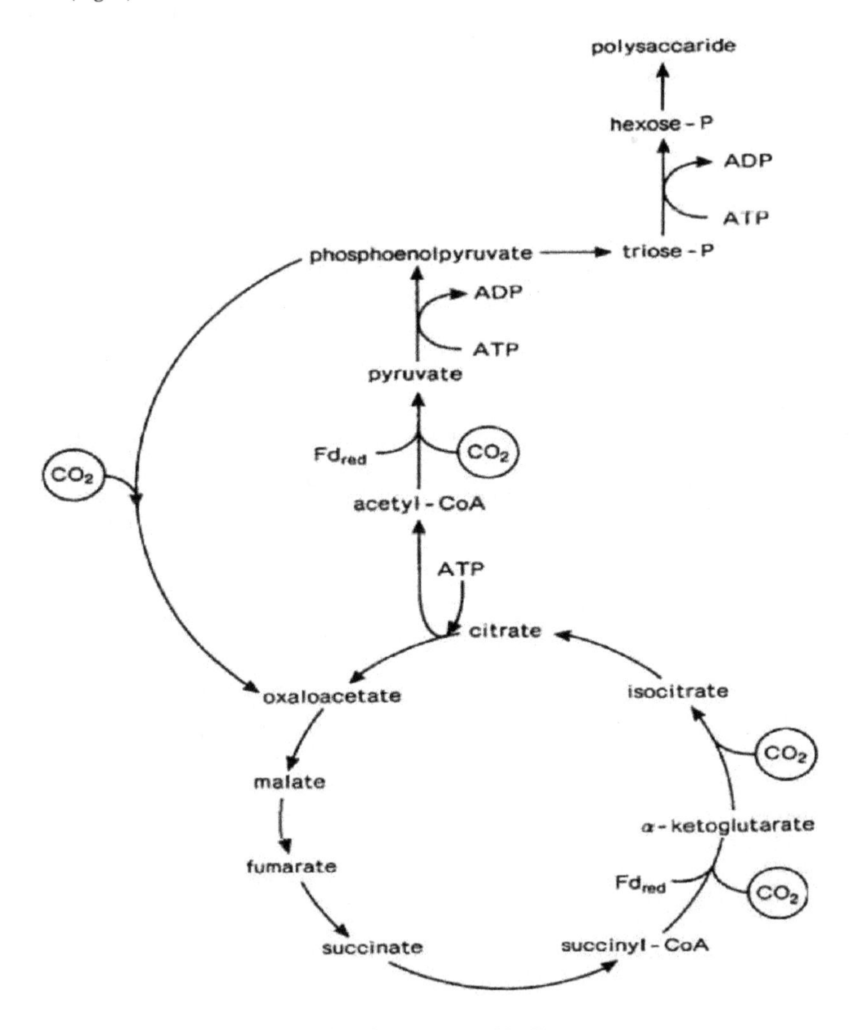

Figure 1. The revers tricarboxylic acid cycle (RTAC) in *Chlorobium* .

First, revers tricarboxylic acid cycle (RTAC) considered an additional mechanism for better functioning of rehabilitation Calvin cycle of the genus *Chlorobium.* Assumed that its main function is the formation of precursors for the synthesis of amino acids, lipids and porphyrins, while restorative Calvin cycle given the main role in the synthesis of carbohydrates. However, the lack of activity in cells rubisco put the availability of restorative Calvin cycle in cells of *Chlorobium limicola* doubt. Confirmation of operation of restorative tricarboxylic acid cycle in cells of green sulfur bacteria was discovered in them a key enzyme of this cycle. Using tracer and fractionation of isotopes of carbon have shown that (RTAC) is the only recovery mechanism for fixation of CO_2 in green sulfur bacteria, and the product cycle acetyl - CoA directly used for the synthesis of carbohydrates. It was also found that the genes of cells rubisco *Rhodospirillum rubrum*, not related of DNA isolated from cells of bacteria genus *Chlorobium*. Similar negative results were obtained using genes to cells rubisco *Anacystis nidulans*.

The study of restorative (RTAC) can explain the inability of green sulfur bacteria photoheterotroph. Simultaneously with the operation of the mechanism fixation of CO_2 cycle intermediates also provide cells needed organic matter for the synthesis of fatty acids (from acetyl-CoA), amino acids (from pyruvate, α-ketoglutarat acid) and carbohydrates (with pyruvate). However, since the activity of α-ketoglutaratdehidrohenaz not found in species of the genus *Chlorobium*, this cycle can operate only in recreation and hence organic compounds can not oxidate with the formation of reduction equivalents .

Recovery (RTAC) provides fixation of CO_2, to be based on restorative carboxylation reaction of organic acids. Fixation of carbon dioxide occurs in three enzymatic reactions, two of which occur with photochemically reduced ferredoksyn, and one - the same way formed provided with (H^+). As a result of a turnover cycle of four molecules of CO_2 and 10 [H^+] using the energy of three molecules of ATP synthesized molecule oxaloacetat acid is the end product cycle.

Described as "short" version of the cycle, in which 2 molecules of CO_2 are fixed using for their restoration 8 [H^+] and the energy of ATP. The final product in this case is acetyl-CoA, which is used to build components of cells. Addition of acetate in the culture medium promotes the accumulation of biomass and stimulates the formation of reserve polysacharides in the cells of green sulfur bacteria. Representatives of the family *Chlorobiaceae*, including *Chlorobium limicola* and *C. thiosulfatophilum*, often accumulate in cells poliglycose and / or glycogen. Accumulation of polysacharides increases in carbon dioxide assimilation by cells under conditions of deficiency of nitrogen and phosphorus. Under certain conditions of cultivation the level of glycogen in the cells can exceed 12% of dry weight of cells. Formed spare polysaccharides play an important role in changing the conditions of cultivation of green sulfur bacteria, especially when ingested bacteria in the extreme conditions of growth.

Larsen and collaborators found that the bacteria *C. thiosulfatophilum* not grow on media containing traces of hydrogen sulfide (0.01%) and various organic compounds: alcohols, sugars, organic acids. Only media with acetic, lactic or pyruvatic acid was seen a slight increase in biomass under conditions of hydrogen sulfide and carbon dioxide in the environment. Regarding the nature of organic compounds green sulfur bacteria similar to purple sulfur bacteria.

Larsen found that washed suspensions of cells *C. thiosulfatophilum* the light can absorb only small amounts of CO_2. The greatest amount of carbon dioxide *C. thiosulfatophilum* assimilates in an atmosphere containing 86% N_2, 9,2% CO_2, 3,9% H_2S and 0,5% H_2. Most data on how metabolic carbon dioxide and other carbon compounds obtained in experiments using *C. thiosulfatophilum* and *C. phaeobacteroides*.

Found that in cells *C thiosulfatophilum* not Calvin cycle enzyme activity. Important role in the assimilation of CO_2 is open in this group of microorganisms revers Crebs cycle, which was named Arnon cycle. This series provides a record of CO_2 through renewable carboxylation of organic acids. As a result of the work cycle in cells of green sulfur bacteria in the process of photosynthesis, glucose is formed, which is the first product of photosynthesis, carbohydrate nature. Ways to transform cells in green sulfur bacteria studied not enough. According to it becomes poliglucose, other authors believe that the glucose immediately polymerizes to form glycogen.

To detect sugars that accumulate in cells of *C. limicola* IMB- K-8 in the process of photosynthesis they were grown under illumination and in the presence of electron donor, which served as H_2S. After 10 days culturing cells destroy bacteria and cell less extract analyzed for the presence in it is reduced sugars. The total number of cell less extracts was determined by Shomodi-Nelson and for determination of glucose using enzymatic set "Diaglyc- 2". It turned out that the total number is reduced sugars determined by the method of Shomodi-Nelson did not differ from the rate obtained specifically for glucose.

It follows that the sugar is reduced *C. limicola* IMB- K-8 represented only glucose, which is the first carbohydrate, which is formed during photosynthesis.To test whether glucose in cells is in free or bound state spent acid hydrolysis cell less extract. It found that the content is reduced sugars increased approximately two fold. It follows that the glucose in the cells located in the free and in a bound state. In these experiments investigated the dynamics of accumulation of intracellular glucose bacterium *C. limicola* IMB- K-8 in the process of growth. It was found that the formation of glucose in the cells is observed throughout the period of growth and completed the transition culture stationary phase.

Growth of *C. limicola* IMB- K-8 and glucose in cells growing in culture in the light in a mineral medium with $NaHCO_3$ and Na_2S. We investigated the growth and accumulation of glucose in the cells of *C. limicola* IMB- K-8 for varying light intensity.

It was found that light intensity plays an important role in CO_2 assimilation in *C. limicola* IMB- K-8. More intensive process proceeded in low light, which does not exceed 40 lux. Reduction or increase in illumination intensity was accompanied by reduced productivity of photosynthesis.

On the intensity of photosynthesis reveals a significant influence of mineral nutrition of bacteria We shows the influence of different sources nitrogen and phosphorus supply of glucose in the cells *C. limicola* IMB- K-8.

Simultaneous limitation of growth of culture nitrogen and phosphorus accompanied by increase in glucose in the cells. Her level of these compounds for the deficit grew by about 60%. Separately salts of nitrogen and phosphorus showed much less effect In these experiments investigated how bacteria use glucose under various conditions of cultivation. This used washed cells were incubated under light and dark. When incubation of cells at the light in the presence of CO_2 and H_2S levels of glucose in the cells practically did not change while under these conditions in the dark glucose concetration in the cells decreased about 2.5 times.

Obviously, in the dark using glucose as an energy source, turning towards Embdena-Meyerhof-Parnas. The level of intracellular glucose is reduced and the conditions of incubation of cells at the light in the environment without hydrogen sulfide, indicating that the use of glucose under these conditions as the sole source of renewable equivalents. In the dark, without glucose hydrogen sulfide is the only source of energy. Thus, the glucose formed by cells plays an important role in the life of cell *C. limicola* IMB - K-8. When staying in the light cells in the process of photosynthesis observed formation of glucose in the cells, and in darkness it is used to maintain cell viability.

2. Isolation, identification and patterns of accumulation poliglucose *C. limicola* IMB- K-8

Nature poliglucose formed in the cells of *C. limicola* IMB- K-8. As already mentioned above green sulfur bacteria in the process of growth can form glucose, a small part of which the cells are in a free state, and the part becomes glycogen. To test the ability of *C. limicola* IMB-K-8 form glycogen was held their extraction from the cells by the method. Grown under light conditions cells in acetic acid. Polisacharide precipitate obtained by adding to the supernatant concentrated ethanol. The obtained precipitate distroy 10M H_2SO_4 obtained hydrolyzate were separated by chromatography. As witnesses used the glucose and galactose, and atzer. After manifestation chromatography revealed only one spot, which is slowly moving in the system butanol - water and Rf value was identical to glucose. It follows that polisaharide that piled up in the cells *C. limicola* IMB- K-8, is polisaharide. As in the literature found allegations that members of *Chlorobium* form glycogen, we extracted polisaharide by Zacharova-Kosenko, which is specific for bacterial glycogen deposition. The formation of glycogen in the cells is only the lighting conditions and the presence of carbon dioxide and hydrogen sulfide in the culture medium. In the absence of H_2S and CO_2 accumulation of glycogen in the cells was observed. In microsections of cells grown under different light, the presence of carbon dioxide and hydrogen sulfide, clearly visible rozet not surrounded by a membrane, glycogen granules (Fig. 2).

Comparative analysis of selected polisaharide and glycogen company "Sigma" showed that the resulting sample shows identical chemical and physical properties: white crystalline powder soluble in water, not soluble alcohol, hydrolyzed in acidic medium to form glucose. Infrared spectroscopy etylaceton extract the studied sample and glycogen "Sigma" has shown that these substances are characterized by the presence of identical functional groups, O-H bonds (the interval 3608 - 3056 cm^{-1}), revealed specific absorption in the carbonyl group (1656 cm^{-1}),- CH2-group (2932 cm^{-1}), and -C-O-H groups (1048 cm^{-1}) and others, indicating the identity of the investigated sample of bovine liver glycogen (the drug company "Sigma") (Fig. 3.) Therefore, we first selected polisacharide of cells *C. limicola* IMB-K-8, which by the nature of the infrared spectrum identical with glycogen "bovine liver".

Accumulation of glycogen in the cells may be an indicator of flow speed. Therefore, we first selected polisacharide of cells *C. limicola* IMB- K-8, which by the nature of the infrared spectrum identical with glycogen "bovine liver". Accumulation of glycogen in the cells may be an indicator of flow speed.

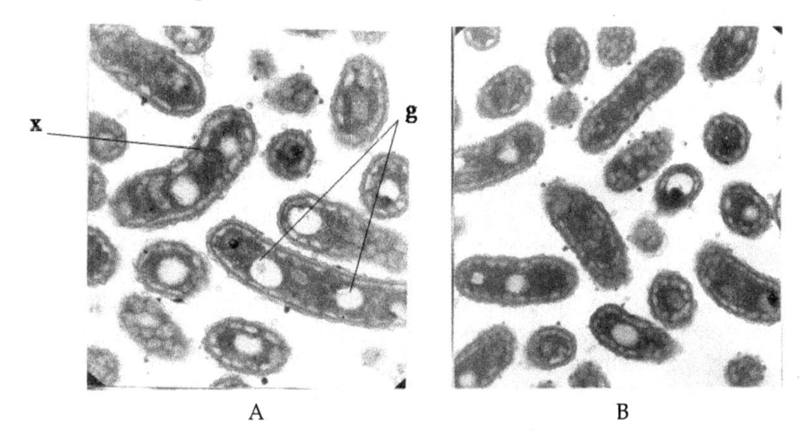

Figure 2. Microsections of cells *C. limicola* IMB- K-8, grown under different light intensity (A - 40lk, B - 100lk): g – glycogen granules, x – chlorosomu.

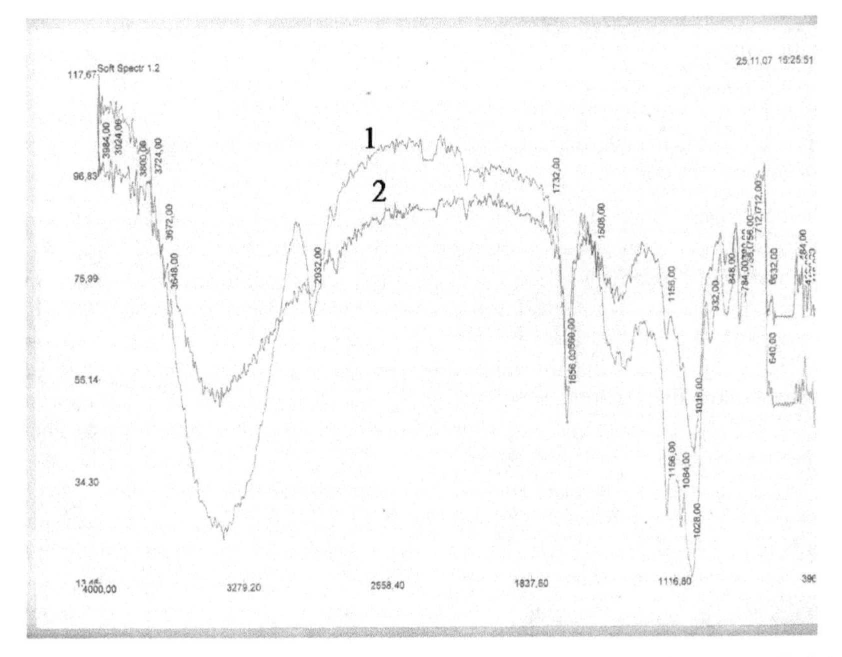

Figure 3. Infrared spectrum of glycogen company "Sigma" (1) and glycogen cells of *C. limicola* IMB- K-8 (2)

The laws of accumulation and utilization of glycogen *C. limicola*. In the presence of light *C. limicola* IMB- K-8 can use organic compounds only, subject to the availability of hydrogen sulfide as an additional source of carbon and continue in their presence actively assimilate carbon dioxide. Assimilation green sulfur bacteria carbon dioxide and organic compounds leads not only to form cells of substances necessary for their growth, but can also affect the synthesis of glycogen. Assuming in these experiments, we investigated the influence of organic carbon sources of power in the process of accumulation of this compound. It turned out that adding to the medium glucose, sucrose, maltose, lactate, not accompanied by changes in intracellular glucose and glycogen.

Only adding to the environment pyruvate and acetate stimulated the growth of glycogen content in cells of *C. limicola* IMB- K-8 which clearly explains the functioning of the studied bacteria cycle Arnon, in the process which produced acetate. Notably, cells with elevated levels of glycogen synthesis that is caused by the addition of pyruvate and acetate, in contrast to cells grown in the presence of other sources of carbon, used almost entirely endogenous glucose. Only adding to the environment pyruvate and acetate stimulated the growth of glycogen content in cells of *C. limicola* IMB- K-8 which clearly explains the functioning of the studied bacteria cycle Arnon, in the process which produced acetate. Notably, cells with elevated levels of glycogen synthesis that is caused by the addition of pyruvate and acetate, in contrast to cells grown in the presence of other sources of carbon, used almost entirely endogenous glucose.

The results obtained give grounds to assert that *C. limicola* IMB- K-8 the most effective use as an additional source of carbon nutrition acetate. It is used only in the presence of hydrogen sulfide and carbon dioxide in the environment and occurs through the inclusion of this compound in the Arnon cycle with the formation of cell components and glycogen. In the presence of pyruvate and acetate in the environment there are some differences in photoreduction CO_2 cells.

So when the concentration of CO_2 in the atmosphere 60mM observed maximum cell growth and increased by 50% the level of glycogen. A slight reduction of carbon dioxide in the environment (20%) accompanied by a reduction in biomass, while increasing the level of glycogen in the cells by about 30%. Further reduction of CO_2 was accompanied by decrease in the intensity of photosynthesis. Increase in glycogen levels in cells with the shortage of carbon dioxide in the atmosphere, apparently, can be explained by inhibition of pyruvate carboxylation reaction and its conversion in to oksaloatsetat and then using it in a constructive metabolism. Note that formed in the process of photosynthesis annoxy carbohydrates not allocated to the environment and stockpiled exclusively in the cell. As evidenced by a negative test for glucose and other sugars is reduced before and after hydrolysis of culture broth. To find ways of further use of glycogen in these experiments, free cells of *C. limicola* IMB K-8 with a high content of polisacharide, incubated in light and in darkness, and then determined the level of glycogen in the cells and analyzed the nature of the organic matter accumulating in the environment. It turned out that the absence of light and presence of CO_2 and H_2S in the medium, cells of *C. limicola* IMB K-8 used a significant amount of glycogen, which testified to a significant reduction of its level in cells.

Analysis of the products of glucose catabolism, obtained after deposition of the mixture (acetone - petroleum ether) from the environment showed that the cells incubated in the dark in the environment accumulate organic compounds as evidenced by their elemental analysis (C-40.25%, H-4.5 %, N-0%). Infrared spectrometry etylatseton hoods showed that these substances are characterized by the presence of O-H bonds (the interval 3608 - 3056cm^{-1}), CH_3-CH_2 bonds (1456sm^{-1}) and specific absorption in the carbonyl (1656 cm^{-1}) and methyl group (2920 cm^{-1}) and R-COOH groups (2700 cm^{-1}) and others that indicate the presence in culture fluid of carboxylic acids (Fig. 4).

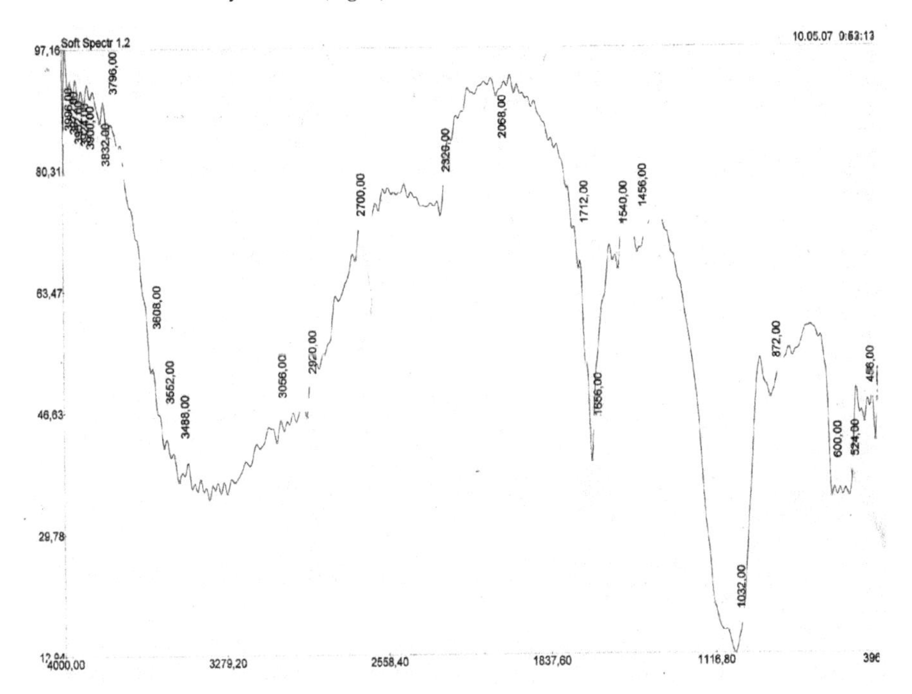

Figure 4. Infrared spectrum of culture fluid components *C. limicola* IMB K-8 cells under incubation in the dark

These results are consistent with data Sirevag, under which the cells incubated with *C. thiosylfatophillum* in the dark in culture fluid accumulated carboxylic acids: acetate (the main component that makes up 70%), propionate and succinate. They are the authors produced by reactions of glycolysis, pyruvate decarboxylation and other reactions.

Under the conditions of incubation, washed cells *C. limicola* IMB K-8 in light of the formation of organic compounds in the culture fluid was observed, and the total content of glucose after hydrolysis of glycogen, not significantly different from control. We shows the dynamic changes in the concentration of glucose (after hydrolysis of glycogen) and the accumulation of carboxylic acids during incubation of cells *C. limicola* IMB K-8 in the dark.

As seen from for 40h incubation, the contents of glycogen (for glucose) in the incubation mixture decreased almost three times while there was accumulation of carboxylic acids in the environment. Thus, synthesized by cells during photosynthesis *C. limicola* IMB K-8 - glucose and glycogen play an important role in the life of these bacteria during their stay on the light and in darkness. In the first case in the photoreduction CO_2 observed accumulation of glucose and its conversion into glycogen. In the darkness degradation glycogen to glucose, catabolism of which provides energy and constructive metabolism of green sulfur bacteria

In addition to the family *Chlorobiaceae* green bacteria carry the family *Chloroflexaceae*, which is called green nonsulfur bacteria. Green nonsulfur bacteria form filaments capable of sliding movement, optional anaerobes that can use organic compounds as sources of carbon and energy. By type of metabolism they phototropy under anaerobic conditions and under aerobic heterotrophs. Their cells contain bakteriohlorophily and carotenoids. Some molecules of green pigments nonsulfur bacteria contained directly in the cytoplasmic membrane, and part of chlorosom. Protein membrane chlorosom similar representatives of families *Chlorobiaceae* and *Chloroflexaceae*. Slow growth of photoavtotroph on the environment of sulphide was first described Median of employes in 1974. The representatives of green nonsulfur bacteria detected actively functioning oxidative tricarboxylic acid cycle. Like most photobacter bacteria genus *Chloroflexus* can grow using CO_2 as the sole carbon source. Found that green nonsulfur bacteria can use hydrogen sulfide as electron donor for photosynthesis and *Chloroflexus aurantiacus* may molecular hydrogen in the process reduce CO_2.

Found that one of the key enzymes - piruvatsyntaza that catalyzes the formation of pyruvate from acetyl-CoA and CO_2 detects activity in *Ch. aurantiacus*. The activity of other specific enzymes that are restorative (RTAC) was absent. On this basis it was concluded that cells of these bacteria, acetyl-CoA is synthesized from CO_2. The mechanism of this synthesis is different from what is in *C. limicola*.

Holo and Grace in 1987 found that in autotrophic conditions is inhibiting the tricarboxylic acid cycle and gliocsilate shunt, and in the cells is a new metabolic pathways in which acetyl-CoA is an intermediate product. Later Holo found that in autotrophic conditions *Ch. aurantiacus* converts acetyl-CoA in 3-hidroksypropionat, which is an intermediate product in the fixation of CO_2. Further to its transformation leads to the formation of malate and succinate. The results were confirmed by Strauss in 1992, which showed that the autotrophic cell growth *Ch. aurantiacus* isolated succinate and many 3-hidroksypropionat in the period from the late exponential phase to early stationary.

When culture *Ch. aurantiacus* was placed in an environment of ^{13}C labeled succinate and analyzed the different components of cells using ^{13}C spectroscopy, which determines the distribution of ^{13}C isotope in various compounds of cells, the results confirmed the role as an intermediate metabolite 3-hidroksypropionatu in CO2 fixation.

Hidrokspropionat role as intermediate in the fixation of CO_2 was investigated Fuchs and Staff in experiments using ^{13}C. The relative amount of ^{13}C after growth *Ch. aurantiacus* in the presence of ^{13}C and ^{13}C 3-hidroksypropionatu acetate. From the samples were labeled ^{13}C marker central intermediate metabolite as trioz and dicarboxylic acids. These experiments showed that cell growth *Ch. aurantiacus* was determined by adding ^{13}C 3-hidroksypropionat for several generations of cells, where it was concluded that this substance is a precursor of all cellular compounds *Ch. aurantiacus*. Thus, 3-hidroksypropionat functions in the body of *Ch. aurantiacus* as a central intermediate metabolite. The data obtained with labeled acetate, also confirmed the key role 3-hidroksypropionat as intermediate in the cyclic mechanism of CO_2 fixation (Fig. 5).

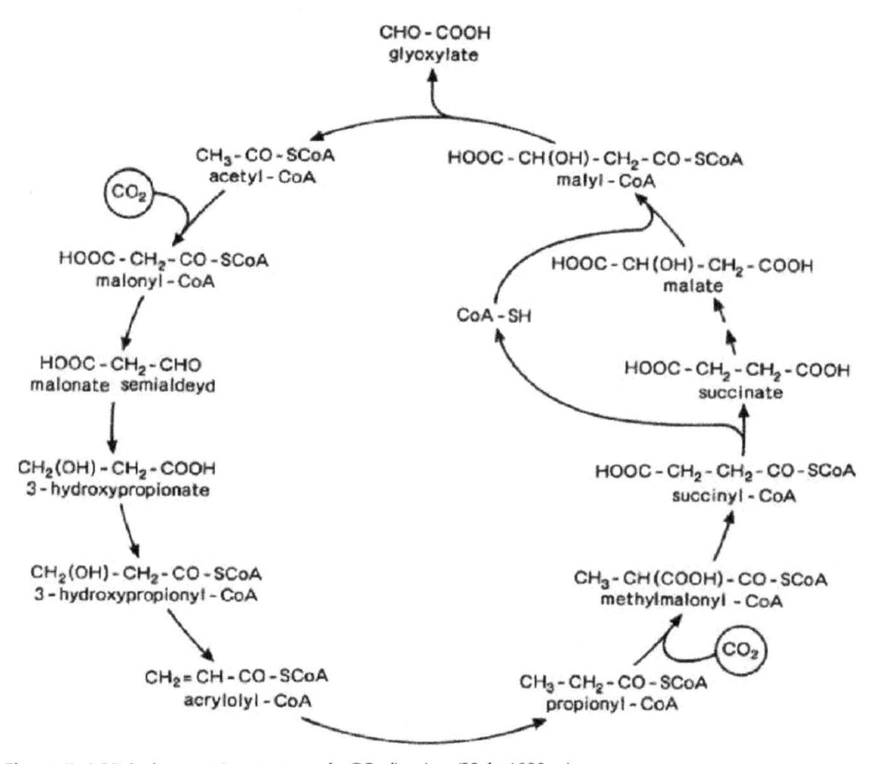

Figure 5. 3-Hidroksypropionatnyy cycle CO_2 fixation (Holo 1989 p.)

For a final check of the cycle Strauss and Fuchs had enzymatic studies and showed that the cells of green bacteria is nonsulfur activity of all enzymes required for assimilation cycle 3-hidroksypropionat reduction of carbon dioxide. In this cycle acetyl - CoA in malonil - CoA and then, reducing turns through 3-hidroksypropionat to propionil - CoA.

Thus, in green bacteria nonsulfur *Ch. aurantiacus* operating mechanism of autotrophic fixation of CO_2, the key intermediates which are 3-hidroksypropionat.. The final product of this cycle is glyoxylate, who fotoheterotrofiv becomes a backup compound poli -*β*-hidrocsybutyrat.

Thus, green bacteria families *Chlorobiaceae* and *Chloroflexaceae,* despite the similarity of their photosintetic system, assimilation of CO_2 reduction carried out in different ways. In the family *Chlorobiaceae* CO_2 fixation reactions proceeding with revers tricarboxylic acid cycle. Carbohydrates - products of photosynthesis, they lay in store as glycogen, which is used in extreme conditions for energy and carbon. Green nonsulfur bacteria family *Chloroflexaceae* used for CO_2 fixation reaction 3-hidroksypropionat way. Under these conditions produced a poli-β-hidroksybutyrat, which, like glycogen in the family *Chlorobiaceae,* is used in the energy and constructive exchanges.

Author details

M. B. Gorishniy and S. P. Gudz
National University of Lviv Ivan Franko, Lviv, Ukraine

3. References

[1] Bergstein T., Henis Y., Cavari B. Investigation of the photosynthetic bacterium *Chlorobium phaeobacteroides* causing seasonal blooms in Lake Kinneret // Canada J. Microbiol. – 1979. – Vol. 25. – P. 999–1007.

[2] Bergstein T., Henis Y., Cavari B. Uptake and metabolism of organic compounds by Chlorobium isolated from Lake Kinneret // Microbiol. – 1981. – Vol. 27. – P. 1087–1091.

[3] Castenholz R. W., Pierson B. K. The prokaryotes. – New York: Springer, 1978. – P. 290–298.

[4] Cork D., CarunasR., Sajjad A. *Chlorobium limicola* forma *thiosulfatophilum* biocatalyst in the production of sulfur and organic carbon from gas stream containing H_2S and CO_2 // Appl. And Envir. Microbiol. – 1983. – Vol. 45. – P. 913–918.

[5] Gemerden H. Physiological ecology of green sulfur bacteria // Ann. Microbiol. – 1983. – Vol. 134. – P. 73–92.

[6] Gorishniy M. B. Ecological significance of green sulfur bacteria in the utilization of hydrogen sulfide / / Thesis for PhD degree in specialty 03.00.16. - Ecology - Institute of Agroecology biochemistry of Ukraine, Kyiv, 2008.

[7] Gorishni y M. B, Gudz S. P, Hnatush S. O The metabolism of glucose and glycogen in the cells of green sulfur bacteria fotosyntezuvalnyh / / Herald of Lviv. Univ. Biological Series. - 2008. - Vol. 46. S. 129-136.

[8] Gorishniy M. B., Gudz. S. P., Hnatush S. O. Metabolism of carbohydrates in the cells of green sulfur bacteria *Chlorobium limicola* Ya-2002 "Ukrainian Biochemical Journal" - 2009. - N. 5. T-81. - S. 26-33.

[9] Gudz S. P, Gorishniy M. B., Hnatush S. O. Bacterial photosynthesis - Lions: Ivan Franko Lviv National University, 2011. – 180 p.

[10] Herter S., Farfsing J., Gad'On N. et al. Autotrophic CO_2 fixation by *Chloroflexus aurantiacus*: study of glyoxylate formation and assimilation via the 3-hydroxypropionate cycle // J. Biol. Chem. – 2001. – Vol. 267. – P. 20256–20273.

[11] Mas J., Gemerden H. Storage products in purple and green sulfur bacteria. In: Anoxygenic photosynthetic – D.: KmwerAcad. Pub. (Netherlands), 1995. – P. 973–990.
[12] Overmann J. Green Sulfur Bacteria. In: Bergey's Manual Systematic Bacteriology. – 2002. – 2nd ed. – Vol. 1. – P. 601–605.
[13] Pfennig N. Photoprophic green bacteria: a comparative, systematic survey // Ann. Rev. Microbiol. – 1977. – Vol. 31. – P. 275–290.
[14] Pfennig N. The phototrophic bacteria and their role in the sulfur cycle // Plant and Soil. – 1975. – Vol. 43. – P. 1–16.
[15] Pfennig N., Widdel F. The bacteria of sulfur cycle // Phil. Trans. R. Soc. Lond. – 1982. – Vol. 298. – P. 433–441.
[16] Repeta D. J., Simpson D.J., Jorgensen B. B., Janasch H. W. Evidence for anoxygenic photosynthesis from the distribution of bacteriochlorophylls in the Black Sea // Nature. – 1989. – Vol. 12. – P. 69–72.
[17] Ugolkova N. V., Ivanovsky R.N. On the mechanism of autotrophic fixation of CO_2 by Chloroflexus aurantiacus // Microbiology. – 2000. – Vol. 5. – P. 139–142.
[18] Van Niel C. B. Natural selection in the microbial world // Gen. Microbiol. – 1955. – Vol. 13, N 1. – P. 201.
[19] Overmann J., Garcia-Pichel F. The Phototrophic Way of Life. The Prokaryotes. – New York: Springer, 2000. – 887 p.
[20] Blankenship M. T. Madigan C. E. Anoxygenic Photosynthetic Bacteria. – Boston: Kluwer Academic Publishers Dordrecht, 1997. – P. 49–89.107.
[21] Casamayor E., Mas J., Pedro-Alio C. In situ assessment on the physiological state of Purple and Green Sulfur Bacteria through the analyses of pigment and 5S rRNA content // Microbiol. Ecol. – 2001. – Vol. 42. – P. 427–437.
[22] Gemerden H. Physiological ecology of green sulfur bacteria // Ann. Microbiol. – 1983. – Vol. 134. – P. 73–92.
[23] Gest H., Favinger J. *Heliobacterium chlorum,* an anoxygenie brownish-green photosyn thetic bacterium, containing a new form of bacteriochlorophill // Arch. Microbiol. – 2002. – Vol. 136 – P. 11-16.
[24] Gorlenko V. M., Pivovarova T. A. On the belonging of blue-green alga *Oscillatoria coerulescens,* to a new genus of chlorobacteria *Oscillochloris* nov. gen. // Izd. Acad. Nauk SSSR. – Ser. Biol. – 1977. – P. 396–409.
[25] Guerrero R., Pedros-Alio C, Esteve I., Mas J. Communities of phototrophic sulfur bacteria in lakes of the Spanish Mediterranean region // Acta Acad. Abo. – 1987. – Vol. 47. – P. 125–151.
[26] Gusev M. V., Shenderova L. V., Kondratieva E. N. The relation to molecular oxygen in different species of photoprophic bacteria // Microbiol. – 1969. – Vol. 38. – P. 787–792.
[27] Hanson T. E., Tabita F. R. A ribulose-1, 5-bisphosphate carboxylase / oxygenase (Rubisco)-like protein from *Chlorobium tepidum*that is involved with sulfur metabolism and the response to oxidative stress. Proc. Natl. Acad. Sci. – 2001. – P. 4397–4402
[28] Holf G., Brattar I. Taxonomic diversity and metabolic activity of microbial communities in the water column of the central Baltic Sea // Limnol. Oceanogr. – 1995. – Vol. 40. – P. 868–874.

[29] IKerryon C. N. Comlex lipids and fatty acids of photosynthetic bacteria. In: The photosynthetic bacteria – New York: Plenum publ. Co, 1978. – P. 281–313.
[30] Kuster E., Dorusch F., Vogt C., Weiss H., Altenburger R. Online biomonitors used as a tool for toxicity reduction evaluation of in situ ground water remediation techniques // Biosens Bioelectron. – 2004. – Vol. 19. – P. 1711–1722.
[31] Larsen H. On the culture and general physiology of the green sulfur bacteria // J. Bacteriol. – 1952. – Vol. 64. – P. 187–196.
[32] Miller M., Liu X., Snyder S. et al. Photosynthetic electron-transferreactions in the green sulphur bacterium *Chlorobium vibrioforme.* Evidence forthe functional involvement of iron-sulfur redox centers on the acceptorside of the reaction center // Biochemistry. – 1992. – Vol. 31. – P. 4354–4363.
[33] Okkels J., Kjaer B., Hansson O., Svendsen I., Moller B., Scheller H. A membrane bound monogeme *cyt. c*551 of a nowel type is the immediate electron donor to P_{840} of the *Chlorobium vibrioforme* photosynthetic reaction centre complex // J. Biol. Chem. – 1992. – Vol. 267. – P. 21139–21145.
[34] Otte S. C., van de Meent E. J., van Veelen P. A. et al. Identification of the major chromosomal bacteriochlorophills of the green sulfur bacteria *Chlorobium vibrioforme* and *Chlorobium phaeobacteroides;* their function in lateral energy transfer // Photosynt. Res. – 1993. – Vol. 35. – P. 159–169.
[35] Overmann J., Cypionka H., Pfennig N. An extremely low-light-adapted prototrophic sulphur bacterium from the Black Sea // Limnol. Oceanogr. – 1992. – Vol. 37. – P. 150–155.
[36] Parkin T. B., Brock T. D. The effects of light quality on the growth of phototrophic in lakes // Arch. Microbiol. – 1980. – Vol. 125. – P. 19–27.
[37] Parkin T. B., Brock T. D. The role of phototrophic in the sulfur cycle of a meromictic lake // Limnol. Oceanogr. – 1981. – Vol. 26. – P. 880–890.
[38] Peschek G., Loffelhardt W. The phototrophic prokaryotes. – New York: Plenum, 1999. – P. 763–774.
[39] Pfennig N. Syntrophic mixed cultures and symbiotic consortia with phototrophic bacteria. In: Anaerobes and Anaerobic Infections. – Studgard: Fischer, 1980. – P. 127-131.
[40] Pfennig N., Widdel F. The bacteria of sulfur cycle // Phil. Trans. R. Soc. Lond. – 1982. – Vol. 298. – P. 433–441.
[41] Pierson B. K., Kenn L. M., Leovy J. G. Isolation of Pigmentation Mutants of the green Filamentous Photosynthetic Bacterium *Chloroflexus aurantiacus* of Bacteriology // Arch. Microbiol. – 1984. – Vol. 159. – P. 222–227.
[42] Powell E.O. The growth rate of microorganism as a function of substrate concentration. In: Microbial physiology and continuous culture, N. Y.: Porton P, 1967. – P. 365–369.
[43] Pringault O., Kühl R. Growth of green sulphur bacteria in experimental benthic oxygen, sulphide, pH and light gradients // Microbiol. – 1998. – Vol. 144. – P. 1051–1061.
[44] Puchkova N. N. Green sulphur bacteria as a component of the sulfureta of shallow saline waters of the Crimea and northern Caucasus // Microbiol. – 1984. – Vol. 53. – P. 324–328.

[45] Puchkova N. N., Gorlenko V. M. A new green sulfur bacterium *Chlorobium chlorovibrioides* spec. // Microbiol. – 1984. – Vol. 51. – P. 118–124.
[46] Rocap G. F., Larimer J. F. Genome divergence in two Prochlorococcus ecotypes reflects oceanic niche differentiation // Nature. – 2003. – Vol. 424. – P. 1042–1047.
[47] Savikhin S., Zhou W., Blankenship R., Struve W. Femtosecond energy transferand spectral equilibration in bacteriochlorophyll α-protein antenna trimers from the green bacterium *Chlorobium tepidum* // Biophys. J. – 1994. – Vol. 66. – P. 110–114.
[48] Scheer H. Structure and occurrence of chlorophylls – Boca Raton.: CRC Press, 1991. – P. 3–30.
[49] Shill D. A., Wood P. M. Light-driven reduction of oxygen as a method forstuduing electron transport in the green photosynrhetic *Chlorobium limicola* // Arch. Microbiol. – 1985. – Vol. 143. – P. 82–87.
[50] Sirevåg R., Buchanan B. Mechanisms of CO_2 fixation in bacterial photosynthesis studied by carbon isotope fractionation technique // Arch. Microbiol. – 1977. – Vol. 16, N 112. – P. 35–38.
[51] Smith H. et al. Nomenclature of the bacteriochlorophills *c, d* and *e* II Photosynth. Res. – 1994. – Vol. 41. – P. 23-26.
[52] Stackebrandt E., Embly M., Weckesser J. Phylogenetic evolutionary, and taxonomic aspects of Phototrophic bacteria // New York: Plenum Press, 1988. – P. 201–215.
[53] Stackebrandt E., Woese C. The evolution of prokaryotes. In: Molecularand cellular aspects of microbioal evolution // Symp. Soc. Gen. Microbiol. – 1981. – Vol. 32. – P. 1–31.
[54] Takaichi S., Wang Z-Y., Umetsu M. et al. New carotenoids from the thermophilic green sulfurbacteria *Chlorobium tepidum*: 1,2-dihydro-carotene, 1,2-dihydrochlorobactene, OH-chlorobactene glucoside ester, and the carotenoid composition of different strains // Arch. Microbiol. – 1997. – Vol. 168. – P. – 270–276.
[55] Tamiaki H. Supramolecuiar structure in extramembraneous antennae of green photosynthetic bacteria // Coord. Chem. Rev. – 1996. – Vol. 148. – P. 183–197.
[56] Thompson R. W., Valentine H. L., Valentine W. N. Cytotoxic mechanisms of hydrosulfide anion and cyanide anion in primaryrat hepatocyte cultures // Toxicology. – 2003. – Vol. 12. – P. 149–159.
[57] Truper H.G. Culture and isolation of phototrophic sulfur bacteria from the marine environment // Helgol. Wiss. Meeresunters. – 1970. – Vol. 20. – P. 6–16.
[58] Truper H., Fisher U. Anaerobic oxidation of sulphur compounds as electron donors for bacterial photosynthesis // Phil. Trans. R. Soc. Lond. – 1982. – Vol. 298. – P. 254–258.
[59] Tuschak C., Glaeser J., Overmann J. Specific detection of green sulfur bacteria by in situ-hybridization with a fluorescently labeled oligonucleotide probe // Arch. Microbiol. – 2003. – Vol. 22. – P. 34–39.
[60] Van Niel C. B. On the nuwphology and physiology of the green sulfur bacteria // Arch. Microbiol. – 1932. – Vol. 3. – P. 11–112.
[61] Van Noort P. I., Zhu Y. R., LoBrutto R. E. Redox effects on the excited-state lifetime in chlorosomes and bacteriochlorophyll c oligomers // Biophys. J. – 1997. – Vol. 72. – P. 316–325

[62] Veldhuis M. J., van Gemerden W. H. Competition between purple and brown phototrophic in stratified lakes: sulfide, acetate, and light as limiting factors // FEMS Microbiol. Ecol. – 1998. – Vol. 38. – P. 31–38.

[63] Vignais P. M., Colbeau J. C. et al. Hydrogenase, nitrogenase, and hydrogen metabolism in photosynthetic bacteria // Adv. Microb. Physiol. – 1985. – Vol. 26. – P. 155–234.

[64] Wahlund T. M., Tabita R. F. The reductive tricarboxylic acid cycle of carbon dioxide assimilation: initial studies and purification of ATP citratlyase from the green sulfur bacteria *Chlorobium* // Bacteriology. – 1997. – Vol. 179. – P. 48–59.

[65] Welsh D. T., Herbert R. A. Identification of organic solutes accumulated by purple and green sulfur bacteria during osmotic stress using natural abundance nuclearmagnetic resonance spectroscopy // FEMS Microbiol. EcoL. – 1993. – Vol. 13. – P. 145–150.

[66] Wu H., RatsepM., Young C. et al. High pressure and stark hole burning studies of chlorosome antennas from green sulfur bacterium *Chlorobium tepidum* // Biophys. J. – 2000. – Vol. 79. – P. 1561–1572.

[67] Xiong J., Inoue K., Nakahara M., Bauer E. Molecularevidence forthe early evolution of photosynthesis // Science. – 2000. – Vol. 289. – P. 1724–1730.

[68] Overmann J., Tuschak C. Phylogeny and molecular fingerprinting of green sulfur bacteria // Arch. Microbiol. – 1997. – Vol. 167. – P. 302–309.

Chapter 9

DC-SIGN Antagonists – A Paradigm of C-Type Lectin Binding Inhibition

Marko Anderluh

Additional information is available at the end of the chapter

1. Introduction

Carbohydrates have rarely been a matter of research by medicinal chemistry-oriented scientists in the past, but has recently gained substantially more publicity [1]. Although equally or even more abundant in nature than peptides/proteins or nucleic acids, they were often simply neglected as potential drug targets and/or drug leads or even drug candidates. There are a number of reasons for this. Glycans – complex carbohydrates linked to proteins or lipids are essential components of every cell surface as all cells are coated with a so-called glycocalyx layer. These cell surface carbohydrates allow for intercellular communication by binding to the carbohydrate-binding proteins (CBPs). Many physiological and even pathophysiological processes like pathogen-cell contact rely upon these interactions. Currently, more than 80 human CBPs have been identified [2,3], so one might immediately recognise CBPs as promising targets for anti-infective therapy. However, only a few of them have been thoroughly studied and as a result, few CBPs have been recognised and validated as drug targets. Furthermore, CBPs' affinity is generally weak per monosaccharide unit, so CBPs usually form strong interactions by binding massive glycans that bear several hundred terminal carbohydrate units, many of which form contact with a single CBP or even a cluster of CBPs on the cellular surface. To inhibit glycan-protein interaction efficiently, one would have to consider designing and synthesizing multivalent carbohydrate ligands. This is by no means an easy task, and makes CBPs relatively unattractive targets in terms of druggability. Even if one would consider producing multivalent carbohydrates with optimal pharmacodynamic properties, they possess unattractive pharmacokinetic properties. Carbohydrates, especially large multivalent ones, share some intrinsic pharmacokinetic shortcomings as drugs; they are rapidly digested in the gut in most cases, and even if they survive the gut metabolism obstacles, they are practically unable to passively diffuse through the enterocyte layer in the small intestine. This inevitably means that no oral application can be guaranteed for a carbohydrate drug, which is the preferred application route. Once in the body, carbohydrates are mostly a source of energy, but are also excreted rapidly with glomerular filtration. Due to these drawbacks, carbohydrates or carbohydrate-derived drugs were often considered unappealing even before their design would take place.

To tackle the first drawback, advances in CBP biochemistry has led to a progressive gain of knowledge on CBPs. A vast international research network named *Consortium for Functional Glycomics* (CFG, http://www.functionalglycomics.org/static/consortium/consortium.shtml) was founded in 2001 with the ultimate goal of enlightening the role of glycans and glycan-binding proteins. To date, hundreds of members of this network as well as outside researchers have revealed the secrets of CBPs and many are now considered potential drug targets: their specific ligands were discovered with screening on large glycoarrays and their binding sites were mostly elucidated [3]. Understanding the 3D structures of CBPs and their binding modes is indispensable for drug design. In order to offer a central repository of knowledge available about CBPs, PROCARB has been constructed [4]. PROCARB is a database of known and modelled carbohydrate-binding protein structures with sequence-based prediction tools, and is a single resource where all the relevant information about a pair of interacting proteins and carbohydrates is available. A similar, although a bit less appealing and informative database on lectin 3D structure is the database of lectins of *Centre de Recherches sur les Macromolécules Végétales* (CERMAV), a French fundamental research centre devoted to glycosciences (http://www.cermav.cnrs.fr/lectines/). Glycoscience.de (http://www.glycosciences.de/) is a German internet portal to support glycomics and glycobiology research. These web sites offer proof that the scientific community has recognised the potential of CBPs. Accordingly, many national and international associations have been involved in deciphering CBPs fundamental roles in human pathophysiological processes, which inevitably leads to on-going cognizance of how CBPs might be exploited to help develop novel therapeutic approaches for treatment of human diseases.

CBPs may be classified primarily into two broad categories: lectins (which are further subdivided into specific types, like C-, H-, I-, L-, P-, R-, and S-type lectins) and sulphated glycosaminoglycan-binding proteins [3]. Extracellular lectins are the most promising drug targets among CBPs and CFG divides them into 4 specific groups: C-type lectins, galectins (or S-type lectins), siglecs (a subclass of I-type lectins that bind sialic acid) and other. Out of 174 total CBPs records on the CFG webpage, 120 of them are classified as C-type lectins, which in itself highlights the importance of this class of CBPs. Like all extracellular lectins, C-type lectins bind terminal carbohydrate epitopes of glycans that originate from either pathogens or other cells. DC-SIGN, (Dendritic Cell-Specific Intercellular adhesion molecule-3-Grabbing Non-integrin) is a type II C-type lectin that functions as an adhesion molecule located exclusively on dendritic cells (DCs). Originally discovered in 1992 [5], it was defined as a C-type lectin capable of binding the HIV-1 envelope glycoprotein gp120, but it took time until Steinman [6] and van Kooyk [7] unravelled its specific role in immunology and pathology and revealed the information to the broader scientific community. Since then, it has been shown to have a major role in primary immune response [8], but it also enhances several pathogens (like HIV, Ebola) infection of T cells and other cells of the immune system

[7,9]. This makes DC-SIGN a very interesting CBP and a target of interest towards novel immunomodulatory and anti-infective therapy. DC-SIGN function can be modulated by small molecules termed DC-SIGN antagonists that bind to DC-SIGN and prevent binding of native DC-SIGN ligands. These molecules could act as novel anti-infectives, and are currently in the early phase of drug development. Latest *in vitro* studies demonstrate that DC-SIGN antagonists effectively block the transmission of pathogens like HIV-1 and Ebola to CD4+ T cells [10]. Although DC-SIGN has not been validated *in vivo* as a drug target yet, DC-SIGN antagonists are a fruitful example of how inhibition of a C-type lectin function might be achieved by small synthetic molecules. As exposed before, CBPs' affinity is generally weak per carbohydrate unit, so the design of molecules that bind to DC-SIGN with high affinity is a demanding assignment. In general, the affinity issue and pharmacokinetic drawbacks of carbohydrates may be overcome by two predominant strategies implied in the design of DC-SIGN antagonists: screening of non-carbohydrate compounds to obtain ligands that are completely devoid of carbohydrate nature [11] and the design of glycomimetics, the compounds designed based on carbohydrate leads which usually still retain some or even a significant degree of carbohydrate nature [3,12]. DC-SIGN binds mannose- and fucose-based oligo- and polysaccharides, so their glycomimetics have been designed and proved to inhibit pathogen-DC-SIGN interaction potently. The author foresees that the approach used to design glycomimetic DC-SIGN antagonists is of general applicability when designing lectin antagonists and will be the thoroughly presented in the present chapter.

2.1. C-type lectins

Probably the largest type or family of lectins is the C-type found on animal (and human) cells [13]. This family includes several endocytic receptors and proteoglycans, and all collectins and selectins identified to date. They are of paramount importance for normal function of the immune system, as they mediate innate immunity, inflammation and immunity to tumour and virally infected cells. Although these CBPs vary greatly in structure among themselves, they have in common a domain, named C-type lectin-like domain (CTLD) or carbohydrate recognition domain (CRD). It is a compact globular structure responsible for selective binding of terminal units found in large carbohydrates [14]. The unique structural hallmark of such a domain is that it binds a carbohydrate by Ca^{2+} ion, which acts as a bridge between the protein and the "core monosaccharide" unit through complex interactions with sugar hydroxyl groups [15]. Namely, the binding with Ca^{2+} ion involves just one terminal saccharide unit – the "core monosaccharide" (as illustrated in Figure 1 for two distinct C-type lectins, the mannose-binding protein and DC-SIGN), while other ligand carbohydrate units (if present) form structural and bonding complementarity with the CRD. Several amino acid residues of the CRD offer 6 coordinate bonds for a Ca^{2+} ion and the carbohydrate donates 2 coordinate bonds with its hydroxyls, so that the Ca^{2+} ion is octacoordinated [16]. Distinct CRD aminoacid residues form hydrogen bonds with other hydroxyl groups on the carbohydrate directly or through water bridges. Changes in the amino acid residues that interact with the "core monosaccharide" modify the carbohydrate-binding specificity of the lectin, so that specific carbohydrate is recognised.

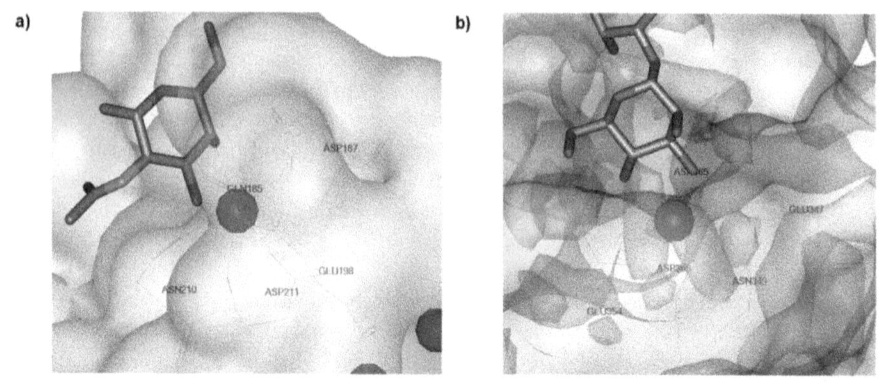

Figure 1. Binding mode of 2 C-type lectin CRDs with the "core monosaccharide". a) Crystal structure of mannose-binding protein with *N*-acetylgalactosamine (GalNAc) in the binding site (PDB code: 1AFB) [17]. b) Crystal structure of DC-SIGN in complex with Man_4 tetrasaccharide (PDB code: 1SL4) [18]. The 3- and 4-OH groups of the "core monosaccharide" (sticks rendering, without hydrogens) directly coordinate Ca^{2+} (blue sphere) and form hydrogen bonds with amino acids that also serve as Ca^{2+} ligands. Ca^{2+} is octacoordinated (violet broken line). For the sake of clarity, proteins are presented as transparent surfaces with amino acid residues (thin sticks) that coordinate Ca^{2+}.

The free energy of such interactions is relatively weak per carbohydrate unit, as we have to take into account high desolvatation penalties of numerous hydroxyls upon carbohydrate binding. The C-type lectin family however has means to obtain high binding affinity, or better, avidity; some C-type lectins oligomerize in order to promote high avidity for specific glycans [19]. The clustering of the CRDs influences not only avidity, but also the lectin selectivity, since each individual CRD can act independently to bind end mono- or oligosaccharide moiety [20]. Although the majority of lectins contain a single C-type CRD, the macrophage mannose receptor has multiple independent CRDs in a single polypeptide. The adjacent CRDs in the mannose receptor may help to direct its binding to specific multivalent, mannose-containing glycans [21].

2.2. DC-SIGN function and structure

DC-SIGN (Dendritic Cell-Specific Intercellular adhesion molecule-3-Grabbing Non-integrin) is a C-type lectin that functions as an adhesion molecule expressed specifically by dendritic cells (DCs), a class of professional antigen presenting cells (APCs). The intrinsic role of DCs is guidance of adaptive immune responses, since they are the major APCs that capture, process and present antigens [22]. DC-SIGN is a specific pattern recognition receptor (PRR) that recognizes distinct molecular patterns (PAMPs – Pathogen-Associated Molecular Patterns) of a number of pathogens [23]. It induces intracellular signalling pathways and triggers DC maturation upon pathogen binding [8,9]. To promote efficient transport of DCs towards effector cells – T cells, DC-SIGN binds human adhesion molecules ICAM-2 on vascular and lymphoid endothelium and enables cell interactions during DC migration [24]. Furthermore, DC-SIGN binding to ICAM-3 allows early antigen-nonspecific contact between DC and T cells in the lymph nodes, enabling T cell receptor engagement by stabilizing the DC-T-cell interaction [25]. Thus, we may conclude that DC-SIGN enables some of the normal DC functions by binding endogenous ligands, but also modulates immune responses to diverse pathogens via its ability to induce antigen-specific intracellular signalling.

DC-SIGN is a transmembrane C-type lectin that consists of one CRD, which defines the ligand specificity of the receptor, a neck region composed of seven complete and one incomplete tandem repeats, and a transmembrane region followed by a cytoplasmic tail containing recycling, internalization and intracellular signalling motifs (Figure 2) [9].

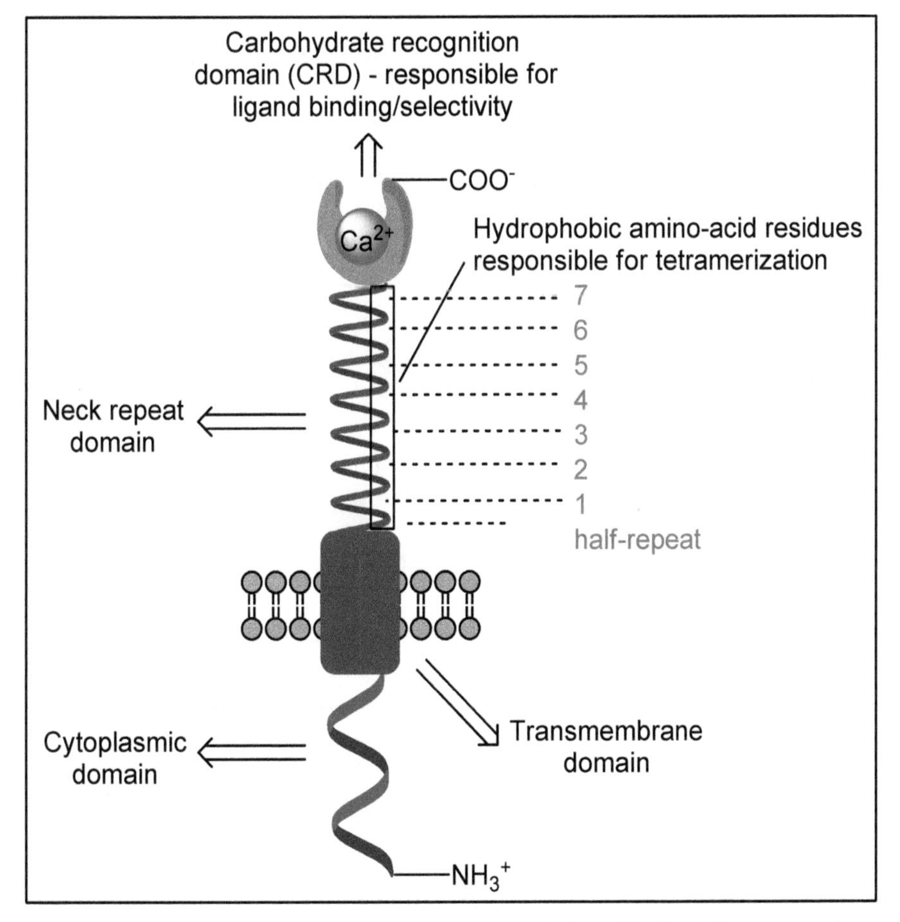

Figure 2. DC-SIGN schematic structure: Ca^{2+} CRD responsible for ligand binding/selectivity, a neck region composed of seven complete and one incomplete repeats, and a transmembrane region followed by a cytoplasmic tail (modified from Švajger et al.) [9].

Like many other C-type lectins, DC-SIGN tetramerizes to increase binding affinity and oligomerization occurs through association of the DC-SIGN neck domain (Figure 3) [26]. The oligomerization status of the DC-SIGN and related C-type lectins depends on the number of helical repeats of the neck region; at least 6 repeats are needed for tetramerization [27]. The stacked CRDs of tetramerized DC-SIGN provides the means of increasing the specificity for multiple repeating (oligo)saccharide units on host molecules, thereby defining the set of pathogens or endogenous molecules that are recognized by DC-SIGN. DC-SIGN neck domain, while allowing stacking and tetramerization, plays a central role as a pH-sensor that balances the equilibrium between the monomeric and tetrameric states of DC-SIGN [28]. In this sense, affinity for carbohydrates may be changed markedly by changing the pH, which helps DC-SIGN to realize its native function. Namely, upon binding, pathogen particles are internalized and further degraded into smaller particles and conjugated to MHC class-II proteins. As degradation proceeds in an acidic endosomal environment, the acid-triggered pathogen release from DC-SIGN is needed for successful degradation. Apart from tetramerization, DC-SIGN forms clusters that organize in membrane microdomains [29]. This organization on the plasma membrane is important for the binding and internalization, suggesting that clustered assemblies act as functional docking sites for pathogens (Figure 3).

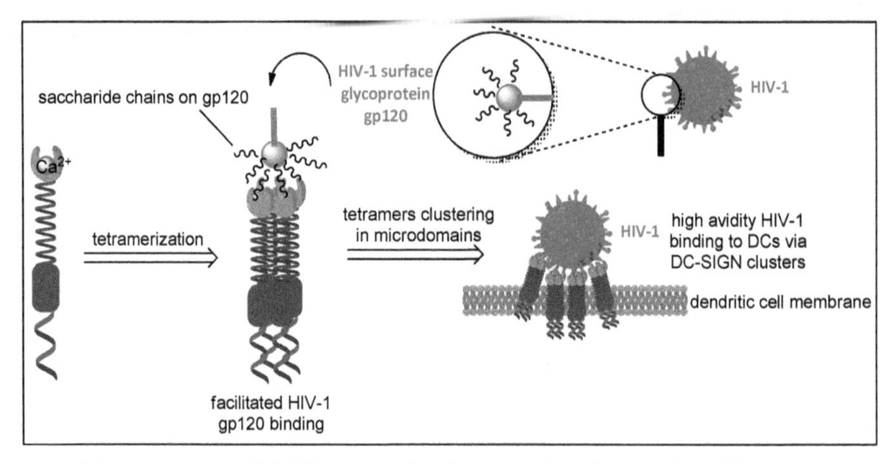

Figure 3. Tetramerization of DC-SIGN and further clustering allows high binding avidity and influences pathogen binding, as depicted with the example of HIV-1 binding to DCs: tetramerization increases avidity to PAMPs like gp120 on HIV-1 surface, while DC-SIGN clusters act as functional docking site for HIV-1 (modified from Švajger et al.) [9].

Apart from allowing oligomerization, DC-SIGN neck repeat domains separate CRDs from the cell surface to enable multivalent interaction with glycans. DC-SIGN binds particularly well to viral and bacterial glycans with closely spaced terminal oligo- or monosaccharides with roughly 5 nm between the units [18,30,31]. DC-SIGN is not a totally rigid macromolecule and shows a degree of flexibility upon ligand binding [31]. Namely, DC-

SIGN adapts to an arrangement of terminal oligo- and monosaccharides, so all CDRs are allowed to interact with their ligands. Taken together, the tetrameric form of DC-SIGN and its conformational flexibility enable effective and selective binding of various glycans. Furthermore, due to the nature of the receptor-mediated cellular signalling, tetramerization of DC-SIGN could contribute to signal transduction after ligand binding.

2.3. DC-SIGN, a target for anti-infective therapy

Probably the most important feature of DC-SIGN is its ability to bind a great number of highly virulent pathogens [32]. Accordingly, it has been recognised as a potential new target for anti-infective therapy that perpetuates basic research of considerable importance. Apart from the already mentioned CFG, the importance of DC-SIGN as a new drug target is reflected in several other projects aimed to thoroughly clarify its function in pathogen infection. For example, several prominent European research teams have joined to form CARMUSYS, a collaborative training project aimed at designing and synthesizing carbohydrate multivalent systems to be used as inhibitors for pathogen attachment and penetration into target cells (CARMUSYS - *Carbohydrate Multivalent System as Tools to Study Pathogen Interactions with DC-SIGN*).

DC-SIGN modulates the outcome of the immune response of DCs by binding and recognizing a variety of microorganisms, including viruses (HIV-1, HCV, CMV, Dengue, Ebola, SARS-CoV, HSV, coronaviruses, H5N1, West Nile virus, measles virus), bacteria (*H. pylori, M. tuberculosis, L. interrogans*), fungi (*C. albicans, A. fumigatus*) and parasites (Leishmania, *S. mansoni*) [9]. On mucosal surfaces of the body where immature dendritic cells sample pathogens, DC-SIGN serves as one of the very first pathogen attachment points and the usual result of this interaction is pathogen internalization, degradation and exposure of the pathogen PAMPs to recruit CD4+ T-cells and to start the humoral immune response [6,22]. HIV-1 exploits native DC-SIGN functions to enslave DCs as carriers to boost T cell infection with or without becoming infected themselves [33]. The very first interaction between HIV-1 and DCs occurs via HIV-1 envelope glycoprotein gp120 with DC-SIGN on immature DCs (Figure 3). The HIV-1-DC-SIGN complex is rapidly endocyted into early endosomes, where the acidic medium causes ligands to dissociate from DC-SIGN [34]. Upon dissociation, most DC-SIGN-bound ligands are lyzed and processed as a normal degradation pathway. Some HIV-1 particles however remain bound to DC-SIGN and thus protected from the host immune system, so a fraction of HIV-1 retains its infectiveness [35]. HIV-1 may rest in DCs in an infectious state for days, hidden in undefined bodies that differ from both endosomes and lysosomes [36]. N-glycan composition of surface proteins governs the viral faith upon interaction of viral envelope with DC-SIGN [37]. By altering the N-linked glycan composition from mixed to oligomannose-enriched, one increases the affinity of HIV-1 for DC-SIGN, which enhances the viral degradation and reduces virus transfer to target cells. On the contrary, HIV-1 with oligomannose-enriched N-glycans is presented to viral envelope-specific CD4+ T cells more efficiently. In the alternative scenario, HIV-1 binds to DC-SIGN and facilitates lateral binding of HIV-1 to CD4 and CCR5 receptors expressed on the same immature DCs [33]. The direct consequence of this interaction is HIV-1 infection of DCs [38]. To conclude, HIV-1 uses DC-SIGN as an entry mode to DCs, and DCs may be regarded as a Trojan horse that takes HIV-1 to CD4+ T-cells while protecting it from the host immune system [39]. A similar entry mode has been observed for some other pathogens like viruses (Ebola) and bacteria (*M. tuberculosis, H. pylori*) [32]. Apart from this straightforward infection pathway, HIV-1-infected DCs are able to mediate transmission of HIV-1 to CD4+ cells by means of immune response modulation and infectious synapse formation. Namely, HIV-1 causes both inhibition of DC maturation while inducing formation of viral synapse, a process previously attributed to mature DCs only [40]. As only several pathogens are able to modulate DCs maturation process, it may be concluded that the DC maturation depends upon selective ligand recognition, possibly also by DC-SIGN.

The fact that DC-SIGN acts as an entry point and a mediator of pathogen infections points out the possible therapeutic usefulness of DC-SIGN. DC-SIGN antagonists work by inhibiting pathogen interaction with DC-SIGN, so the very first phase of pathogen infection can be inhibited, as proven in *in vitro* experiments [10,41,42]. DC-SIGN antagonists may be designed as monovalent glycomimetics based on the DC-SIGN-binding oligosaccharides and their multimeric presentation [43]. Alternatively, screening of compound libraries has been successful in obtaining non-carbohydrate DC-SIGN antagonists [44]. As mentioned, potential carbohydrate-derived drugs have poor pharmacokinetic properties in general and this might raise some concern as to whether there is potential for therapeutically useful glycomimetic DC-SIGN antagonists. In the case of HIV-1, the major initial contact site between HIV-1 and DC-SIGN is in the vaginal mucosa, so a DC-SIGN antagonist could be administered topically to prevent HIV-1 transmission without systemic application [45]. It has to be stressed that no clinically proven therapy based on inhibition of DC-SIGN-mediated pathogen infection has been presented [43].

2.4. How does DC-SIGN choose among terminal monosaccharides?

DC-SIGN has a highly regulated recognition of its ligands as it selectively binds glycans with terminal D-mannose- and L-fucose expressed on a number of bacteria, parasites, fungi and viruses [46]. However, the mere presence of D-mannose and L-fucose does not assure binding selectivity itself. Namely, monosaccharide binding to DC-SIGN CRD is generally very weak, with Ki(D-mannose) = 13.1 mM and Ki(L-fucose) = 6,7 mM being the strongest binders among monosaccharides [26]. DC-SIGN CRD forms a 1-to-1 complex with terminal mono- or oligosaccharides, which relies upon already mentioned octacoordination of Ca^{2+} ion in the binding site by the "core monosaccharide"; the equatorial 3- and 4-hydroxyls each form coordination bonds with the Ca^{2+} in the binding site common to all C-type lectins, but also offer hydrogen bonds with amino acids that also serve as Ca^{2+} ligands [47]. A crucial structural feature of a mannose residue is an axial position of the 2-hydroxyl group; this allows tight surface complementarity of the core mannose with the binding site. Equatorial position of the 2-hydroxyl group would probably prevent this tight binding due to steric clash, so hexopyranoses with equatorial 2-hydroxyl groups do not form strong interactions. To increase binding affinity, other saccharide units form additional interactions with the binding site, while binding specificity is based on spatial constraints. An excellent description of selectivity/specificity mechanism may be found in the work of Guo et al., who demonstrated that DC-SIGN selectively binds high-mannose and fucosylated oligosaccharides [18]. The difference in the affinity of each oligosaccharide results from a different spatial arrangement of the mannose- and fucose-based ligands demonstrated in two crystal structures of fucose-based tetrasaccharide LNFP III and Man_4 tetrasaccharide bound to DC-SIGN CRD; the Man3 mannose moiety of Man_4 (the "core monosaccharide") inclines the rest of the molecule towards Phe313 with high surface complementarity, while the fucose moiety (the "core monosaccharide" of LNFP III) makes hydrophobic contact with Val351 and positions the second saccharide in a vertical manner, away from the protein surface (Figure 4). The binding mode presented for Man_4 tetrasaccharide (PDB code: 1SL4) in which Phe313 forms a steric hindrance is a prevalent one for a mannose-containing oligosaccharides. However several other PDB structures of DC-SIGN CRD (PDB code: 1K9I, 1K9J, 2IT5 and 2IT6) show that Phe313 residue is rather flexible and allows 2 distinct binding modes, both including coordination of the Ca2+ by one mannose residue. One mode is shown in Figure 4, while in the second, the binding orientation of the mannose at the principal Ca^{2+} site is reversed, and creates different interactions between the terminal mannose and the region around Phe313 [47,48]. Thus we conclude that the binding mode of a specific oligosaccharide does not depend solely on the "core monosaccharide" involved, but is highly sensitive to the substitution pattern and 3D structure of adjacent monosacccharides. Furthermore, the mannose- and fucose-based ligands have overlapping, but different binding modes, which might offer a rational explanation for the different biological effects of mannose- and fucose-based ligands [49]. Consequently, the two observed binding modes offer a solid basis upon which DC-SIGN antagonists with either fucose or mannose anchors can be designed [50].

On the supramolecular level, DC-SIGN tetramerization has a major impact on binding affinity. By forming tetramers, binding affinity for glycans with repetitive sugar motifs with high-mannose or fucose N-linked oligosaccharides increases markedly [46]. This simple observation drives us to the conclusion that only polyvalent ligands could efficiently bind to DC-SIGN and offer a rationale for design of multivalent carbohydrate systems as DC-SIGN antagonists [51].

DC-SIGN was first regarded as the major binding lectin for various mannose-glycosylated PAMPS including HIV-1 gp120. However, other C-type lectins bind both D-mannose and L-fucose-containing glycans, which brings under question DC-SIGN's exclusive role in pathogen binding, but above all, raises the concern of binding selectivity when designing DC-SIGN antagonists. In particular, two C-type lectins - Langerin on epithelial Langerhans cells and a mannose receptor on dermal DCs - bind high-mannose oligosaccharides and HIV-1 gp120 with high affinity [52]. Their function in HIV-1 (and possibly other pathogen) infection differs: interaction of HIV-1 with DC-SIGN enables HIV-1 to survive a host immune system, while Langerin helps Langerhans cells to eradicate HIV-1 [53,54]. On the other hand, mannose receptor facilitates HIV-1 infection of DCs by the CD4/CCR5 entry pathway [53]. Therefore, blockade of both DC-SIGN and mannose receptors seems the right strategy for prevention of HIV-1 entry into DCs while Langerin inhibition suppresses viral

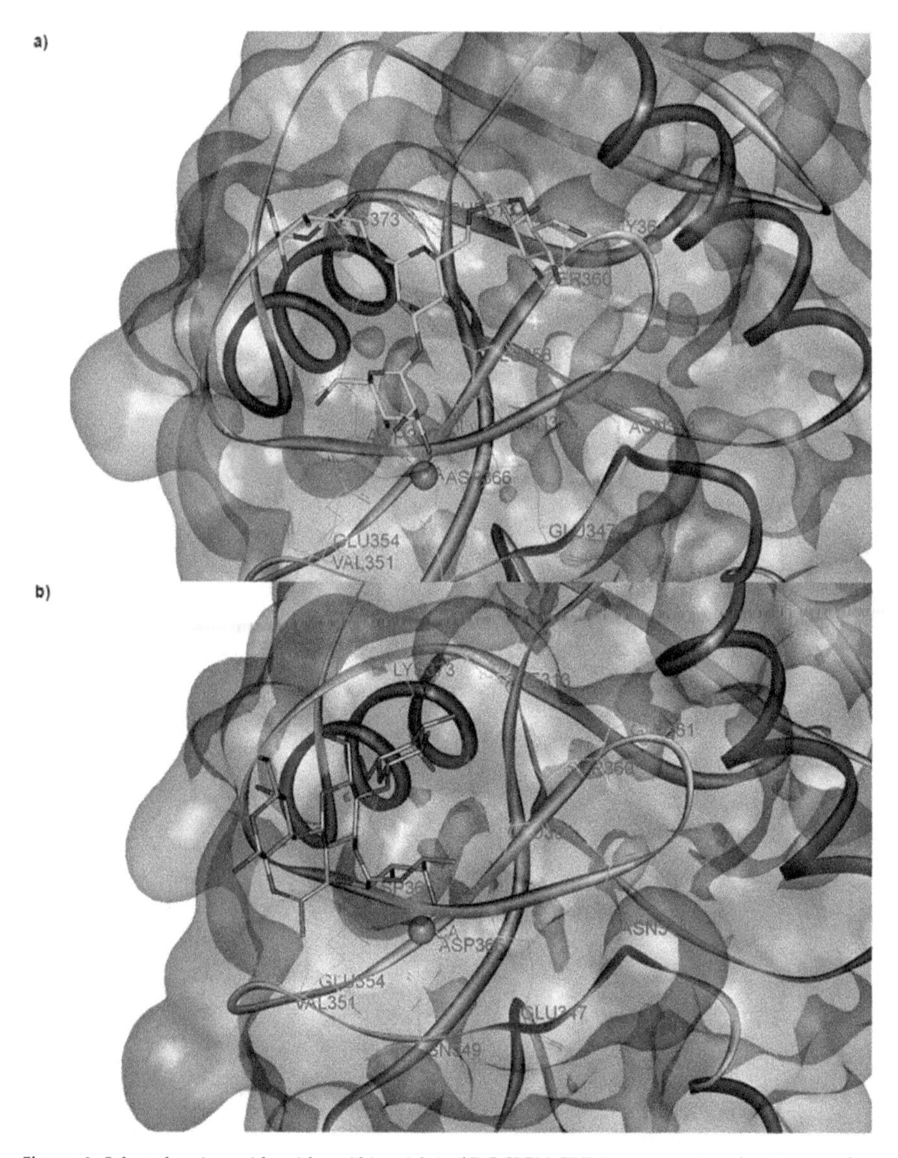

Figure 4. Selected amino acid residues (thin sticks) of DC-SIGN CRD (transparent surface) in complex with: a) Man_4 tetrasaccharide (solid sticks, PDB code: 1SL4), and b) fucose-based tetrasaccharide LNFP III (solid sticks, PDB code: 1SL5) [18]. The Man3 mannose moiety of Man_4 or the "core monosaccharide" inclines the rest of the molecule towards Phe313 with high surface complementarity, while the fucose moiety or the "core monosaccharide" of LNFP III makes hydrophobic contact with Val351 and positions the second saccharide in a vertical manner, away from the protein surface, so only one galactose moiety makes additional contacts with DC-SIGN CRD.

clearance thus allowing a boost of HIV-1 infection. When designing an efficient HIV-1 entry inhibitor based on C-type lectin antagonism, an agent of choice should bind to both DC-SIGN and mannose receptors, while having no or marginal affinity to Langerin. Although all three receptors bind to virtually the same ligands, selectivity against Langerin is an achievable objective since Langerin binding sites differ from that of DC-SIGN [55]. As expected, selectivity versus Langerin does not necessarily rely on the "core monosaccharide" involved in the binding process, but on spatial constraints formed by adjacent glycan monosaccharide units [56].

2.5. Binding of monovalent DC-SIGN antagonists

As mentioned earlier, binding of DC-SIGN natural ligands depends upon the presence of an L-fucose or D-mannose hexopyranose unit. Their inherent disadvantages in terms of low activity and/or insufficient drug-like properties can be modified by the design of glycomimetics – compounds that mimic the bioactive function of carbohydrates but have far better drug-like properties [3]. This concept has been successfully used in the design of monovalent DC-SIGN antagonists. The term "monovalent" is used here to describe low-molecular weight molecules that can occupy only one DC-SIGN CRD at a time, so it incorporates mono- and oligosaccharide structures and their mimetics. The design of these monovalent glycomimetics can be structurally divided into the following sections, as depicted in figure 5:

- the choice of a monosaccharide unit,
- the choice of glycosidic bond surrogate,
- the choice of adjacent saccharides or structures that contribute to overall binding affinity.

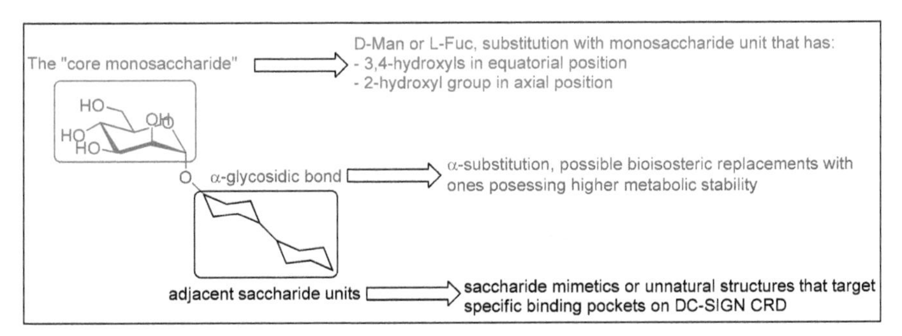

Figure 5. Schematic presentation of the design of monovalent glycomimetics that act as DC-SIGN antagonists: systematic replacements in structure that lead to efficient DC-SIGN ligands.

2.5.1. *The choice of a monosaccharide unit*

The most extensive work regarding the choice of the monosaccharide unit comes from the work of Bernardi and Rojo's groups from CARMUSYS [41,50,57]. In particular, they have shown that L-fucose can be incorporated in glycomimetic surrogates of Lewis-X trisaccharide to obtain even better affinity than the native trisaccharide. The full Lewis-X mimetic **1** was shown to inhibit DC-SIGN binding of mannosylated BSA in the upper micromolar range (IC_{50}=350 μM), but the second generation of analogous compounds failed to give any significant improvement over **1** (Figure 6) [50,58]. STD-NMR studies of **1** with DC-SIGN ECD have shown that only fucose residue makes strong contact with the DC-SIGN CRD. A reasonable explanation for this observation might be the before mentioned binding mode of the L-fucose moiety: it positions the second saccharide in a vertical manner away from the protein surface, and thus the rest of the molecule fails to form tight interactions with protein. Just a glance at figure 4 reveals that DC-SIGN CRD is quite flat and high structural complementarity is one of the requirements for strong binding. In the absence of functional groups that would allow ionic interactions which are not highly dependent on the distance, the rest of the Lewis-X mimetic **1** and its analogues probably form the majority of interactions with the solvent. Lewis-X mimetics however share one important figure; their affinity for Langerin is insignificant, so they are selective DC-SIGN antagonists [58].

The L-fucose monosaccharide has the highest affinity for DC-SIGN among monosaccharides, and L-fucose should be the logical choice as the "core monosaccharide" when designing DC-SIGN antagonists. On the contrary, D-mannose has received most of the attention as the majority of mono- and polyvalent DC-SIGN antagonists incorporate D-mannose as the "core monosaccharide".

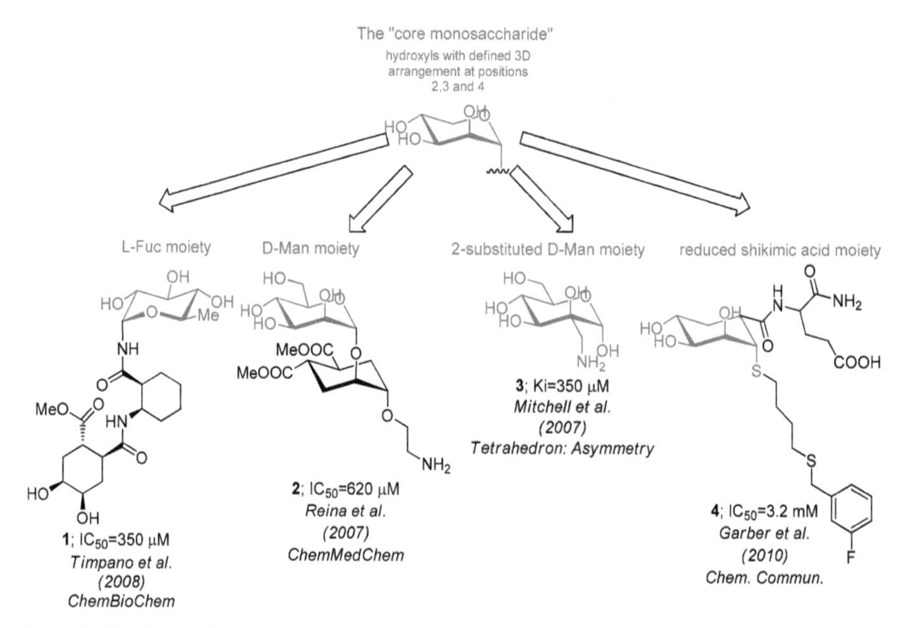

Figure 6. The choice of "core monosaccharide" of DC-SIGN antagonists; L-fucose (L-Fuc), D-mannose (D-Man), 2-substituted D-mannose and reduced shikimic acid are useful "core monosaccharides".

Pseudo-1,2-mannobioside **2** and its analogues of Reina et al. (Figure 6) contain a D-mannose unit substituted at the anomeric position with conformationally constrained cyclohexanediol [41]. The STD-NMR of DC-SIGN ECD with an azido derivative of **2** shows that the compound makes close contact with the protein, which is in agreement with the binding mode of Man_4 tetrasaccharide. The inhibitory activity of **2** on Ebola virus entry into DC-SIGN expressing Jurkat cells was quite high (IC_{50}=0.62 mM) and this was the first functional assay showing that DC-SIGN antagonism with small molecules might be used to inhibit viral transfection mediated by DC-SIGN. Another example of substituted D-mannose as the "core monosaccharide" might be found in the work of Mitchell et al.: they have synthesized a small library of 2-C-substituted branched D-mannose analogues of which compound **3** exhibited a 48-fold stronger binding to DC-SIGN (Ki=0.35 mM, Ki(mannose)=17.1 mM) [59].

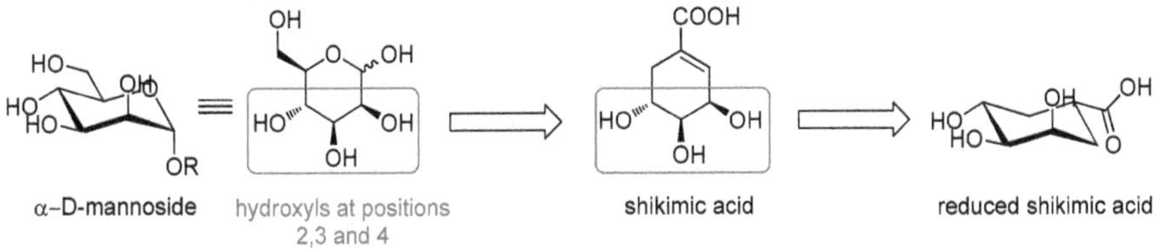

Figure 7. Design of reduced shikimic acid "core monosaccharide" as D-mannose mimetic [60].

An innovative approach was used by Garber et al.: taking D-mannose as the lead structure, they have concentrated on the spatial relationship of hydroxyls at positions 2,3 and 4 and concluded that reduced shikimic acid should enable the same spatial relationship of hydroxyls (Figure 7) [60]. They have synthesized 192 derivatives of reduced shikimic acid and compound **4** (Figure 6) was found to be the most potent hit of this focused library.

2.5.2. *The choice of glycosidic bond surrogate*

The metabolic instability of glycosidic bonds renders it inappropriate for the design of drugs, and hence requires an appropriate surrogate when designing mimetics of oligosaccharides. Numerous changes in carbohydrate structures have been successful in the design of glycosidase inhibitors, and exactly these structures could be used to modify glycosidic bond instability [61]. However, unwanted inhibition of glycosidases may as well provoke side effects of potential drugs, so careful choice of glycosidic bond surrogates has to be made if only its stability is the ultimate goal. The α-glycosidic bond found in both L-fucose and D-mannose containing oligosaccharides that bind to DC-SIGN was successfully replaced by Timpano et al. by a stable α-fucosylamide structure (compound **1**, Figure 6) and shown not to affect the binding affinity in a detrimental sense [50]. The derivatives of reduced shikimic acid (compound **4**, Figure 6) have 2 features that influence glycosidic bond stability: first, they have a thioglycosidic bond, which is proven to be metabolically more stable towards glycosidases [62], and second, they are a constitute of reduced shikimic acid which is a carbasugar. Carbasugars lack anomeric reactivity, which implies their metabolic stability towards glycosidases and glycosyltransferases. Although glycosidic bond surrogates were often not used in the design of DC-SIGN antagonists, the stability of glycosidic bonds has to be challenged when designing stable glycomimetics. According to the latest literature, a large number of alternatives already exist [63].

2.5.3. *The choice of adjacent saccharides or structures that contribute to overall binding affinity*

Monosaccharide moieties other than "core monosaccharide" in the structure of DC-SIGN oligosaccharide ligands form not only additional interactions with the binding site, but also influence the binding specificity with spatial constraints and point other monosaccharide units towards or away from the protein surface, as seen in the Figure 4. Notable quality of the adjacent monosaccharide units is that they do not form the same interactions like the "core monosaccharide", but instead form a network of H-bonds, possibly through water molecules. For example, Man$_4$ tetrasaccharide makes contact with DC-SIGN through at least two water molecules, while one stabilizes its binding conformation [18,64]. Alternatively, both "core" and adjacent monosaccharides make surface complementarity and hydrophobic interactions. In particular, Van Liempt et al. demonstrated that Val351 in DC-SIGN creates a hydrophobic pocket that strongly interacts with the Fucalpha1,3/4-GlcNAc moiety of the Lewis antigens [65]. So, although highly polar in nature, the adjacent monosaccharide units contribute significantly to overall binding free energy also by hydrophobic interactions apart from being mere "linker" to other structures. This implies that altering hydrophilic character of adjacent monosaccharide units to more hydrophobic surrogates should increase free binding energy by increasing hydrophobic interactions. With this in mind, Timpano et al. have used the (1*S*,2*R*)-2-aminocyclohexanecarboxylic acid as a scaffold/linker to attach D-galactose mimetic into the structure of compound **1** and its derivatives (Figure 8) [50]. The molecule was carefully chosen to mimic Lewis-X structure while minimizing the hydrophilic ballast of the Lewis-X, and galactose mimetic was incorporated based on the observation, that galactose residue makes contact with the DC-SIGN CRD surface and is thus important in binding [18].

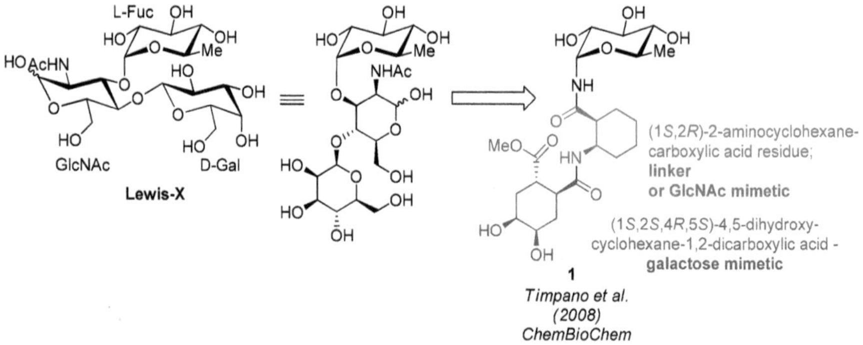

Figure 8. (1*S*,2*R*)-2-aminocyclohexanecarboxylic acid as a scaffold/linker to attach D-galactose mimetic into the structure of compound **1** and its derivatives, all mimic Lewis-X [50].

In the series of mannose-based DC-SIGN antagonists, Reina et al. and Sattin et al. (groups of Rojo and Bernardi) have incorporated (1*S*,2*S*,4*S*,5*S*)-dimethyl 4,5-dihydroxycyclohexane-1,2-dicarboxylate as adjacent monosaccharide mimicking "trans" conformation of 1,2-hydroxyls in D-mannose (Figure 9, compounds **2** and **5**) [10,41]. Again, the rational for this change was to imitate the 3D relationship of key hydroxyls in the D-mannose moiety, while lowering the overall hydrophilicity. Furthermore, the cyclohexane saccharide mimic lacks glycosidic bond and is thus metabolically stable.

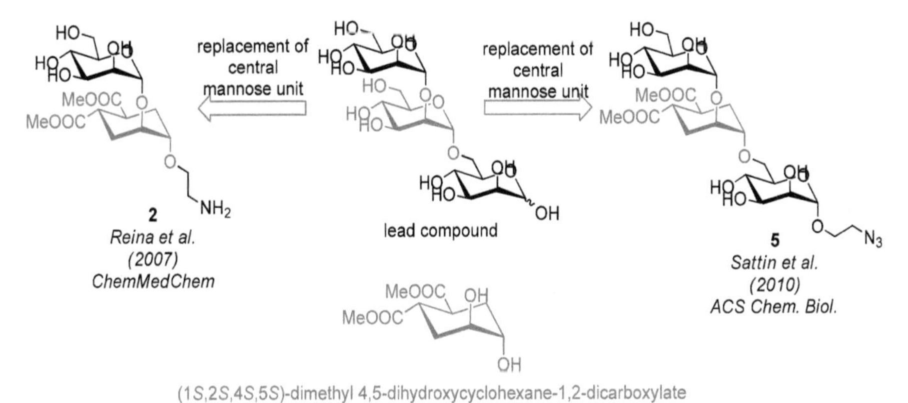

Figure 9. (1*S*,2*S*,4*S*,5*S*)-dimethyl 4,5-dihydroxycyclohexane-1,2-dicarboxylate as central monosaccharide surrogate that mimics "trans" conformation of 1,2-hydroxyls in D-mannose.

Starting from compound **2** or its azido derivative, the group of Bernardi (Obermajer et al.) continued to pursue the idea of increasing the binding affinity by identifying two binding areas around Phe313 in the DC-SIGN binding site that were only partially occupied by co-crystallized tetramannoside Man4 [64]. These hydrophobic areas were targeted by attaching different hydrophobic moieties to deprotected carboxylates of pseudo-1,2-mannobioside **2** (Figure 10). A number of mannose-based DC-SIGN antagonists were synthesized (an illustrative example is compound **6**), and the majority of them inhibited DC adhesion at low micromolar concentrations improving the potency of the starting compound **2** by two orders of magnitude. Probably the same hydrophobic areas have contributed to the affinity of **4**.

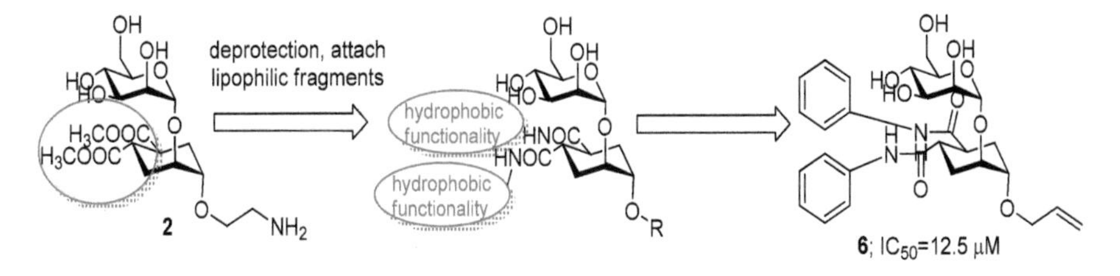

Figure 10. Increasing potency of DC-SIGN antagonists by attaching hydrophobic moieties to deprotected carboxylates of **2** to afford **6** and its derivatives [64].

2.6. Increasing affinity and/or avidity?

The evident disadvantage of monovalent DC-SIGN antagonists – their low affinity for DC-SIGN and thus low potency – can be overcome by conjugating monovalent units to various scaffolds for polyvalent presentation (Figure 11) [51]. Namely, even the most potent monovalent ligands have inhibitory constants in low micromolar concentrations, while therapeutically useful compounds have the same effect in nanomolar or even picomolar concentrations. The polyvalent antagonists are believed to interact with possibly all DC-SIGN CRDs on tetramers, or perhaps, they might interact and collate DC-SIGN clusters on the cell surface in the same manner as HIV-1 increases its avidity to DC surface (depicted in Figure 3). The avidity observed for polyvalent ligands in general originates from multiple binding: the polyvalent molecule with reversible mechanism of binding has higher possibility for being bound to at least one of receptor binding sites (i.e. one of CRDs), so that dissociation rate constant significantly decreases [66]. The other rationale may be derived from the observation of Andrews et al., who concluded that the average loss of overall rotational and translational entropy accompanying drug-receptor interaction is a constant for relatively small molecules and was estimated to be approx. 14 kcal/mol at 310 K [67]. In other words, more favourable free binding energy is obtained by combining binding epitopes/structures into one larger molecule, and this is also true for uniting same monovalent structures into a large polyvalent one because all rotational and translational entropic losses that occur during binding of individual monovalent molecules are reduced to one entropic loss for a larger molecule. According to both theories, when larger polyvalent molecules are employed, binding avidity is observed instead of just a linear increase in binding affinity. The main concern when choosing appropriate polyvalent support is the spatial relationship between individual monovalent ligands: monovalent ligands should be appropriately spaced to allow binding to at least two binding sites (or CRDs in case of C-type lectins), otherwise avidity cannot be achieved.

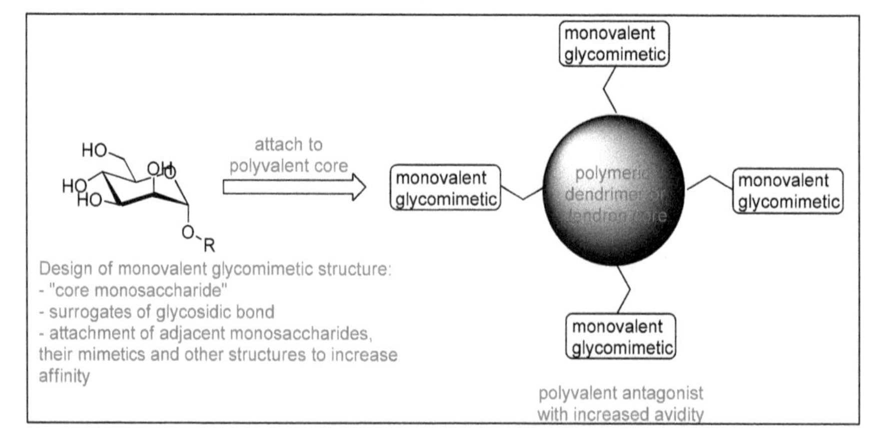

Figure 11. Strategy for increasing binding affinity/avidity of DC-SIGN antagonists: monosaccharides or monovalent glycomimetics are attached to polyvalent dendrimer or dendron core (modified from Anderluh et al.) [43].

The first polyvalent DC-SIGN antagonists were designed, synthesized and assayed in the laboratories of Rojo and Delgado [68]. A simple glycodendritic core bearing 32 mannose residues conjugated to BoltornH30 dendrimer through a succinic acid spacer (**7**, Figure 12) hindered DC-SIGN mediated Ebola virus infection at nanomolar concentrations (IC_{50}=337 nM). In comparison, the same inhibition was achieved with only millimolar concentrations of α-methyl-D-mannopyranoside (IC_{50}=1.27 mM); the core bears only 32 mannose residues and the difference in binding affinity is more than 3000-fold, which clearly indicates high binding avidity.

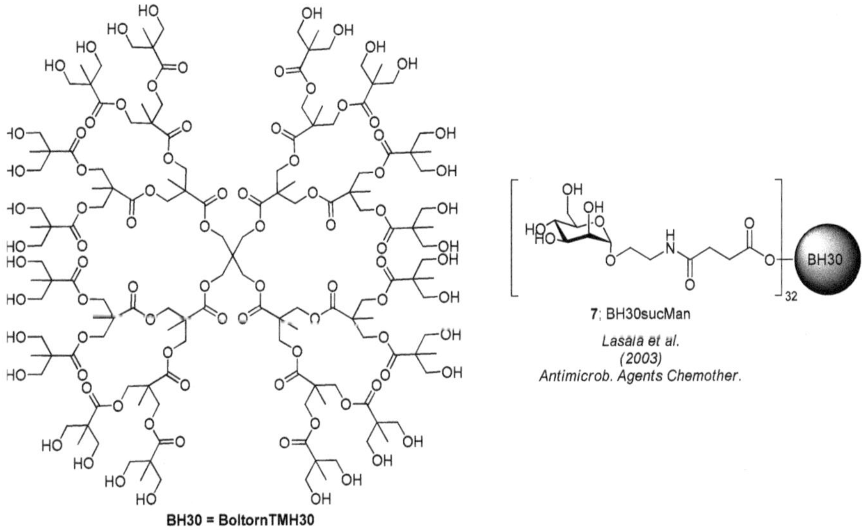

BH30 = BoltornTMH30

7; BH30sucMan

Lasala et al. (2003) Antimicrob. Agents Chemother.

Figure 12. Glycodendritic structure bearing 32 mannose residues conjugated to BoltornH30 dendrimer as DC-SIGN antagonist [68].

Relatively simple glycopolymers with a different load of combined α-mannose and β-galactose in ratios were synthesized by Becer et al. and evaluated in inhibition of DC-SIGN-gp120 binding (**8**, Figure 13) [69]. Glycopolymer with 25% mannose failed to inhibit gp120 binding potently (IC_{50} of 1.45 μM), while glycopolymer with 100% mannose was far more efficient (IC_{50} of 37 nM) even when calculated per mannose unit, as it had approx. 40 times higher affinity compared with 4-times higher load of the mannose unit.

Polyvalent DC-SIGN antagonists described before relied on the use of "core monosaccharide" only. The design and choice of a potent monovalent DC-SIGN ligand might reduce the requirement for huge dendrimeric presentation while retaining desired effect. Designed monovalent glycomimetic **4** (Figure 6) of Garber et al. was loaded onto a carefully selected multivalent core designed to link at least 2 CRDs of DC-SIGN tetramer as monovalent units were roughly 40Å apart, which is exactly the width of DC-SIGN CRD (**8**, Figure 13) [60]. Glycopolymer **8** with 29 units of **4** was found to have 1000-fold higher affinity for DC-SIGN (IC_{50}=2.9 μM) than **4**, showing that high avidity binding was obtained.

It is hard though to assess the relevance of designed monovalent ligand versus D-mannose, as the analogous polyvalent DC-SIGN antagonist bearing D-mannose or reduced shikimic acid was not synthesized. The direct evidence however came from the work of Sattin et al [10]. They have attached four copies of monovalent trimannoside mimetic **5** (Figure 9) to a tetravalent dendron **10** (Figure 14), which potently inhibited HIV-1 transfection to CD4+ T lymphocytes (>94 % inhibition at 100 μM). For comparison only, a tetrameric dendron bearing 4 copies of simple D-mannose (**11**, Figure 14) failed to inhibit HIV-1 transfection with even comparable potency (65 % inhibition at 100 μM). Furthermore, Dendron **10** inhibited Ebola *cis* infection of Jurkat T cells one order of magnitude more potently than dendron **11** [70]. This data clearly demonstrates that potent monovalent DC-SIGN antagonists reduce the need for high polyvalency number and influences binding affinity markedly.

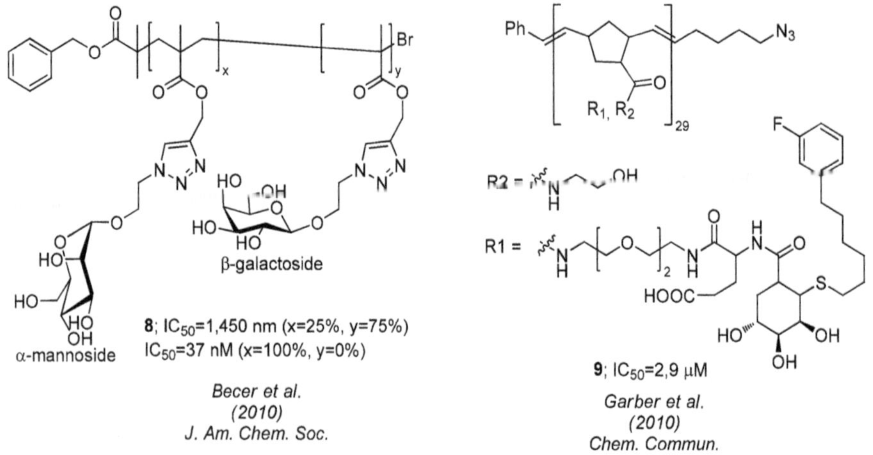

Figure 13. Glycopolymers of Becer et al. and Garber et al. with high binding avidity [60,69].

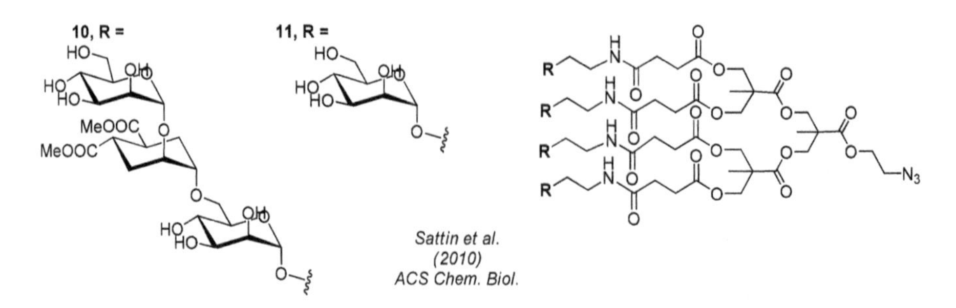

Figure 14. Tetravalent dendrons of Sattin et al. bearing four copies of monovalent trimannoside mimetic **5** (**10**) or D-mannose residue (**11**) [10].

3. Conclusion; could it work for all C-type lectins?

As mentioned earlier, all C-type lectins share a structural feature, namely C-type lectin-like domain (CTLD) or carbohydrate recognition domain (CRD) responsible for Ca^{2+}-dependent selective binding of terminal mono- or oligosaccharide units of large carbohydrates. This implies that only the "core monosaccharide" makes contact with Ca^{2+} ion while other ligand carbohydrate units (if present) form structural and bonding complementarity with the CRD. From this point of view, the systematic approach presented herein could be of general applicability when designing glycomimetic C-type lectin antagonists. It consists of designing the monovalent ligand based on three distinct steps: the choice/design of a "core" monosaccharide unit, the choice/design of glycosidic bond surrogate and the choice/design of adjacent saccharides or structures that contribute to overall binding affinity. Still, the evident drawback of monovalent glycomimetics is their low affinity not only to DC-SIGN, but to lectins in general [3]. The other characteristic of C-type lectins is their ability to oligomerize, and further make clusters of functional oligomers. The tactics of polyvalent presentation targets exactly the oligomerized or even clustered structures and is a prerequisite of polyvalent structures as C-type lectin antagonists. Accordingly, high avidity for DC-SIGN and other C-type lectins can be achieved with high loading of monovalent ligands to various polyvalent systems. Taken together, the results of several groups presented in this chapter clearly demonstrate that this general procedure for designing glycomimetic DC-SIGN antagonists gives notable results, and according to the structural resemblance of diverse C-type lectin CRDs, there is a high probability of its applicability to distinct C-type lectins.

Author details

Marko Anderluh
Department of Medicinal Chemistry, University of Ljubljana,
Faculty of Pharmacy, Ljubljana, Slovenia

Acknowledgement

This work is dedicated to Prof. Dr. Slavko Pečar, a great mentor and teacher, who taught me not only pure science, but above all how to nourish my enthusiasm for science.

4. References

[1] Finkelstein J (2007) Insight - Glycochemistry & glycobiology. Nature 446: 999-999.
[2] Ernst JD (1998) Macrophage receptors for Mycobacterium tuberculosis. Infect. Immun. 66: 1277-1281.
[3] Ernst B, Magnani JL (2009) From carbohydrate leads to glycomimetic drugs. Nat. Rev. Drug Discover. 8: 661-677.

[4] Malik A, Firoz A, Jha V, Ahmad S (2010) PROCARB: A Database of Known and Modelled Carbohydrate-Binding Protein Structures with Sequence-Based Prediction Tools. Adv. Bioinformatics. 436036.

[5] Curtis BM, Scharnowske S, Watson AJ (1992) Sequence And Expression Of A Membrane-Associated C-Type Lectin That Exhibits CD4-Independent Binding Of Human-Immunodeficiency-Virus Envelope Glycoprotein-Gp120. Proc. Natl. Acad. Sci. USA 89: 8356-8360.

[6] Steinman RM (2000) DC-SIGN: A guide to some mysteries of dendritic cells. Cell. 100: 491-494.

[7] Geijtenbeek TBH, Kwon DS, Torensma R, van Vliet SJ, van Duijnhoven GCF, Middel J, Cornelissen I, Nottet H, KewalRamani VN, Littman DR (2000) DC-SIGN, a dendritic cell-specific HIV-1-binding protein that enhances trans-infection of T cells. Cell.100: 587-597.

[8] Geijtenbeek TBH, Torensma R, van Vliet SJ, van Duijnhoven GCF, Adema GJ, van Kooyk Y, Figdor CG (2000) Identification of DC-SIGN, a novel dendritic cell-specific ICAM-3 receptor that supports primary immune responses. Cell 100: 575-585.

[9] Svajger U, Anderluh M, Jeras M, Obermajer N (2010) C-type lectin DC-SIGN: An adhesion, signalling and antigen-uptake molecule that guides dendritic cells in immunity. Cell. Signall. 22: 1397-1405.

[10] Sattin S, Daghetti A, Thepaut M, Berzi A, Sanchez-Navarro M, Tabarani G, Rojo J, Fieschi F, Clerici M, Bernardi A (2010) Inhibition of DC-SIGN-Mediated HIV Infection by a Linear Trimannoside Mimic in a Tetravalent Presentation. ACS Chem. Biol. 5: 301-312.

[11] Komath SS, Kavitha M, Swamy MJ (2006) Beyond carbohydrate binding: new directions in plant lectin research. Org. Biomol. Chem. 4: 973-988.

[12] Becker B, Condie GC, Le GT, Meutermans W (2006) Carbohydrate-based scaffolds in drug discovery. Mini-Rev. Med. Chem. 6: 1299-1309.

[13] Varki A (1999) Essentials of glycobiology. Cold Spring Harbor, NY: Cold Spring Harbor Laboratory Press; 1999. xvii, 653 p.

[14] Zelensky AN, Gready JE (2005) The C-type lectin-like domain superfamily. FEBS J. 272: 6179-6217.

[15] Drickamer K (1993) Ca2+-Dependent Carbohydrate-Recognition Domains In Animal Proteins. Curr. Opin. Struct. Biol. 3: 393-400.

[16] Harding MM (2006) Small revisions to predicted distances around metal sites in proteins. Acta Crystallogr. Sect. D-Biol. Crystallogr. 62: 678-682.

[17] Kolatkar AR, Weis WI (1996) Structural basis of galactose recognition by C-type animal lectins. J. Biol. Chem. 271: 6679-6685.

[18] Guo Y, Feinberg H, Conroy E, Mitchell DA, Alvarez R, Blixt O, Taylor ME, Weis WI, Drickamer K (2004) Structural basis for distinct ligand-binding and targeting properties of the receptors DC-SIGN and DC-SIGNR. Nat. Struct. Mol. Biol. 11: 591-598.

[19] Teillet F, Dublet B, Andrieu JP, Gaboriaud C, Arland GJ, Thielens NM (2005) The two major oligomeric forms of human mannan-binding lectin: Chemical characterization,

carbohydrate-binding properties, and interaction with MBL-associated serine proteases. J. Immunol. 174: 2870-2877.

[20] Rini JM (1995) Lectin Structure. Annu. Rev. Biophys.Biomol. Struct. 24: 551-577.

[21] Martinez-Pomares L, Linehan SA, Taylor PR, Gordon S (2001) Binding properties of the mannose receptor. Immunobiology 204: 527-535.

[22] Banchereau J, Steinman RM (1998) Dendritic cells and the control of immunity. Nature 392: 245-252.

[23] Janeway CA, Medzhitov R (2002) Innate immune recognition. Annu. Rev. Immunol. 20: 197-216.

[24] Geijtenbeek TB, Krooshoop DJ, Bleijs DA, van Vliet SJ, van Duijnhoven GC, Grabovsky V, Alon R, Figdor CG, van Kooyk Y (2000) DC-SIGN-ICAM-2 interaction mediates dendritic cell trafficking. Nat. Immunol. 1: 353-357.

[25] Geijtenbeek TBH, Torensma R, van Vliet SJ, van Duijnhoven GCF, Middel J, Cornelissen ILMHA, Adema GJ, Nottet HSLM, Figdor CG, van Kooyk Y (1999) DC-SIGN, a novel dendritic cell-specific adhesion receptor for ICAM-3 mediates DC-T cell interactions and HIV-1 infection of DC. Blood 94: 434a-434a.

[26] Mitchell DA, Fadden AJ, Drickamer K (2001) A novel mechanism of carbohydrate recognition by the C-type lectins DC-SIGN and DC-SIGNR. Subunit organization and binding to multivalent ligands. J. Biol. Chem. 276: 28939-28945.

[27] Feinberg H, Guo Y, Mitchell DA, Drickamer K, Weis WI (2005) Extended neck regions stabilize tetramers of the receptors DC-SIGN and DC-SIGNR. J. Biol. Chem. 280: 1327-1335.

[28] Tabarani G, Thepaut M, Stroebel D, Ebel C, Vives C, Vachette P, Durand D, Fieschi F (2009) DC-SIGN Neck Domain Is a pH-sensor Controlling Oligomerization: Saxs and Hydrodynamic Studies of Extracellular Domain. J. Biol. Chem. 284: 21229-21240.

[29] Cambi A, de Lange F, van Maarseveen NM, Nijhuis M, Joosten B, van Dijk EM, de Bakker BI, Fransen JA, Bovee-Geurts PH, van Leeuwen FN and others (2004) Microdomains of the C-type lectin DC-SIGN are portals for virus entry into dendritic cells. J. Cell Biol. 164: 145-155.

[30] Cambi A, Koopman M, Figdor CG (2005) How C-type lectins detect pathogens. Cell. Microbiol. 7: 481-488.

[31] Menon S, Rosenberg K, Graham SA, Ward EM, Taylor ME, Drickamer K, Leckband DE (2009) Binding-site geometry and flexibility in DC-SIGN demonstrated with surface force measurements. Proc. Natl. Acad. Sci. USA 106: 11524-11529.

[32] Khoo US, Chan KY, Chan VS, Lin CL. DC-SIGN and L-SIGN: the SIGNs for infection (2008) J. Mol. Med. 86: 861-874.

[33] Tsegaye TS, Pohlmann S (2010) The multiple facets of HIV attachment to dendritic cell lectins. Cell. Microbiol. 12: 1553-1561.

[34] Cambi A, Beeren I, Joosten B, Fransen JA, Figdor CG (2009) The C-type lectin DC-SIGN internalizes soluble antigens and HIV-1 virions via a clathrin-dependent mechanism. Eur. J. Immunol. 39: 1923-1928.

[35] Kwon DS, Gregorio G, Bitton N, Hendrickson WA, Littman DR (2002) DC-SIGN-mediated internalization of HIV is required for trans-enhancement of T cell infection. Immunity 16: 135-144.

[36] Trumpfheller C, Park CG, Finke J, Steinman RM, Granelli-Piperno A (2003) Cell type-dependent retention and transmission of HIV-1 by DC-SIGN. Int. Immunol. 15: 289-298.

[37] van Montfort T, Eggink D, Boot M, Tuen M, Hioe CE, Berkhout B, Sanders RW (2011) HIV-1 N-Glycan Composition Governs a Balance between Dendritic Cell-Mediated Viral Transmission and Antigen Presentation. J. Immunol. 187: 4676-4685.

[38] Sodhi A, Montaner S, Gutkind JS (2004) Viral hijacking of G-protein-coupled-receptor signalling networks. Nat. Rev. Mol. Cell. Biol. 5: 998-1012.

[39] van Kooyk Y, Appelmelk B, Geijtenbeek TB (2003) A fatal attraction: Mycobacterium tuberculosis and HIV-1 target DC-SIGN to escape immune surveillance. Trends Mol. Med. 9: 153-159.

[40] Hodges A, Sharrocks K, Edelmann M, Baban D, Moris A, Schwartz O, Drakesmith H, Davies K, Kessler B, McMichael A, Simmons A (2007). Activation of the lectin DC-SIGN induces an immature dendritic cell phenotype triggering Rho-GTPase activity required for HIV-1 replication. Nat Immunol 2007;8:569-77.

[41] Reina JJ, Sattin S, Invernizzi D, Mari S, Martinez-Prats L, Tabarani G, Fieschi F, Delgado R, Nieto PM, Rojo J, Bernardi A (2007) 1,2-Mannobioside mimic: synthesis, DC-SIGN interaction by NMR and docking, and antiviral activity. ChemMedChem 2: 1030-1036.

[42] Alen MMF, Kaptein SJF, De Burghgraeve T, Balzarini J, Neyts J, Schols D (2009) Antiviral activity of carbohydrate-binding agents and the role of DC-SIGN in dengue virus infection. Virology 387: 67-75.

[43] Anderluh M, Jug G, Svajger U, Obermajer N (2012) DC-SIGN Antagonists, a Potential New Class of Anti-Infectives. Curr. Med. Chem. 19: 992-1007.

[44] Borrok MJ, Kiessling LL (2007) Non-carbohydrate inhibitors of the lectin DC-SIGN. J. Am. Chem. Soc. 129: 12780-12785.

[45] Lederman MM, Offord RE, Hartley O (2006) Microbicides and other topical strategies to prevent vaginal transmission of HIV. Nat. Rev. Immunol. 6: 371-382.

[46] Appelmelk BJ, van Die I, van Vliet SJ, Vandenbroucke-Grauls CM, Geijtenbeek TB, van Kooyk Y (2003) Cutting edge: carbohydrate profiling identifies new pathogens that interact with dendritic cell-specific ICAM-3-grabbing nonintegrin on dendritic cells. J. Immunol. 170: 1635-1639.

[47] Feinberg H, Mitchell DA, Drickamer K, Weis WI (2001) Structural basis for selective recognition of oligosaccharides by DC-SIGN and DC-SIGNR. Science 294: 2163-2166.

[48] Feinberg H, Castelli R, Drickamer K, Seeberger PH, Weis WI (2007) Multiple modes of binding enhance the affinity of DC-SIGN for high mannose N-linked glycans found on viral glycoproteins. J. Biol. Chem. 282: 4202-4209.

[49] Gringhuis SI, den Dunnen J, Litjens M, van der Vlist M, Geijtenbeek TB (2009) Carbohydrate-specific signaling through the DC-SIGN signalosome tailors immunity to Mycobacterium tuberculosis, HIV-1 and Helicobacter pylori. Nat. Immunol. 10: 1081-1088.

[50] Timpano G, Tabarani G, Anderluh M, Invernizzi D, Vasile F, Potenza D, Nieto PM, Rojo J, Fieschi F, Bernardi A (2008) Synthesis of novel DC-SIGN ligands with an alpha-fucosylamide anchor. Chembiochem 9: 1921-1930.

[51] Rojo J, Delgado R (2004) Glycodendritic structures: promising new antiviral drugs. J. Antimicrob. Chemother. 54: 579-581.

[52] Cunningham AL, Turville S, Wilkinson J, Cameron P, Dable J (2003) The role of dendritic cell C-type lectin receptors in HIV pathogenesis. J. Leukoc. Biol. 74: 710-718.

[53] Turville SG, Cameron PU, Handley A, Lin G, Pohlmann S, Doms RW, Cunningham AL (2002) Diversity of receptors binding HIV on dendritic cell subsets. Nat. Immunol. 3: 975-983.

[54] de Witte L, Nabatov A, Pion M, Fluitsma D, de Jong M, de Gruijl T, Piguet V, van Kooyk Y, Geijtenbeek TBH (2007) Langerin is a natural barrier to HIV-1 transmission by Langerhans cells. Nat. Med. 13: 367-371.

[55] Chatwell L, Holla A, Kaufer BB, Skerra A (2008) The caxbohydrate recognition domain of Langerin reveals high structural similarity with the one of DC-SIGN but an additional, calcium-independent sugar-binding site. Mol. Immunol. 45: 1981-1994.

[56] Holla A, Skerra A (2011) Comparative analysis reveals selective recognition of glycans by the dendritic cell receptors DC-SIGN and Langerin. Prot. Eng. Des. & Sel. 24: 659-669.

[57] Reina JJ, Diaz I, Nieto PM, Campillo NE, Paez JA, Tabarani G, Fieschi F, Rojo J (2008) Docking, synthesis, and NMR studies of mannosyl trisaccharide ligands for DC-SIGN lectin. Org. Biomol. Chem. 6: 2743-2754.

[58] Andreini M, Doknic D, Sutkeviciute I, Reina JJ, Duan JX, Chabrol E, Thepaut M, Moroni E, Doro F, Belvisi L, Weiser J, Rojo J, Fieschi F, Bernardi A (2011) Second generation of fucose-based DC-SIGN ligands: affinity improvement and specificity versus Langerin. Org. Biomol. Chem. 9: 5778-5786.

[59] Mitchell DA, Jones NA, Hunter SJ, Cook JMD, Jenkinson SF, Wormald MR, Dwek RA, Fleet GWJ (2007) Synthesis of 2-C-branched derivatives of D-mannose: 2-C-aminomethyl-D-mannose binds to the human C-type lectin DC-SIGN with affinity greater than an order of magnitude compared to that of D-mannose. Tetrahedron-Asymmetry 18: 1502-1510.

[60] Garber KCA, Wangkanont K, Carlson EE, Kiessling LL (2010) A general glycomimetic strategy yields non-carbohydrate inhibitors of DC-SIGN. Chem. Commun. 46: 6747-6749.

[61] Gloster TM, Davies GJ (2010) Glycosidase inhibition: assessing mimicry of the transition state. Org. Biomol. Chem. 8: 305-320.

[62] Almendros M, Danalev D, Francois-Heude M, Loyer P, Legentil L, Nugier-Chauvin C, Daniellou R, Ferrieres V (2011) Exploring the synthetic potency of the first furanothioglycoligase through original remote activation. Org. Biomol. Chem. 9: 8371-8378.

[63] Cipolla L, La Ferla B, Airoldi C, Zona C, Orsato A, Shaikh N, Russo L, Nicotra F (2010) Carbohydrate mimetics and scaffolds: sweet spots in medicinal chemistry. Future Med. Chem. 2:587-599.

[64] Obermajer N, Sattin S, Colombo C, Bruno M, Svajger U, Anderluh M, Bernardi A (2011) Design, synthesis and activity evaluation of mannose-based DC-SIGN antagonists. Mol. Divers. 15: 347-360.

[65] Van Liempt E, Imberty A, Bank CMC, Van Vliet SJ, Van Kooyk Y, Geijtenbeek TBH, Van Die I (2004) Molecular basis of the differences in binding properties of the highly related C-type lectins DC-SIGN and L-SIGN to Lewis x trisaccharide and Schistosoma mansoni egg antigens. J. Biol. Chem. 279: 33161-33167.

[66] Shinohara Y, Hasegawa Y, Kaku H, Shibuya N (1997) Elucidation of the mechanism enhancing the avidity of lectin with oligosaccharides on the solid phase surface. Glycobiology 7: 1201-1208.

[67] Andrews PR, Craik DJ, Martin JL (1984) Functional-Group Contributions to Drug Receptor Interactions. J. Med. Chem. 27: 1648-1657.

[68] Lasala F, Arce E, Otero JR, Rojo J, Delgado R (2003) Mannosyl glycodendritic structure inhibits DC-SIGN-mediated Ebola virus infection in cis and in trans. Antimicrob. Agents Chemother. 47: 3970-3972.

[69] Becer CR, Gibson MI, Geng J, Ilyas R, Wallis R, Mitchell DA, Haddleton DM (2010) High-Affinity Glycopolymer Binding to Human DC-SIGN and Disruption of DC-SIGN Interactions with HIV Envelope Glycoprotein. J. Am. Chem. Soc. 132: 15130-15132.

[70] Luczkowiak J, Sattin S, Sutkeviciute I, Reina JJ, Sanchez-Navarro M, Thepaut M, Martinez-Prats L, Daghetti A, Fieschi F, Delgado R Bernardi A, Rojo J (2011) Pseudosaccharide Functionalized Dendrimers as Potent Inhibitors of DC-SIGN Dependent Ebola Pseudotyped Viral Infection. Bioconjug. Chem. 22: 1354-1365.

Chapter 10

Broad Antiviral Activity of Carbohydrate-Binding Agents Against Dengue Virus Infection

Marijke Alen and Dominique Schols

Additional information is available at the end of the chapter

1. Introduction

1.1. Origin and epidemiology

Dengue virus (DENV) is a member of the *Flavivirus* genus within the family of the *Flaviviridae* and is the most common mosquito-borne viral disease. *Flaviviruses* derived from a common viral ancestor 10,000 years ago. DENV has a relative recent revolutionary history originating 1000 years ago and establishes transmission in humans since a few hundred years. There is strong evidence that DENV was originally a monkey virus in non-human primates in Africa and Asia. Cross-species transmission to humans has occurred independently for all four DENV serotypes [1,2]. Each serotype shares around 65% of the genome and despite of the differences, each serotype causes nearly identical syndromes in humans and circulates in the same ecological niche [3]. First clinical symptoms of dengue infections date from the 10th century but it is not for sure that this was a dengue epidemic. The first large dengue epidemics were in 1779 in Asia, Africa and North America. The first reported epidemic of dengue hemorrhagic fever (DHF) was in Manilla, Philippines, in 1953 after World War II. It was suggested that he movement of the troops during World War II has led to the spread of the virus. It has been shown that in the 19th and 20th century, the virus was widespread in the tropics and subtropics where nowadays 3.6 billion people are at risk of getting infected with DENV (Figure 1). Every year, 50 million infections occur, including 500,000 hospitalizations for DHF, mainly among children, with a case fatality rate exceeding 15% in some areas [4,5]. In 40 years of time, DENV became endemic in more than 100 countries because of the increase in human population, international transport and the lack of vector control.

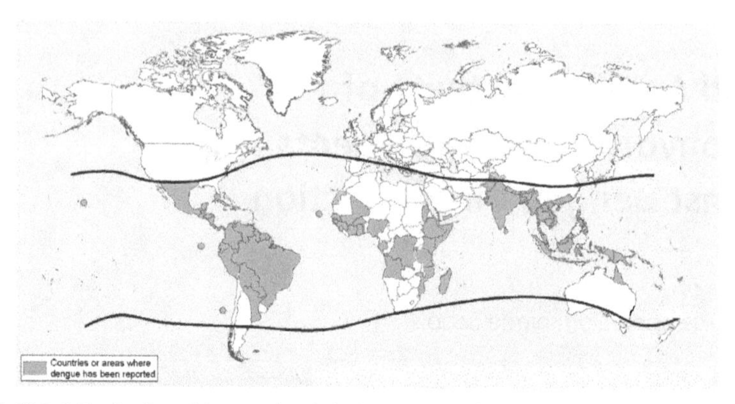

Figure 1. Global distribution of dengue virus infections in 2011. Contour lines represent the areas at risk (Source: WHO, 2012).

1.2. Transmission

The transmission of DENV can only occur by the bite of an infected female mosquito, the *Aedes aegypti* and the *Aedes albopictus. Aedes aegypti* originated, and is still present, in the rainforests of Africa feeding on non-human primates (sylvatic cycle, Figure 2). DENV infection in non-human primates occurs asymptomatically. However, the mosquito became domesticated due to massive deforestation and breeds in artificial water holdings, like automobile tires, discarded bottles and buckets that collect rainwater [4]. On one hand *Aedes aegypti* is not an efficient vector because it has a low susceptibility for oral infection with virus in human blood. Since mosquitoes ingest 1 μl of blood, the virus titer in human blood has to be 10^5-10^7 per ml for transmission to be sustained. After 7-14 days the virus has passed the intestinal tract to the salivary glands and can be transmitted by the infected mosquito to a new host. On the other hand, *Aedes aegypti* is an efficient vector because it has adapted to humans and they repeatedly feed themselves in daylight on different hosts. After a blood meal, the oviposition can be stimulated and the virus can be passed transovarially to the next generation of mosquitoes (vertical transmission, Figure 2) [6].

The tiger mosquito *Aedes albopictus* is spreading his region from Asia to Europe and the United States of America (USA). In the 1980s, infected *Aedes Albopictus* larvae were transported in truck tires from Asia to the United States. Dengue viruses were introduced into port cities, resulting in major epidemics [6].

Because there is no vaccine available, the only efficient way to prevent DENV infection is eradication of the mosquito. In the 1950s and 1960s there was a successful vector control program in the USA organized by the Pan American Health Organization. They eradicated the mosquito from 19 countries. Unfortunately, the vector control program was stopped in 1972 because the government thought that DENV was not important anymore [2,4]. This resulted in a re-emergence of the mosquito and DENV infections in the USA. Both demographic and ecological changes contributed to the world wide spread of DENV infections.

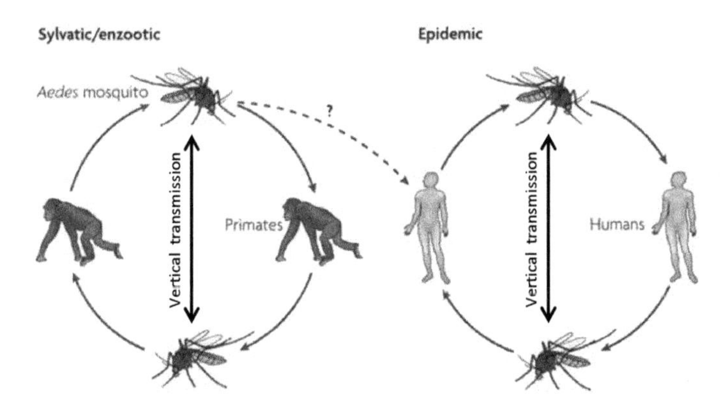

Figure 2. Transmission of DENV by Aedes mosquitoes. Modified from Whitehead et al. [7].

Very recently, another approach to attack the vector has been documented [8,9]. There was a mosquito made resistant to DENV infection after trans infection with the endosymbiont *Wolbachia* bacterium which can infect a lot of insects. A certain strain of the *Wolbachia* bacteria was trans infected in *Aedes* mosquitoes and was reported to inhibit the replication and dissemination of several RNA viruses, such as DENV. Embryos of a *Wolbachia* uninfected female die if the female has bred with a *Wolbachia* infected male. This means that *Wolbachia* infected mosquitoes can take over the natural population. This was recently tested in Australia, where dengue is endemic, and after 2 months the *Wolbachia* infected mosquitoes resistant to DENV had taken over the natural mosquito population. Thus, this indicates the beginning of a new area in vector control efforts with a high potential to succeed.

1.3. Pathogenesis

Although DENV infections have a high prevalence, the pathogenesis of the disease is not well understood. The disease spectrum can range from an asymptomatic or flu like illness to a lethal disease. After a bite of an infected mosquito, there is an incubation period of 3 to 8 days. Then there is on acute onset of fever (≥ 39°C) accompanied by nonspecific symptoms like severe headache, nausea, vomiting, muscle and joint pain (dengue fever). Clinical findings alone are not helpful to distinguish dengue fever from other febrile illnesses such as malaria or measles. Half of the infected patients report a rash and is most commonly seen on the trunk and the insides of arms and thighs. Skin hemorrhages, including petechiae and purpura, are very common. Liver enzyme levels of alanine aminotransferase and aspartate aminotransferase can be elevated. Dengue fever is generally self-limiting and is rarely fatal [5,10,11].

The disease can escalate into dengue hemorrhagic fever (DHF) or dengue shock syndrome (DSS). DHF is primarily a children's disease and is characterized as an acute febrile illness with thrombocytopenia (≤ 100,000 cells/mm^3). This results in an increased vascular permeability and plasma leakage from the blood vessels into the tissue. Plasma leakage has been documented by an increased hematocrit and a progressive decrease in platelet count. Petechiae and subcutaneous bleedings are very common [12].

DSS is defined when the plasma leakage becomes critical resulting in circulatory failure, weak pulse and hypotension. Plasma volume studies have shown a reduction of more than 20% in severe cases. A progressively decreasing platelet count, a rising hematocrit, sustained abdominal pain, persistent vomiting, restlessness and lethargy may be all signs for DSS. Prevention of shock can only be established after volume replacement with intravenous fluids [5,11]. When experienced clinicians and nursing staff are available in endemic areas, the case fatality rate is < 1%.

DHF and DSS occur during a secondary infection with a heterologous serotype. The first infection with one of the four serotypes provides lifelong immunity to the homologous virus. During a second infection with a heterologous serotype, non-neutralizing IgG antibodies can enhance disease severity. This phenomenon is called antibody-dependent enhancement (ADE). The pre-existing non-neutralizing heterotypic antibodies can form a complex with DENV and enhance the access to Fc-receptor bearing cells such as monocytes and macrophages [13,14] (Figure 3).

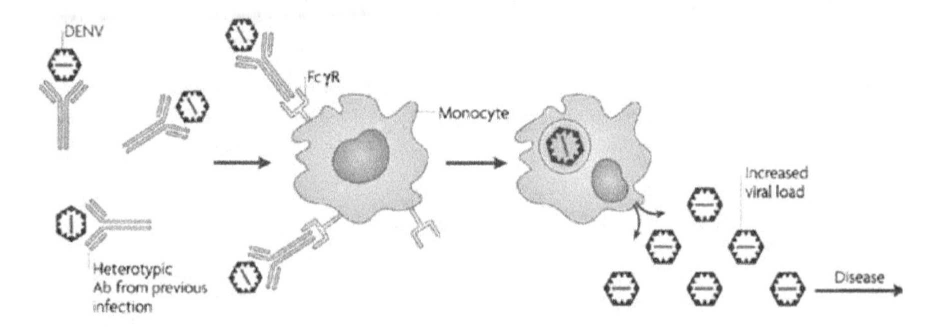

Figure 3. Mechanism of antibody-dependent enhancement (ADE). During a secondary infection caused by a heterologous virus, the pre-existing heterotypic antibodies can cross-react with the other DENV serotypes. The non-neutralizing antibody-virus complex can interact with the Fc-receptor on monocytes or macrophages. This will lead to an increased viral load and a more severe disease. Figure derived from Whitehead et al. [7].

This will lead to an increase in viral load and a more severe disease. These non-neutralizing antibodies can cross-react with all four virus serotypes, as well as with other flaviviruses. This phenomenon explains why young infants born to dengue immune mothers often experience a more severe disease due to transplacental transfer of DENV-specific antibodies [15]. Another approach to assist this phenomenon is the observation of increased viremia in non-human primates which received passive immunization with antibodies against DENV [16].

A second mechanism to explain ADE of flaviviruses is the involvement of the complement system. It has been shown that monoclonal antibodies against complement receptor 3 inhibit ADE of West Nile virus *in vitro* [14]. But Fc-receptor-dependent ADE is believed to be the most common mechanism of ADE.

2. DENV entry

2.1. Entry process

The infectious entry of DENV into its target cells, mainly dendritic cells [17], monocytes and macrophages, is mediated by the viral envelope glycoprotein E via receptor-mediated endocytosis [18]. The E-glycoprotein is the major component (53 kDa) of the virion surface and is arranged as 90 homodimers in mature virions [19]. Recent reports demonstrated also that DENV enters its host cell via clathrin-mediated endocytosis [20,21], as observed with other types of flaviviruses [22,23]. Evidence for flavivirus entry via this pathway is based on the use of inhibitors of clathrin-mediated uptake, such as chlorpromazine. However, DENV entry via a non-classical endocytic pathway independent from clathrin has also been described [24]. It seems that the entry pathway chosen by DENV is highly dependent on the cell type and viral strain. In case of the classical endocytic pathway, there is an uptake of the receptor-bound virus by clathrin coated vesicles. These vesicles fuse with early endosomes to deliver the viral RNA into the cytoplasm. The E-protein responds to the reduced pH of the endosome with a large conformational rearrangement [25,26]. The low pH triggers dissociation of the E-homodimer, which then leads to the insertion of the fusion peptide into the target cell membrane forming a bridge between the virus and the host. Next, a stable trimer of the E-protein is folded into a hairpin-like structure and forces the target membrane to bend towards the viral membrane and eventually fusion takes place [25,27,28]. The fusion results in the release of viral RNA into the cytoplasm for initiation of replication and translation (Figure 4).

2.2. The DENV envelope

The DENV E-glycoprotein induces protective immunity and flavivirus serological classification is based on its antigenic variation. During replication the virion assumes three conformational states: the immature, mature and fusion-activated form. In the immature state, the E-protein is arranged as a heterodimer and generates a "spiky" surface because the premembrane protein (prM) covers the fusion peptide. In the Golgi apparatus, the virion maturates after a rearrangement of the E-protein. The E-heterodimer transforms to an E-homodimer and results in a "smoothy" virion surface. After a furin cleavage of the prM to pr and M, the virion is fully maturated and can be released from the host cell. Upon fusion, the low endosomal pH triggers the rearrangement of the E-homodimer into a trimer [29].

The E-protein monomer is composed out of β-barrels organized in three structural domains (Figure 5).

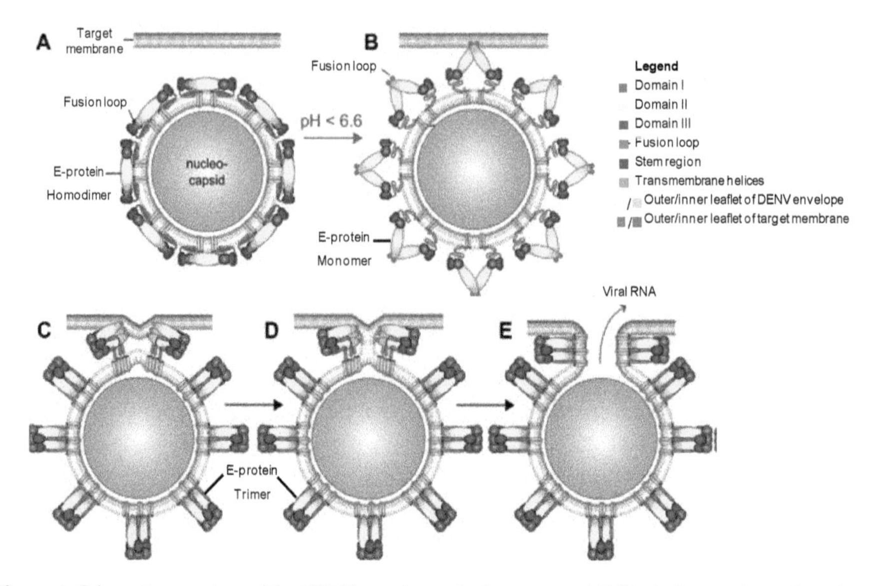

Figure 4. Schematic overview of the DENV membrane fusion process. (A) Pre-fusion conformation of the E-protein consists of homodimers on the virus surface. (B) Low endosomal pH triggers dissociation of the E-dimers into monomers which leads to the insertion of the fusion peptide with the endosomal target membrane. (C) A stable E-protein trimer is folded in a hairpin-like structure. (D) Hemifusion intermediate in which only the outer leaflets of viral and target cellular membranes have fused. (E) Formation of the post-fusion E-protein trimer and opening of the fusion pore allows the release of the viral RNA into the cytoplasm. Modified from Stiasny et al. [26].

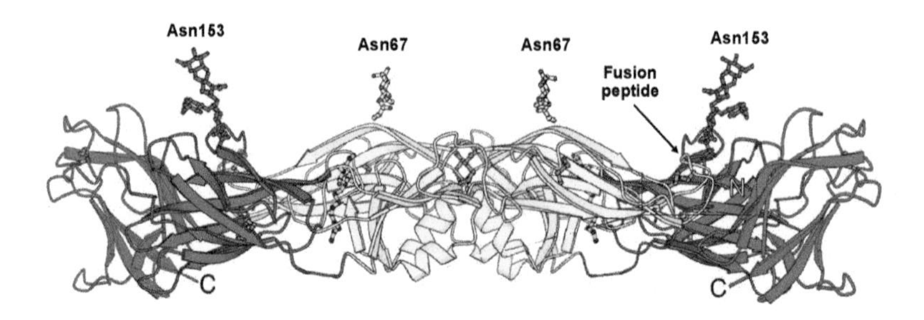

Figure 5. Location of the 2 N-glycans on the envelope protein of DENV. The DENV E-protein dimer carries 2 N-glycans on each monomer at Asn67 and Asn153. β-strands are shown as ribbons with arrows, α-helices are shown as coiled ribbons. Thin tubes represent connecting loops. Domain I is shown in red, domain II is shown in yellow and contains the fusion peptide near Asn153. Domain III is shown in blue. Disulfide bridges are shown in orange. In green, the ligand N-octyl-D-glucoside is shown, which interacts with the hydrophobic pocket between domain I and II. Modified from Modis et al. [30].

The central domain I contains the aminoterminus and contains two disulfide bridges. Domain II is an extended finger-like domain that bears the fusion peptide and stabilizes the dimer. This sequence contains three disulfide bridges and is rich in glycine. Between domain I and domain II is a binding pocket that can interact with a hydrophobic ligand, the detergent β-N-octyl-glucoside. This pocket is an important target for antiviral therapy because mutations in this region can alter virulence and the pH necessary for the induction of conformational changes. The immunoglobulin-like domain III contains the receptor binding motif, the C-terminal domain and one disulfide bond [30,31]. Monoclonal antibodies recognizing domain III are the most efficient of blocking DENV [32,33] and this domain is therefore an interesting target for antiviral therapy.

Because dendritic cell-specific intercellular adhesion molecule 3-grabbing non-integrin (DC-SIGN) (See 1.3.1) is identified as an important receptor for DENV in primary DC in the skin and DC-SIGN recognizes high-mannose sugars, carbohydrates present on the E-protein of DENV could be important for viral attachment. The E-protein has two potential glycosylation sites: asparagines 67 (Asn67) and Asn153. Glycosylation at Asn153 is conserved in flaviviruses, with the exception of Kunjin virus, a subtype of West Nile virus [34] and is located near the fusion peptide in domain II [30,31] (Figure 5). Glycosylation at Asn67 is unique for DENV [31].

3. Role of DC-SIGN in DENV infection

Prior to fusion, DENV needs to attach to specific cellular receptors. Because DENV can infect a variety of different cell types isolated from different hosts (human, insect, monkey and even hamster), the virus must interact with a wide variety of cellular receptors. In the last decade, several candidate attachment factor/receptors are identified. DC-SIGN is described as the most important human cellular receptor for DENV.

Since 1977, monocytes are considered to be permissive for DENV infection [35]. More recent, phenotyping of peripheral blood mononuclear cells (PBMCs) from pediatric DF and DHF cases resulted in the identification of monocytes as DENV target cells [36].

First, it was believed that monocytes are important during secondary DENV infections during the ADE process, because of their Fc-receptor expression. The complex formed between the non-neutralizing antibody and the virus can bind to Fc-receptors and enhance infection in neighboring susceptible cells [14,18,37]. However, *in vitro*, monocytes isolated from PBMCs, apparently have a very low susceptibility for DENV infection for reasons that remain to be elucidated.

More detailed observation of the natural DENV infection, changes the idea of monocytes being the first target cells. Following intradermal injection of DENV-2 in mice, representing the bite of an infected mosquito, DENV occurs to replicate in the skin [38]. The primary DENV target cells in the skin are believed to be immature dendritic cells (DC) or Langerhans cells [17,39-41]. Immature DC are very efficient in capturing pathogens whereas mature DC are relatively resistant to infection. The search for cellular receptors responsible for DENV capture leads to the identification of cell-surface C-type lectin DC-specific intercellular adhesion molecule 3-grabbing non-integrin (DC-SIGN; CD209) [42-45]. DC-SIGN is mainly expressed by immature DC, but also alveolar macrophages and interstitial DC in the lungs, intestine, placenta and in lymph nodes express DC-SIGN [46]. DC-SIGN is a tetrameric transmembrane receptor and is a member of the calcium-dependent C-type lectin family. The receptor is composed of four domains: a cytoplasmic domain responsible for signaling and internalization due to the presence of a dileucine motif, a transmembrane domain, seven to eight extracellular neck repeats implicated in the tetramerization of DC-SIGN and a carbohydrate recognition domain (CRD) (Figure 6) [47].

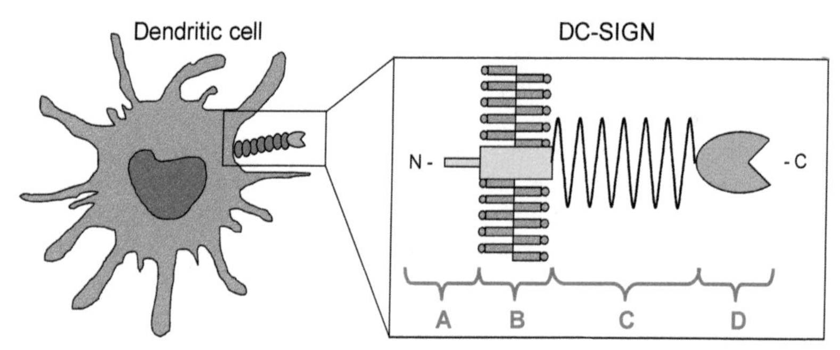

Figure 6. Structure of DC-SIGN. DC-SIGN, mainly expressed by human dendritic cells in the skin, is composed out of four domains. (A) cytoplasmic domain containing internalization signals, (B) transmembrane domain, (C) 7 or 8 extracellular neck repeats implicated in the oligomerization of DC-SIGN and (D) carbohydrate recognition domain which can interact calcium-dependent with a variety of pathogens.

Alen *et al.* [42] investigated the importance of DC-SIGN receptor in DENV infections using DC-SIGN transfected Raji cells versus Raji/0 cells. A strong contrast in DENV susceptibility was observed between Raji/DC-SIGN$^+$ cells and Raji/0 cells. DC-SIGN expression renders cells susceptible for DENV infection. Also in other cell lines, the T-cell line CEM and the astroglioma cell line U87, expression of DC-SIGN renders the cells permissive for DENV infection. To evaluate the importance of DC-SIGN, Raji/DC-SIGN$^+$ cells were incubated with a specific anti-DC-SIGN antibody prior to DENV infection. This resulted in an inhibition of the DENV replication by ~90%, indicating that DC-SIGN is indeed an important receptor for DENV. Also 2 mg/ml of mannan inhibited the DENV infection in Raji/DC-SIGN$^+$ cells by more than 80%. This data indicate that the interaction between DC-SIGN and DENV is dependent on mannose-containing N-glycans present on the DENV envelope [42].

Thus, the CRD of DC-SIGN recognizes high-mannose N-glycans and also fucose-containing blood group antigens [48,49]. Importantly, DC-SIGN can bind a variety of pathogens like human immunodeficiency virus (HIV) [50], hepatitis C virus (HCV) [51], ebola virus [52] and several bacteria, parasites and yeasts [46]. Many of these pathogens have developed strategies to manipulate DC-SIGN signaling to escape from an immune response [46]. Following antigen capture in the periphery, DC maturate by up regulation of the costimulatory molecules and down regulation of DC-SIGN. By the interaction with ICAM-2 on the vascular endothelial cells, DC can migrate to secondary lymphoid organs [53]. Next, the activated DC interact with ICAM-3 on naïve T-cells. This results in the stimulation of the T-cells and subsequently in the production of cytokines and chemokines [54]. Inhibition of the initial interaction between DENV and DC could prevent an immune response. DC-SIGN could be considered as a target for antiviral therapy by interrupting the viral entry process. But caution must be taken into account as the DC-SIGN receptor has also an important role in the activation of protective immune responses instead of promoting the viral dissemination. However, several DC-SIGN antagonists have been developed such as small interfering RNAs (siRNA) silencing DC-SIGN expression [55], specific anti-DC-SIGN antibodies [56] and glycomimetics interacting with DC-SIGN [57]. The *in vivo* effects of DC-SIGN antagonists remain to be elucidated.

Besides DC, macrophages play a key role in the immune pathogenesis of DENV infection as a source of immune modulatory cytokines [58]. Recently, Miller *et al.* showed that the mannose receptor (MR; CD206) mediates DENV infection in macrophages by recognition of the glycoproteins on the viral envelope [59]. Monocyte-derived DC (MDDC) can be generated out of monocytes isolated from fresh donor blood incubated IL-4 and GM-CSF. After a differentiation process MDDC were generated highly expressing DC-SIGN (Figure 7A, B) and showing a significantly decrease in CD14 expression in contrast to monocytes [59,60]. Again, DC-SIGN expression on MDDC renders cells susceptible for DENV in contrast to monocytes (Figure 7A, B). MR is also present on monocyte-derived DC (MDDC) and anti-MR antibodies can inhibit DENV infection, although to a lesser extent than anti-DC-SIGN antibodies do (Figure 7C) [61]. Furthermore, the combination of anti-DC-SIGN and anti-MR antibodies was even more effective in inhibiting DENV infection. Yet, complete inhibition of DENV infection was not achieved, indicating that other entry pathways are potentially involved. Two other receptors on DC reported to be responsible for HIV attachment are syndecan-3 (a member of the heparan sulfate proteoglycan family) [62] and the DC immune receptor [63]. Since DENV interacts with heparan sulfate, syndecan-3 may be a possible (co)-receptor on DC. It has been hypothesized that DENV needs DC-SIGN for attachment and enhancing infection of DC *in cis* and needs MR for internalization [59]. In fact, cells expressing mutant DC-SIGN, lacking the internalization domain, are still susceptible for DENV infection because DC-SIGN can capture the pathogen [43].

Another C-type lectin, CLEC5A (C-type lectin domain family 5, member A) expressed by human macrophages can also interact with DENV and acts as a signaling receptor for the release of proinflammatory cytokines [64]. However, whereas the DC-SIGN-DENV interaction is calcium-dependent, CLEC5A binding to its ligand is not dependent on calcium. Mannan and fucose can inhibit the interaction between CLEC5A and DENV, indicating that the interaction is carbohydrate-dependent [64]. However, a glycan array demonstrated no binding signal between CLEC5A and N-glycans of mammals or insects [65]. The molecular interaction between CLEC5A and DENV remains to be elucidated.

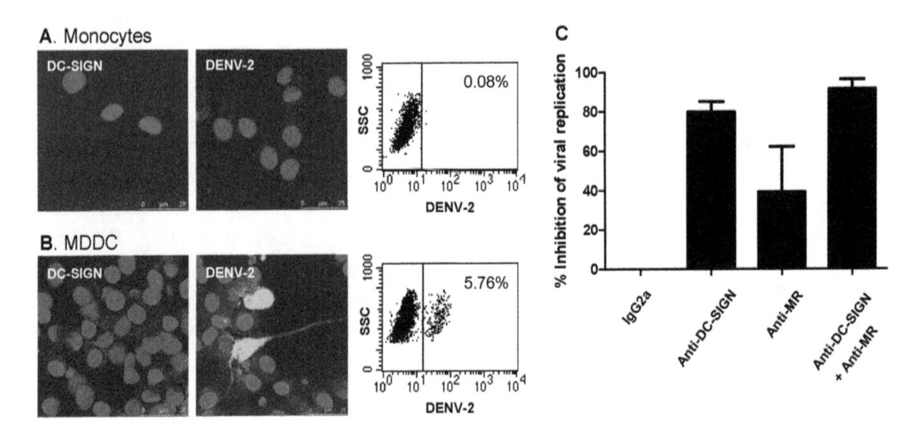

Figure 7. Infection of MDDC by DENV. Monocytes isolated from PBMCs were untreated (A) or treated with 25 ng/ml IL-4 and 50 ng/ml GM-CSF (B) for 5 days prior to DENV-2 infection. Two days after infection the cells were permeabilized and analyzed for DC-SIGN expression and DENV infection by confocal microscopy and flow cytometry. Uninfected cells were stained with a PE-labeled monoclonal DC-SIGN-antibody (red). Infected cells were stained with a mixture of antibodies recognizing DENV-2 E-protein and PrM protein (green). Nuclei were stained with DAPI (blue). Infected monocytes (A) and MDDC (B) were analyzed by flow cytometry to detect DENV-2 positive cells. The values indicated in each dot plot represent the % of DENV-2 positive cells. (C) MDDC were preincubated with 10 μg/ml of isotype control IgG_{2a}, anti-DC-SIGN or anti-MR antibody for 30 minutes before DENV-2 infection. Viral replication was analyzed by flow cytometry. % Inhibition of viral replication ± SEM of 4 different blood donors is shown.

Immune cells, in particular dendritic cells, are the most relevant cells in the discovery of specific antiviral drugs against dengue virus, but the isolation of these cells and the characterization is unfortunately labour intensive and time consuming.

Liver/lymph node-specific ICAM-3 grabbing non-integrin (L-SIGN) is a DC-SIGN related transmembrane C-type lectin expressed on endothelial cells in liver, lymph nodes and placenta [66,67]. Similar to DC-SIGN, L-SIGN is a calcium-dependent carbohydrate-binding protein and can interact with HIV [67], HCV [51], Ebola virus [52], West Nile virus [68] and DENV [45]. Zellweger *et al.* observed that during antibody-dependent enhancement in a mouse model that liver sinusoidal endothelial cells (LSEC) are highly permissive for antibody-dependent DENV infection [69]. Given the fact that LSEC express L-SIGN, it is interesting to focus on the role of L-SIGN in DENV infection. L-SIGN expression on LSEC has probably an important role in ADE *in vivo* and therefore it is interesting to find antiviral agents interrupting the DENV-L-SIGN interaction and subsequently prevent the progression to the more severe and lethal disease DHF/DSS. Although endothelial cells [70] and liver endothelial cells [71] are permissive for DENV and L-SIGN-expression renders cells susceptible for DENV infection, the *in vivo* role for L-SIGN in DENV entry remains to be established.

4. Antiviral therapy

At present, diagnosis of dengue virus infection is largely clinical, treatment is supportive through hydration and disease control is limited by eradication of the mosquito. Many efforts have been made in the search for an effective vaccine, but the lack of a suitable animal model, the need for a high immunogenicity vaccine and a low reactogenicity are posing huge challenges in the dengue vaccine development [7,72]. There are five conditions for a dengue vaccine to be effective: (i) the vaccine needs to be protective against all four serotypes without reactogenicity, because of the risk of ADE, (ii) it has to be safe for children, because severe dengue virus infections often affects young children, (iii) the vaccine has to be economical with minimal or no repeat immunizations, because dengue is endemic in many developing countries, (iv) the induction of a long-lasting protective immune response is necessary and finally (v) the vaccine may not infect mosquitoes by the oral route [7,73].

As there is no vaccine available until now, the search for antiviral products is imperative. The traditional antiviral approach often attacks viral enzymes, such as proteases and polymerases [74,75]. Because human cells lack RNA-dependent polymerase, this enzyme is very attractive as antiviral target without cytotoxicity issues. Nucleoside analogues and non-nucleoside compounds have previously been shown to be very effective in anti-HIV therapy and anti-hepatitis B virus therapy. The protease activity is required for polyprotein processing which is necessary for the assembly of the viral replication complex. Thereby, the protease is an interesting target for antiviral therapy. However, the host cellular system has similar protease activities thus cytotoxic effects form a major recurrent problem. Very recently, many efforts have been made in the development of polymerase and protease inhibitors of DENV, but until today, any antiviral product has reached clinical trials. This chapter is focusing on a different step in the virus replication cycle, namely, the viral entry process. In the past few years, progression has been made in unraveling the host cell pathways upon DENV infection. It is proposed that viral epitopes on the surface of DENV can trigger cellular immune responses and subsequently the development of a severe disease. Therefore, these epitopes are potential targets for the development of a new class of antiviral products, DENV entry inhibitors. Inhibition of virus attachment is a valuable antiviral strategy because it forms the first barrier to block infection. Several fusion inhibitors, glycosidase inhibitors and heparin mimetics have been described to inhibit DENV entry in the host cell. Here, specific molecules, the carbohydrate-binding agents (CBAs), preventing the interaction between the host and the N-glycans present on the DENV envelope are discussed.

4.1. Carbohydrate-binding agents (CBAs)

The CBAs form a large group of natural proteins, peptides and even synthetic agents that can interact with glycosylated proteins. CBAs can be isolated from different organisms: algae, prokaryotes, fungi, plants, invertebrates and vertebrates (such as DC-SIGN and L-SIGN) [76,77]. Each CBA will interact in a specific way with monosaccharides, such as mannose, fucose, glucose, N-acetylglucosamine, galactose, N-acetylgalactosamine or sialic acid residues present in the backbone of N-glycan structures. Because a lot of enveloped viruses are glycosylated at the viral surface, such as HIV, HCV and DENV (Figure 5), CBAs could interact with the glycosylated envelope of the virus and subsequently prevent viral entry into the host cell [78,79]. Previously, antiviral activity against HIV and HCV [78-80] was demonstrated of several CBAs isolated from plants (plant lectins) and algae specifically binding mannose and N-acetylglucosamine residues.

Here, we focus on the antiviral activity of three plant lectins, *Hippeastrum* hybrid agglutinin (HHA), *Galanthus nivalis* agglutinin (GNA) and *Urtica dioica* (UDA) isolated from the amaryllis, the snow drop and the stinging nettle, respectively. In general, plant lectins form a large diverse group of proteins, exhibiting a wide variety of monosaccharide-binding properties which can be isolated from different sites within the plant, such as the bulbs, leaves or roots. HHA (50 kDa) and GNA (50 kDa) isolated from the bulbs are tetrameric proteins. For GNA, each monomer contains two carbohydrate-binding sites and a third site is created once if tetramerization had occurred, resulting in a total of 12 carbohydrate-binding sites (Figure 8). HHA specifically interacts with α1-3 and α1-6 mannose residues and GNA only recognizes α1-3 mannose residues. UDA, isolated from the rhizomes of the nettle, is active as a monomer containing 2 carbohydrate-binding sites recognizing N-acetylglucosamine residues (Figure 8). In 1984, UDA was isolated for the first time and with its molecular weight of 8.7 kDa, UDA is the smallest plant lectin ever reported [81]. The plant lectins have been shown to possess both antifungal and insecticidal activities playing a role in plant defense mechanisms. Here, the antiviral activity of the plant lectins against DENV will be further highlighted and discussed in detail.

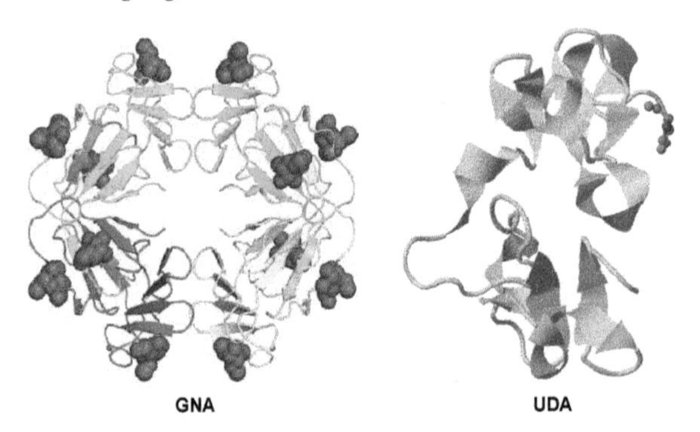

Figure 8. Structure of GNA and UDA. GNA is isolated from the snow drop and is a tetrameric protein. UDA is isolated from the stinging nettle and is a monomeric protein composed out of hevein domains.

Previously, concanavalin A (Con A), isolated from the Jack bean, binding to mannose residues and wheat germ agglutinin binding to N-acetylglucosamine (Glc-NAc) residues, were shown to reduce DENV infection *in vitro*. A competition assay, using mannose, proved that the inhibitory effect of Con A was due to binding α-mannose residues on the viral protein, because mannose successfully competed with Con A [82]. Together with the fact that HHA, GNA and UDA act inhibitory against HIV and HCV, we hypothesized that these plant lectins had antiviral activity against DENV, because DENV has two N-glycosylation sites on the viral envelope and uses DC-SIGN as a cellular receptor to enter DC. The antiviral activity of the three plant lectins was investigated in DC-SIGN+ and L-SIGN+ cells and the infection was analyzed by flow cytometry, RT-qPCR and confocal microscopy.

4.2. Broad antiviral activity of CBAs against DENV

Because DC-SIGN interacts carbohydrate-dependent with DENV, the antiviral activity of the three plant lectins, HHA, GNA and UDA, recognizing monosaccharides present in the backbone of N-glycans on the DENV E-protein, was evaluated. A consistent dose-dependent antiviral activity was observed in DC-SIGN transfected Raji cells against DENV-2 analyzed by flow cytometry (detecting the presence of DENV Ag) and RT-qPCR (detecting viral RNA in the supernatants) [42].

Next, the antiviral potency of the three plant lectins was determined against all four serotypes of DENV, of which DENV-1 and DENV-4 are low-passage clinical virus isolates, in both Raji/DC-SIGN+ cells and in primary immature MDDC. The use of MDDC has much more clinical relevance than using a transfected cell line. MDDC resemble DC in the skin [83] and mimic an *in vivo* DENV infection after a mosquito bite. Moreover, cells of the hematopoietic origin, such as DC, have been shown to play a key role for DENV pathogenesis in a mouse model [84]. A dose-dependent and a DENV serotype-independent antiviral activity of HHA, GNA and UDA in MDDC was demonstrated as analyzed by flow cytometry (Figure 9). These CBAs proved about 100-fold more effective in inhibiting DENV infection in primary MDDC compared to the transfected Raji/DC-SIGN+ cell line.

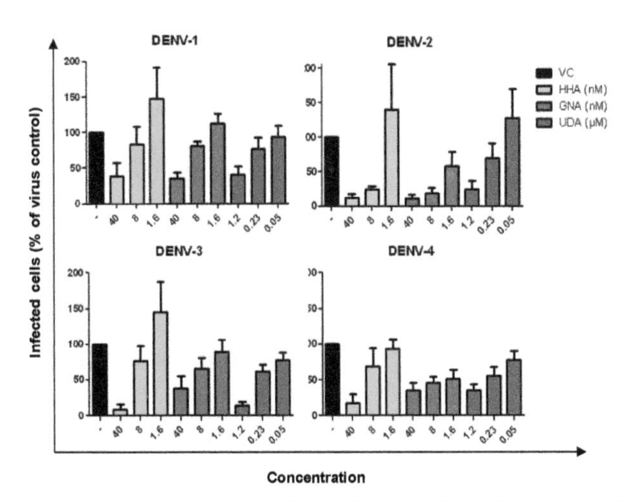

Figure 9. Dose-dependent antiviral activity of HHA, GNA and UDA in DENV-infected MDDC. MDDC were infected with the four serotypes of DENV in the presence or absence of various concentrations of HHA, GNA and UDA. DENV infection was analyzed by flow cytometry using an anti-PrM antibody recognizing all four DENV serotypes (clone 2H2). % of infected cells compared to the positive virus control (VC) ± SEM of 4 to 12 different blood donors is shown. (Adapted from Alen et al. [61]).

When DENV is captured by DC, a maturation and activation process occurs. DC require downregulation of C-type lectin receptors [85], upregulation of costimulatory molecules, chemokine receptors and enhancement of their APC function to migrate to the nodal T-cell areas and to activate the immune system [86]. Cytokines implicated in vascular leakage are produced, the complement system becomes activated and virus-induced antibodies can cause DHF via binding to Fc-receptors. Several research groups demonstrated maturation of DC induced by DENV infection [87,88]. Some groups made segregation in the DC population after DENV infection, the infected DC and the uninfected bystander cells. They found that bystander DC, in contrast to infected DC, upregulate the cell surface expression of costimulatory molecules, HLA and maturation molecules. This activation is induced by TNF-α and IFN-α secreted by DENV-infected DC [40,89,90]. Instead, Alen *et al.* observed an upregulation of the costimulatory molecules CD80 and CD86 and a downregulation of DC-SIGN and MR on the total (uninfected and infected) DC population following DENV infection [61]. This could indicate that the DC are activated and can interact with naive T-cells and subsequently activate the immune system resulting in increased vascular permeability and fever. When the effect of the CBAs was examined on the expression level of the cell surface markers of the total DC population, it was shown that the CBAs are able to inhibit the activation of all DC caused by DENV and can keep the DC in an immature state. Furthermore, DC do not express costimulatory molecules and thus do not interact or significantly activate T cells. An approach to inhibit DENV-induced activation of DC may prevent the immunopathological component of DENV disease.

However, since plant lectins are expensive to produce and not orally bioavailable, the search for non-peptidic small molecules is necessary. PRM-S, a highly soluble non-peptidic small-size carbohydrate-binding antibiotic is a potential new lead compound in HIV therapy, since PRM-S efficiently inhibits HIV replication and prevents capture of HIV to DC-SIGN^{+} cells [91]. PRM-S also inhibited dose-dependently DENV-2 replication in MDDC but had only a weak antiviral activity in Raji/DC-SIGN^{+} cells [61]. Actinohivin (AH), a small prokaryotic peptidic lectin containing 114 amino acids, exhibits also anti-HIV-1 activity by recognizing high-mannose-type glycans on the viral envelope [92]. Although DENV has high mannose-type glycans on the E-protein, there was no antiviral activity of AH against DENV infection. Other CBAs such as microvirin, griffithsin and Banlec have been shown to exhibit potent activity against HIV replication [93-95], but these CBAs did not show antiviral activity against DENV. Previously, it has been shown that the CBAs HHA, GNA and UDA also target the N-glycans of other viruses, such as HIV, HCV [79] and HCMV [80]. This indicates that the CBAs can be used as broad-spectrum antiviral agents against various classes of glycosylated enveloped viruses. Although, the three plant lectins did not act inhibitory against parainfluenza-3, vesicular stomatitis virus, respiratory syncytial virus or herpes simplex virus [79]. Together, these data indicate a unique carbohydrate-specificity, and thus also a specific profile of antiviral activity of the CBAs.

4.3. Molecular target of the CBAs on DENV

It was demonstrated in time of drug addition assays that the mannose binding lectin HHA prevents DENV-2 binding to the host cell and acts less efficiently when the virus had already attached to the host cell. It was shown that HHA interacts with DENV and not with cellular membrane proteins such as DC-SIGN on the target cell. The potency of HHA to inhibit attachment of DENV to Raji/DC-SIGN$^+$ cells is comparable to its inhibitory activity of the capture of HIV and HCV to Raji/DC-SIGN$^+$ cells [79]. CBAs could thus be considered as unique prophylactic agents of DENV infection.

To identify the molecular target of the CBAs on DENV, a resistant DENV to HHA was generated in the mosquito cell line C6/36 by Alen *et al.* (HHA^{res} DENV). Compared with the WT DENV, two highly prevalent mutations were found, namely N67D and T155I, present in 80% of all clones sequenced. Similar mutational patterns destroying both glycosylation motifs (T69I or T69A each in combination with T155I) were present in another 10% of the clones analyzed. The N-glycosylation motif $_{153}$N-D-T$_{155}$ is conserved among the majority of all flaviviruses, while a second N-glycosylation motif $_{67}$N-T-T$_{69}$ is unique among DENV [96]. In the HHA^{res} virus both N-glycosylation motifs were mutated either directly at the actual N-glycan accepting a residue of the first site (Asn67) or at the C-proximal Thr155 being an essential part of the second N-glycosylation site [97], thus both N-glycosylation sites on the viral envelope protein can be considered to be deleted. This indicates that HHA directly targets the N-glycans on the viral E-protein. In fact, all clones sequenced showed the deletion of the N-glycan at Asn153. However, 10% of the clones sequenced had no mutation at the glycosylation motif $_{67}$N-T-T$_{69}$, indicating that this glycosylation motif [96] has a higher genetic barrier compared to $_{153}$N-D-T$_{155}$. Though there are multiple escape pathways to become resistant to HHA, it seems not to be possible to fully escape the selective pressure of favoring a deglycosylation of the viral E-protein. In addition, there were no mutations found either apart from the N-glycosylation sites of the E-protein or in any of the five WT DENV-2 clones passaged in parallel. This is not fully unexpected as flaviviruses replicate with reasonable fidelity and DENV does not necessarily exist as a highly diverse quasispecies neither *in vitro* nor *in vivo* [98,99].

There are some contradictions in terms of necessity of glycosylation of Asn67 and Asn153 during DENV viral progeny. Johnson *et al.* postulated that DENV-1 and DENV-3 have both sites glycosylated and that DENV-2 and DENV-4 have only one N-glycan at Asn-67 [100]. In contrast, a study comparing the number of glycans in multiple isolates of DENV belonging to all four serotypes led to the consensus that all DENV strains have two N-glycans on the E-protein [101]. However, mutant DENV lacking the glycosylation at Asn153 can replicate in mammalian and insect cells, indicating that this glycosylation is not essential for viral replication [96,102]. There is a change in phenotype because ablation of glycosylation at Asn153 in DENV is associated with the induction of smaller plaques in comparison to the wild type virus [96]. Asn153 is proximal to the fusion peptide and therefore deglycosylation at Asn153 showed also an altered pH-dependent fusion activity and displays a lower stability [103,104]. In contrast, Alen *et al.* showed that the mutant virus, HHA^{res} lacking both

N-glycosylation sites, had a similar plaque phenotype in BHK cells (manuscript submitted). It has been shown that DENV lacking the glycosylation at Asn67 resulted in a replication-defective phenotype, because this virus infects mammalian cells weakly and there is a reduced secretion of DENV E-protein. Replication in mosquito cells was not affected, because the mosquito cells restore the N-glycosylation at Asn67 with a compensatory site-mutation (K64N) generating a new glycosylation site [96,105]. These data are in contrast with other published results, where was demonstrated that DENV lacking the Asn67-linked glycosylation can grow efficiently in mammalian cells, depending on the viral strain and the amino acid substitution abolishing the glycosylation process [102]. A compensatory mutation was detected (N124S) to repair the growth defect without creating a new glycosylation site. Thus, the glycan at Asn67 is not necessary for virus growth, but a critical role for this glycan in virion release from mosquito cells was demonstrated [102]. However, HHA resistant virus was found to replicate efficiently in mosquito and insect cells indicating an efficient carbohydrate-independent viral replication in these cell lines. A possible explanation for the differences between our data and data from previous studies could be that the mutant virus has been generated in mosquito C6/36 cells (during replication under antiviral drug pressure) and not in mammalian cells (after introducing the mutations by site-directed mutagenesis). In addition, in previous studies, other amino acid substitutions were generated, resulting in different virus genotypes and subsequently resulting in poorly to predict virus phenotypes.

The glycosylation at Asn67 is demonstrated to be essential for infection of monocyte-derived dendritic cells (MDDC), indicating an interaction between DC-SIGN and the glycan at Asn67 [96,106]. Also the HHA^{res} DENV was not able to infect efficiently $DC\text{-}SIGN^{+}$ cells or cells that express the DC-SIGN-related liver-specific receptor L-SIGN. Interestingly, MDDC are also not susceptible for HHA^{res} DENV infection, indicating the importance of the DC-SIGN-mediated DENV infection in MDDC. Moreover, cells of the hematopoietic origin, such as DC, are described to be necessary for DENV pathogenesis [84]. If the CBA resistant DENV in not able to infect DC anymore, it can be stated that the CBAs interfere with a physiologically highly relevant target. DC-SIGN is postulated as the most important DENV entry receptor until now. The entry process of DENV in Vero, Huh-7, BHK-21 and C6/36 cell lines is DC-SIGN-independent and also carbohydrate-independent. Indeed, HHA^{res} DENV can efficiently enter and replicate in these cell lines. HHA^{res} DENV lacking both N-glycans on the envelope E-glycoprotein is able to replicate efficiently in mammalian cells, with the exception of $DC\text{-}SIGN^{+}$ cells.

The HHA^{res} virus was used as a tool to identify the antiviral target of other classes of compounds as it could replicate in human liver Huh-7 cells. The use of Huh-7 cells has much more clinical relevance than using monkey (Vero) or hamster (BHK) kidney cells. The HHA^{res} DENV was found cross-resistant to GNA, that recognizes like HHA, α-1,3 mannose residues. UDA, which recognizes mainly the N-acetylglucosamine residues of the N-glycans, also lacked antiviral activity against HHA^{res} DENV in Huh-7 cells. This indicates that the entire backbone of the N-glycan is deleted. Likewise, pradimicin-S (PRM-S), a small-size α-1,2-mannose-specific CBA, was also unable to inhibit HHA^{res} DENV. This demonstrates that PRM-S targets also the N-glycans on the DENV envelope. In contrast, ribavirin (RBV), a nucleoside analogue and inhibitor of cellular purine synthesis [74], retained as expected wild-type antiviral activity against HHA^{res} DENV. This argues against that there would be compensatory mutations in the non-structural proteins of DENV which are responsible for an overall enhanced replication of the viral genome [107,108]. SA-17, a novel doxorubicine analogue that inhibits the DENV entry process [109], was equipotent against WT and HHA^{res} DENV. The SA-17 compound is predicted to interact with the hydrophobic binding pocket of the E-glycoprotein which is independent from the N-glycosylation state of the E-glycoprotein [109]. These data confirm the molecular docking experiments of SA-17.

Generally, the function of glycosylation on surface proteins is proper folding of the protein, trafficking in the endoplasmic reticulum, interaction with receptors and influencing virus immunogenicity [110]. Virions produced in the mosquito vector and human host may have structurally different N-linked glycans, because the glycosylation patterns are fundamentally different [101,111]. N-glycosylation in mammalian cells is often of the complex-type because a lot of different processing enzymes could add a diversity of monosaccharides. Glycans produced in insect cells are far less complex, because of less diversity in processing enzymes and usually contain more high-mannose and pauci-mannose-type glycans. DC-SIGN can distinguish between mosquito- and mammalian cell-derived alphaviruses [112] and West Nile virus [68], resulting in a more efficient infection by a mosquito-derived virus, but this was not the case for DENV [101].

Although the CBAs HHA and GNA are not mitogenic and not toxic to mice when administered intravenously [113], caution must be taken in the development of the CBAs to use as antiviral drug in the clinic. First, the natural plant lectins are expensive to produce and hard to scale-up, but efforts have been made to express CBAs in commensal bacteria which provide an easy production process of this class of agents. Second, there can be a systemic reaction against the lectins such as in food allergies against peanut lectin or banana lectin [114,115]. Third, the CBAs can recognize aspecifically cellular glycans and could interfere with host cellular processes. But, DENV glycosylation is of the high-mannose or pauci-mannose type, which is only rare on mammalian proteins. The synthetic production of small non-peptidic molecules, such as PRM-S, with CBA-like activity, could overcome the pharmacological problems of the plant lectins. Therefore, PRM-S forms a potential lead candidate in the development of more potent and specific DENV entry inhibitors.

5. Conclusion

In conclusion, besides active vector control in tropical and subtropical regions, there is an urgent need for antiviral treatment to protect half of the world's population against severe DENV infections. DC-SIGN is thought to be the most important DENV receptor and that the DC-SIGN-DENV envelope protein interaction is an excellent target for viral entry inhibitors such as the CBAs. Resistance against HHA forces the virus to delete its N-glycans and subsequently this mutant virus is not able anymore to infect its most important target cells.

Thus the CBAs act in two different ways: prevention of viral entry by directly binding N-glycans on the viral envelope and indirectly forcing the virus to delete its N-glycans and loose the capability to infect DC. The plant lectins provided more insight into the entry pathway of the virus into the host cell. Hopefully some of these future derivatives with a comparable mode of action will reach clinical trials in the near future.

Author details

Dominique Schols* and Marijke Alen
Department of Microbiology and Immunology, Rega Institute for Medical Research, University of Leuven, Leuven, Belgium

Acknowledgement

We are grateful to S. Claes, B. Provinciael, E. Van Kerckhove and E. Fonteyn for excellent technical assistance. This work was supported by the FWO (G-485-08N and G.0528.12N), the KU Leuven (PF/10/018 and GOA/10/014) and the Dormeur Investment Service Ltd.

6. References

[1] Holmes EC, Twiddy SS (2003) The origin, emergence and evolutionary genetics of dengue virus. Infect Genet Evol 3: 19-28.
[2] Mackenzie JS, Gubler DJ, Petersen LR (2004) Emerging flaviviruses: the spread and resurgence of Japanese encephalitis, West Nile and dengue viruses. Nat Med 10: S98-S109.
[3] Halstead SB (2008) Dengue virus - Mosquito interactions. Annu Rev Entomol 53: 273-291.
[4] Gubler DJ (1998) Dengue and dengue hemorrhagic fever. Clin Microbiol Rev 11: 480-496.
[5] Halstead SB (2007) Dengue. Lancet 370: 1644-1652.
[6] Monath TP (1994) Dengue - the risk to developed and developing countries. Proc Natl Acad Sci U S A 91: 2395-2400.
[7] Whitehead SS, Blaney JE, Durbin AP, Murphy BR (2007) Prospects for a dengue virus vaccine. Nat Rev Microbiol 5: 518-528.
[8] Walker T, Johnson PH, Moreira LA, Iturbe-Ormaetxe I, Frentiu FD, McMeniman CJ et al. (2011) The wMel Wolbachia strain blocks dengue and invades caged Aedes aegypti populations. Nature 476: 450-453.
[9] Hoffmann AA, Montgomery BL, Popovici J, Iturbe-Ormaetxe I, Johnson PH, Muzzi F et al. (2011) Successful establishment of Wolbachia in Aedes populations to suppress dengue transmission. Nature 476: 454-457.

* Corresponding Author

[10] Mairuhu ATA, Wagenaar J, Brandjes DPM, van Gorp ECM (2004) Dengue: an arthropod-borne disease of global importance. Eur J Clin Microbiol Infect Dis 23: 425-433.

[11] Rigau-Pérez JG, Clark GG, Gubler DJ, Reiter P, Sanders RJ, Vorndam AV (1998) Dengue and dengue haemorrhagic fever. Lancet 352: 971-977.

[12] Solomon T, Mallewa M (2001) Dengue and other emerging flaviviruses. J Infect 42: 104-115.

[13] Halstead SB (1988) Pathogenesis of dengue: challenges to molecular biology. Science 239: 476-481.

[14] Takada A, Kawaoka Y (2003) Antibody-dependent enhancement of viral infection: molecular mechanisms and in vivo implications. Rev Med Virol 13: 387-398.

[15] Kliks SC, Nimmanitya S, Nisalak A, Burke DS (1988) Evidence that maternal dengue antibodies are important in the development of dengue hemorrhagic fever in infants. Am J Trop Med Hyg 38: 411-419.

[16] Halstead SB (1979) In vivo enhancement of dengue virus infection in rhesus monkeys by passively transferred antibody. J Infect Dis 140: 527-533.

[17] Wu SJL, Grouard-Vogel G, Sun W, Mascola JR, Brachtel E, Putvatana R et al. (2000) Human skin Langerhans cells are targets of dengue virus infection. Nat Med 6: 816-820.

[18] Kou Z, Quinn M, Chen HY, Rodrigo WW, Rose RC, Schlesinger JJ et al. (2008) Monocytes, but not T or B cells, are the principal target cells for dengue virus (DV) infection among human peripheral blood mononuclear cells. J Med Virol 80: 134-146.

[19] Kuhn RJ, Zhang W, Rossmann MG, Pletnev SV, Corver J, Lenches E et al. (2002) Structure of dengue virus: Implications for flavivirus organization, maturation, and fusion. Cell 108: 717-725.

[20] Acosta EG, Castilla V, Damonte EB (2008) Functional entry of dengue virus into Aedes albopictus mosquito cells is dependent on clathrin-mediated endocytosis. J Gen Virol 89: 474-484.

[21] van der Schaar HM, Rust MJ, Chen C, van der Ende-Metselaar H, Wilschut J, Zhuang XW et al. (2008) Dissecting the cell entry pathway of dengue virus by single-particle tracking in living cells. Plos Pathog 4: e1000244.

[22] Chu JJH, Ng ML (2004) Infectious entry of West Nile virus occurs through a clathrin-mediated endocytic pathway. J Virol 78: 10543-10555.

[23] Nawa M, Takasaki T, Yamada K, Kurane I, Akatsuka T (2003) Interference in Japanese encephalitis virus infection of vero cells by a cationic amphiphilic drug, chlorpromazine. J Gen Virol 84: 1737-1741.

[24] Acosta EG, Castilla V, Damonte EB (2009) Alternative infectious entry pathways for dengue virus serotypes into mammalian cells. Cell Microbiol 11: 1533-1549.

[25] Modis Y, Ogata S, Clements D, Harrison SC (2004) Structure of the dengue virus envelope protein after membrane fusion. Nature 427: 313-319.

[26] Stiasny K, Fritz R, Pangerl K, Heinz FX (2009) Molecular mechanisms of flavivirus membrane fusion. Amino Acids 41: 1159-1163.

[27] Martin CSS, Liu CY, Kielian M (2009) Dealing with low pH: entry and exit of alphaviruses and flaviviruses. Trends Microbiol 17: 514-521.

[28] Zhang Y, Zhang W, Ogata S, Clements D, Strauss JH, Baker TS et al. (2004) Conformational changes of the flavivirus E glycoprotein. Structure 12: 1607-1618.

[29] Perera R, Khaliq M, Kuhn RJ (2008) Closing the door on flaviviruses: Entry as a target for antiviral drug design. Antiviral Res 80: 11-22.

[30] Modis Y, Ogata S, Clements D, Harrison SC (2003) A ligand-binding pocket in the dengue virus envelope glycoprotein. Proc Natl Acad Sci U S A 100: 6986-6991.

[31] Rey FA, Heinz FX, Mandl C, Kunz C, Harrison SC (1995) The envelope glycoprotein from tick-borne encephalitis virus at 2 angstrom resolution. Nature 375: 291-298.

[32] Crill WD, Roehrig JT (2001) Monoclonal antibodies that bind to domain III of dengue virus E glycoprotein are the most efficient blockers of virus adsorption to Vero cells. J Virol 75: 7769-7773.

[33] Rajamanonmani R, Nkenfou C, Clancy P, Yau YH, Shochat SG, Sukupolvi-Petty S et al. (2009) On a mouse monoclonal antibody that neutralizes all four dengue virus serotypes. J Gen Virol 90: 799-809.

[34] Scherret JH, Mackenzie JS, Khromykh AA, Hall RA (2001) Biological significance of glycosylation of the envelope protein of Kunjin virus. Ann N Y Acad Sci 951: 361-363.

[35] Halstead SB, O'rourke EJ, Allison AC (1977) Dengue viruses and mononuclear phagocytes .II. Identity of blood and tissue leukocytes supporting in vitro infection. J Exp Med 146: 218-229.

[36] Durbin AP, Vargas MJ, Wanionek K, Hammond SN, Gordon A, Rocha C et al. (2008) Phenotyping of peripheral blood mononuclear cells during acute dengue illness demonstrates infection and increased activation of monocytes in severe cases compared to classic dengue fever. Virology 376: 429-435.

[37] Kliks SC, Nisalak A, Brandt WE, Wahl L, Burke DS (1989) Antibody-dependent enhancement of dengue virus growth in human-monocytes as a risk factor for dengue hemorrhagic-fever. Am J Trop Med Hyg 40: 444-451.

[38] Taweechaisupapong S, Sriurairatana S, Angsubhakorn S, Yoksan S, Khin MM, Sahaphong S et al. (1996) Langerhans cell density and serological changes following intradermal immunisation of mice with dengue 2 virus. J Med Microbiol 45: 138-145.

[39] Ho LJ, Wang JJ, Shaio MF, Kao CL, Chang DM, Han SW et al. (2001) Infection of human dendritic cells by Dengue virus causes cell maturation and cytokine production. J Immunol 166: 1499-1506.

[40] Libraty DH, Pichyangkul S, Ajariyakhajorn C, Endy TP, Ennis FA (2001) Human dendritic cells are activated by dengue virus infection: Enhancement by gamma interferon and implications for disease pathogenesis. J Virol 75: 3501-3508.

[41] Marovich M, Grouard-Vogel G, Louder M, Eller M, Sun W, Wu SJ et al. (2001) Human dendritic cells as targets of dengue virus infection. J Invest Dermatol 6: 219-224.

[42] Alen MMF, Kaptein SJF, De Burghgraeve T, Balzarini J, Neyts J, Schols D (2009) Antiviral activity of carbohydrate-binding agents and the role of DC-SIGN in dengue virus infection. Virology 387: 67-75.

[43] Lozach PY, Burleigh L, Staropoli I, Navarro-Sanchez E, Harriague J, Virelizier JL et al. (2005) Dendritic cell-specific intercellular adhesion molecule 3-grabbing non-integrin (DC-SIGN)-mediated enhancement of dengue virus infection is independent of DC-SIGN internalization signals. J Biol Chem 280: 23698-23708.

[44] Navarro-Sanchez E, Altmeyer R, Amara A, Schwartz O, Fieschi F, Virelizier JL et al. (2003) Dendritic-cell-specific ICAM3-grabbing non-integrin is essential for the productive infection of human dendritic cells by mosquito-cell-derived dengue viruses. Embo Rep 4: 723-728.

[45] Tassaneetrithep B, Burgess TH, Granelli-Piperno A, Trumpfherer C, Finke J, Sun W et al. (2003) DC-SIGN (CD209) mediates dengue virus infection of human dendritic cells. J Exp Med 197: 823-829.

[46] van Kooyk Y, Geijtenbeek TBH (2003) DC-SIGN: Escape mechanism for pathogens. Nat Rev Immunol 3: 697-709.

[47] Mitchell DA, Fadden AJ, Drickamer K (2001) A novel mechanism of carbohydrate recognition by the C-type lectins DC-SIGN and DC-SIGNR - Subunit organization and binding to multivalent ligands. J Biol Chem 276: 28939-28945.

[48] Appelmelk BJ, van Die I, van Vliet SJ, Vandenbroucke-Grauls CMJE, Geijtenbeek TBH, van Kooyk Y (2003) Cutting edge: Carbohydrate profiling identifies new pathogens that interact with dendritic cell-specific ICAM-3-grabbing nonintegrin on dendritic cells. J Immunol 170: 1635-1639.

[49] Feinberg H, Mitchell DA, Drickamer K, Weis WI (2001) Structural basis for selective recognition of oligosaccharides by DC-SIGN and DC-SIGNR. Science 294: 2163-2166.

[50] Geijtenbeek TBH, Kwon DS, Torensma R, van Vliet SJ, van Duijnhoven GCF, Middel J et al. (2000) DC-SIGN, a dendritic cell-specific HIV-1-binding protein that enhances trans-infection of T cells. Cell 100: 587-597.

[51] Pöhlmann S, Zhang J, Baribaud F, Chen ZW, Leslie G, Lin G et al. (2003) Hepatitis C virus glycoproteins interact with DC-SIGN and DC-SIGNR. J Virol 77: 4070-4080.

[52] Marzi A, Möller P, Hanna SL, Harrer T, Eisemann J, Steinkasserer A et al. (2007) Analysis of the interaction of Ebola virus glycoprotein with DC-SIGN (dendritic cell-specific intercellular adhesion molecule 3-grabbing nonintegrin) and its homologue DC-SIGNR. J Infect Dis 196 (Suppl 2): S237-S246.

[53] Geijtenbeek TBH, Krooshoop DJEB, Bleijs DA, van Vliet SJ, van Duijnhoven GCF, Grabovsky V et al. (2000) DC-SIGN-ICAM-2 interaction mediates dendritic cell trafficking. Nat Immunol 1: 353-357.

[54] Banchereau J, Steinman RM (1998) Dendritic cells and the control of immunity. Nature 392: 245-252.

[55] Nair MPN, Reynolds JL, Mahajan SD, Schwartz SA, Aalinkeel R, Bindukumar B et al. (2005) RNAi-directed inhibition of DC-SIGN by dendritic cells: Prospects for HIV-1 therapy. AAPS J 7: E572-E578.

[56] Dakappagari N, Maruyama T, Renshaw M, Tacken P, Figdor C, Torensma R et al. (2006) Internalizing antibodies to the C-type lectins, L-SIGN and DC-SIGN, inhibit viral glycoprotein binding and deliver antigen to human dendritic cells for the induction of T cell responses. J Immunol 176: 426-440.

[57] Anderluh M, Jug G, Svajger U, Obermajer N (2012) DC-SIGN antagonists, a potential new class of anti-infectives. Curr Med Chem 19: 992-1007.

[58] Chen YC, Wang SY (2002) Activation of terminally differentiated human monocytes/macrophages by dengue virus: Productive infection, hierarchical production of innate cytokines and chemokines, and the synergistic effect of lipopolysaccharide. J Virol 76: 9877-9887.

[59] Miller JL, Dewet BJM, Martinez-Pomares L, Radcliffe CM, Dwek RA, Rudd PM et al. (2008) The mannose receptor mediates dengue virus infection of macrophages. Plos Pathog 4: e17.

[60] Kwan WH, Helt AM, Maranon C, Barbaroux JB, Hosmalin A, Harris E et al. (2005) Dendritic cell precursors are permissive to dengue virus and human immunodeficiency virus infection. J Virol 79: 7291-7299.

[61] Alen MMF, De Burghgraeve T, Kaptein SJF, Balzarini J, Neyts J, Schols D (2011) Broad antiviral activity of carbohydrate-binding agents against the four serotypes of dengue virus in monocyte-derived dendritic cells. Plos One 6: e21658.

[62] de Witte L, Bobardt M, Chatterji U, Degeest G, David G, Geijtenbeek TB et al. (2007) Syndecan-3 is a dendritic cell-specific attachment receptor for HIV-1. Proc Natl Acad Sci U S A 104: 19464-19469.

[63] Lambert AA, Gilbert C, Richard M, Beaulieu AD, Tremblay MJ (2008) The C-type lectin surface receptor DICIR acts as a new attachment factor for HIV-1 in dendritic cells and contributes to trans- and cis-infection pathways. Blood 112: 1299-1307.

[64] Chen ST, Lin YL, Huang MT, Wu MF, Cheng SC, Lei HY et al. (2008) CLEC5A is critical for dengue-virus-induced lethal disease. Nature 453: 672-676.

[65] Watson AA, Lebedev AA, Hall BA, Fenton-May AE, Vagin AA, Dejnirattisai W et al. (2011) Structural flexibility of the macrophage dengue virus receptor CLEC5A implications for ligand binding and signaling. J Biol Chem 286: 24208-24218.

[66] Bashirova AA, Geijtenbeek TBH, van Duijnhoven GCF, van Vliet SJ, Eilering JBG, Martin MP et al. (2001) A dendritic cell-specific intercellular adhesion molecule 3-grabbing nonintegrin (DC-SIGN)-related protein is highly expressed on human liver sinusoidal endothelial cells and promotes HIV-1 infection. J Exp Med 193: 671-678.

[67] Pöhlmann S, Soilleux EJ, Baribaud F, Leslie GJ, Morris LS, Trowsdale J et al. (2001) DC-SIGNR, a DC-SIGN homologue expressed in endothelial cells, binds to human and simian immunodeficiency viruses and activates infection in trans. Proc Natl Acad Sci U S A 98: 2670-2675.

Chapter 11

Digestion in Ruminants

Barbara Niwińska

Additional information is available at the end of the chapter

1. Introduction

Ruminants, cloven-hoofed mammals of the order *Artiodactyla*, obtain their food by browsing or grazing, subsisting on plant material (Hungate, 1966). Today, 193 species of living ruminants exist in 6 families: *Antilocapridae, Bovidae, Cervidae, Giraffidae, Moschidae* and *Tragulidae* (Nowak, 1999). The number of wild ruminants is about 75 million and of domesticated about 3.6 billion (Hackmann and Spain, 2010). Approximately 95% of the population of domesticated ruminants constitute species: cattle, sheep and goats, all of them belong to the *Bovidae* family. Cattle and sheep are the two most numerous species and cattle is of the most economic importance. The economic value of milk and beef production in the EU is almost 125 billion Euro per year and accounts for 40% of total agricultural production (FAIP, 2003). The dairy cows is unique among all other mammalian species because of the intense artificial transgenerational genetic selection for milk production during the last 50 yr, so that annual averages of more than 12,500 kg/cow of milk per lactation are not uncommon (Eastridge, 2006). The selection has increased their peak energy yield by about 250% (20 Mcal × d^{-1} observed vs. 7.76 Mcal × d^{-1} expected) (Hackmann and Spain, 2010). Genetic improvement is accompanied by increasing metabolic demands for energy. The efficient use of energy of the feed resources is the main reason for the numerous and multilateral studies on carbohydrates digestion processes in cattle.

2. Digestive tract

Ruminants digestive system is characterized by functional and anatomical adaptations that allowed them to unlock otherwise unavailable food energy in fibrous plant material, mainly in cellulose and others recalcitrant carbohydrates (Van Soest, 1994). This property gives them an advantage over nonruminants. An important characteristic of ruminants digestive system is the occurrence of the microbial fermentation prior to the gastric and intestinal digestion activity. Their unique digestive system integrates a large microbial population with the animal's own system in the symbiotic relationship. The microbial fermentation occurs mainly in the rumen, the first chamber of the four-compartment stomach, which consists also of the reticulum and omasum (act as filters), and the abomasum (the true enzymatic stomach).

3. Rumen function

The feedstuffs consumed by ruminants are all initially exposed to the fermentative activity in the rumen, the place of more or less complete microbial fermentation of dietary components. Ruminal fermentation initially results in the degradation of carbohydrates and protein to short-term intermediates such as sugars and amino acids. The products of this initial degradation are readily metabolized to microbial mass and carbon dioxide, methane, ammonia and volatile fatty acids (VFA): primarily acetate, propionate and butyrate and to a lesser degree branched chain VFA and occasionally lactate. The rate and extent of fermentation are important parameters that determine protein, vitamins, and short-chain organic acids supply to the animal (Koenig et al., 2003; Hall, 2003). The host ruminant animal absorbs VFA (mostly through the rumen wall) and digests proteins, lipids, and carbohydrate constituents of microbes and feed residues entering the small intestine to supply its maintenance needs and for the production of meat and milk. Ruminant animals derive about 70% of their metabolic energy from microbial fermentation of feed particles and microbial protein accounts for as much as 90% of the amino acids reaching the small intestine (Nocek and Russell, 1988; Bergman, 1990).

The rumen has a complex environment composed of microbes, feed at various stages of digestion, gases, and rumen fluid. Rumen microorganisms usually adhere to feed particles and form biofilms to degrades plant material. The efficiency of ruminants to utilize of feeds is due to highly diversified rumen microbial ecosystem consisting of bacteria (10^{10}–10^{11} cells/ml, more than 50 genera), ciliate protozoa (10^4–10^6/ml, 25 genera), anaerobic fungi (10^3–10^5 zoospores/ml, 5 genera) and bacteriophages (10^8–10^9/ml) (Hobson, 1989). The synergism and antagonism among the different groups of microbes is so diverse and complicated that it is difficult to quantify the role played by any particular group of microbes present in the rumen (Kamra, 2005). Bacterial numbers in the rumen are the highest and bacteria play a dominant role in all facets of ruminal fermentation. They are adopted to live at acidities between pH 5.5 and 7.0, in the absence of oxygen, at the temperature of 39-40°C, in the presence of moderate concentration of fermentation products, and at the expense of the ingesta provided by ruminant (Hungate, 1966). Rumen digesta volume accounts for 8-14% of body weight of cows and is characterized by dry matter content about 15% (Dado and Allen, 1995; Reynolds et al., 2004; Kamra, 2005).

4. Techniques for estimating rumen digestibility

The rumen digestibility of feeds can be estimated by biological methods. The "basic model" which gives the value utilized for defining the nutritive value of a feed is the *in vivo* digestibility, which represents the entire process occurring in the gastro-intestinal tract. *In vitro* methods which simulate the digestion process, have being less expensive and less time-consuming, and they allow to maintain experimental conditions more precisely than do *in vivo* trials. Three major *in vitro* digestion techniques currently available to determine the nutritive value of ruminant feeds are: digestion with rumen microorganisms (Tilley and Terry, 1963; Menke et al., 1979), digestion with enzymes (De Boever et al., 1986), and *in situ* the nylon bag technique (Mehrez and Ørskov, 1977). The nylon bag technique *(in sacco)* has been used for many years to provide estimates of both the rate and the extent of disappearance of feed constituents. Those characteristics are measured by placing feedstuffs in fabric bag and then incubating the bag by certain time intervals in the rumen of animal. However, the single technique does not provide accurate estimation of *in vivo* digestion. Judkins et al. (1990) compared 11 techniques for estimating diet dry matter digestibility across six different diets in experiment with rams. Authors found, that the rumen digestibility of feeds nutrients was influenced by diets composition, feeding conditions and physiological status of animals. It therefore seems appropriate that the developments and use of various modification of mentioned experimental techniques have enabled much progress in rumen studies.

5. Carbohydrates classification in ruminants feeds

Carbohydrates constitute the highest proportion of diets and are important for meeting the energy needs of animals and of rumen microbes, and are important for maintaining the health of the gastrointestinal tract. Typically, carbohydrates make up 70 to 80% of the diets fed to dairy cattle and are composed of mixture of numerous monomers and polymers (Nocek and Russell, 1988). The carbohydrates fraction of feeds are defined according the chemical or enzymatic methods used for their analysis and availability to the ruminants. Broadly, carbohydrates are classified as nonstructural that are found inside the cells of plants or structural that are found in plant cell walls, but these fractions are not chemically uniform (Van Soest et al., 1991).

Fraction of nonstructural carbohydrates (NSC) includes organic acid, mono- di- and oligosaccharides, starches, and other reserve carbohydrates. Total NSC includes pectin is referred as nonfibrous carbohydrates (NFC), calculated as 100−(CP+ether extract+ash+NDF) (Mertens, 1992). NFC are the highly digestible and are the major source of energy for high producing cattle. Fraction of structural carbohydrates is characterized by neutral detergent fiber (NDF) and acid detergent fiber (ADF) contents. NDF includes the crosslinked matrix of the plant cell wall with cellulose, hemicellulose, and lignin as the major components and ADF does not include hemicelluloses (Van Soest, 1963). NDF, ADF, and cellulose content are measured according to methods described by Van Soest et al. (1991). The content of hemicellulose was calculated as NDF – ADF (Mertens, 1992).

Fractions of carbohydrates described above are subdivided by chemical composition, physical characteristics, ruminal degradation, and postruminal digestibility characteristics, because of these factors, various modifications of the analytical methods have been proposed (Hall et al., 1999; Nie et al., 2009).

6. Degradation and utilization of carbohydrates by rumen microbial ecosystem

Dietary carbohydrates are the main rumen microbial fermentation substrates. Microbial yields are related primarily to the growth rate that carbohydrate permits. The individual carbohydrates characterized by faster rumen degradation rates result in greater microbial yield (Hall and Herejk, 2001). The enzyme systems produced by microorganisms for carbohydrates hydrolysis are complex; they usually comprise hydrolases from several families, and there may be multiple enzymes hydrolysing each polysaccharide. Nearly all carbohydrate digestion occurs (>90%) within the rumen, but under certain circumstances (e.g., high rate of passage), a significant amount of carbohydrate digestion can occur in the small and large intestine.

7. Nonfibrous carbohydrates

Nonfiber carbohydrates may provide 30 to 45% of the diet on a dry matter basis (Hall et al., 2010). The NFC fraction is considered a source of readily available energy for microbial growth (Ariza et al., 2001).

7.1. Mono- di- and oligosaccharides

The concentration of monosaccharides, glucose and fructose was estimated from 1% to 3% (in grasses and herbage) and of sucrose from 2% to 8% (Smith, 1973). Sucrose formed from α-D-glucose and β-d-fructose linked by 1, 2 glycosidic linkage is digested by enzyme sucrose phosphorylase (EC 2.4.1.7, according to the IUB-MB enzyme nomenclature; Stan-Glasek et al., 2010). Maltose formed from two units of glucose joined with an α(1–4) bond is digested by enzyme α–glucosidase (EC 3.2.1.20). Oligosaccharides concentration in the different plants ranges between 0.3% and 6% and represent a wide diversity of biomolecules (including stachyose and raffinose), they are chains of monosaccharides that are two to approximately 20 units long. The enzymes belonging to the group of polysaccharide hydrolases (EC 3.2.1.-) which hydrolyse the glycosidic bond between two or more carbohydrates utilize oligosaccharides (Courtois, 2009).

Mono-and disaccharidesare rapidly fermented within the rumen to yield VFA. The rate of glucose fermentation after glucose dosing varied from 422 to 738% h^{-1} and the rate of fermentation of monosaccharides originating from disaccharide hydrolysis was 300 to 700% h^{-1}(Wejsberg et al., 1998).

Ruminal bacteria that ferment sucrose include *Streptococcus bovis, Lachnospira multiparus, Lactobacillus ruminis, Lactobacillus vitulinis, Clostridium longisporum, Eubacterium cellulosolvens,* and some strains of *Eubacterium ruminantium, Butyrivibrio fibrisolvens, Ruminococcus albus, Ruminococcus flavefaciens, Megaspaera elsdenii, Prevotella spp., Selenomonas ruminantium, Pseudobutyrivibrio ruminis* strain A and *Succinivibrio dextrinosolvens* (Stewart et al., 1997, Martin and Russell, 1987, Stan-Glasek et al., 2010). Maltose utilize *Ruminobacter amylophilus*) and oligosaccharides *Actinomyces ruminicola* as a sources of energy (Anderson, 1995; An et al., 2006).

7.2. Pectic substances

Pectic substances are a group of galacturonan polymers with neutral sugars (largely arabinose and galactose) substitutions (Jung, 1997). Pectic substances are found in the middle lamella and other cell wall layers (Van Soest, 1994). The most important pectinolytic activity represents pectin lyase (EC 4.2.2.10) (Wojciechowicz, 1982).

Grasses contain from 3 to 4% of pectin in the dry matter, leguminous plants from 5 to 12%, and sugar beet pulp 25% (Aspinall, 1970; Van Soest, 1983; Cassida et al., 2007). The utilization of pectin varied from 79.4 to 95.9% (Marounek and Duškovă, 1999). Most of pectin degrades at a rate of 13% h^{-1} (Hall et al., 1998).

Pectin-utilizing bacteria include *Butyrivibrio fibrisolvens* and *Prevotella ssp.*(the principal rumen pectin-utilizing bacteria) and *Fibrobacter succinogenes, S. bovis* and *Lachnospira multiparus* (Czerkawski and Breckenridge, 1969; Gradel and Dehority, 1972; Baldwin and Allison, 1983). Pectin is reported to ferment primarily to acetate (Czerkawski and Breckenridge, 1969; Marounek and Duškovă, 1999).

7.3. Starch

Starch is a complex of two structurally distinct polymers: amylose and amylopectin (Chesson, 1997). Amylose is chemically composed of α-*1,4*- linked polymers of glucose. It is degraded by α-amylases (EC 3.2.1.1), which releases oligosaccharides maltodextrins, and β-amylases (EC 3.2.1.2), which remove maltose units. Amylopectin is a highly branched molecule of α-*1,4*- linked polymers of glucose joined *1,6* at intervals along the backbone moleculr. Amylopectin is degraded to maltose by β-amylases action (in 50%), glucanohydrolases (EC 3.2.1.3 and EC 3.2.1.41) and isoamylase (EC 3.2.1.68). Maltose and maltodextrins are degraded to glucose by α-glucosidase (EC 3.2.1.20) (Hobson, 1989). Starch can be degraded by ruminal microbial enzymes as well as enzymes in the small intestine of the ruminats.

Of the nonfibrous fraction, starches are the highest proportion in the diet, and cereal grains is the major source of starch in ruminants diet. The cereal grains differ in their starch content, with wheat containing (on dry matter basis) 77% starch, corn 72%, and barley and oats 57 to 58% (Huntington, 1997). Differences exist among cereal grains in their extents and rates of ruminal starch degradation. In the rumen has been digested from 55 to 70% of corn starch, 80 to 90% of barley and wheat starch, and 92 to 94% of oats starch (Huntington, 1997). The degradation rates is estimated from 4.0 to 6.4% h^{-1} for corn starch and from 14.7 to 24.5% h^{-1} for barley starch (Herrera-Saldana et al., 1990; Tamminga et al., 1990). On average, 5 to 20% of starch consumed is digested postruminally, mainly in the small intestine (from 45 to 85% of starch entering the duodenum), this capacity is limited by the supply of pancreatic amylase (Hunhington, 1997).

The bacteria *Ruminobacter amylophilus, Prevotela ruminicola, Streptococcus bovis, Succinimonas amylolytica* and many strains of *Selenomonas ruminantium, Butyrivibrio fibrisolvens, Eubacterium ruminantium* and *Clostridium* ssp., all of the entodiniomorph protozoa and the chytrid fungi are amylolitic (Chesson, 1997). The high-starch concentrate diets favor the development of propionate producing bacteria species (Ørskov, 1986; France and Dijkstra, 2005). The fermentation of starch in the rumen depending on factors such as structure (amylose/amylopectin ratio), plant source, mechanical alterations (grain processing, chewing), diet composition, amount of feed consumed per unit time and degree of adaptation of ruminal microbiota to the diet (Piva and Masoero, 1996, Huntington, 1997; Eastridge, 2006).

8. Structural carbohydrates

Structural carbohydrates is less digestible than NFC and is negatively correlated with energy concentration in the diet for ruminants, but is important for rumination, saliva flow, ruminal buffering, and health of the rumen wall, however, high dietary concentrations can limit dry matter intake by increased rumen fill. The retention time of plant fiber in the rumen is sufficiently long (48 h or more in some species) to allow the growth of a fibrolytic microbial population whose extensive fiber utilization contributes a major portion of the energy for the animal (Van Soest, 1994). The cellulose fibers are embedded in a matrix of other structural biopolymers, primarily hemicelluloses and lignin (Lynd, 1999, Marchessault and Sundararajan, 1993; Van Soest, 1994). The high-fibre forage diets encourage the grown of acetate producing bacterial species, the acetate : propionate : butyrate molar proportion would typically be in region 70:20:10 (France and Dijkstra, 2005). Fiber digestion may be reduced due to decreased rumen pH (the fiber digesters are most active at a pH of 6.2 to 6.8), and the availability of surface area for colonization (Sutton et al., 1987; Chesson and Forsberg, 1997). Fungi plays an active and positive role in fiber degradation (Williams and Orpin, 1987).

8.1. Cellulose

Cellulose content is in the range from 35 to 50% of plant dry weight (Lynd et al., 1999). It is chemically composed of a homogenous polymers of β-1,4-D-glucose linked through β-1,4-glycosidic bonds. Native cellulose exists as fibrils which are composed of amorphous and crystalline regions formed from cellulose chains, each of them contains between 500 and 14,000 b-1,4D-glucose units (Bazooyar et al., 1012). The digestion of cellulose necessitates a combination of many classes of cellulases. The digestion process including activity of endoglucanases (EC 3.2.1.4), that cut randomly at internal amorphous sites in the cellulose chain; exoglucanases (cellodextrinases EC 3.2.1.74 and cellobiohydrolases EC 3.2.1.91), that act processively on the reducing or non-reducing ends of cellulose chains, releasing either cellobiose or glucose as major products; and glucosidases (EC 3.2.1.21) that hydrolyze soluble cellodextrins and cellobiose to glucose (Lynd et al., 2002).

Fibrobacter succinogenes, Ruminococcus flavefaciens and *R. albus* are considered to be the predominant cellulolytic bacteria present in the rumens, these species gain selective advantage in the rumen is by optimizing two catabolic activities: cellulose hydrolysis (depolymerization) and efficient utilization of the hydrolytic products (cellodextrins) (Weimer, 1996; Koike and Kobayashi, 2001). *F. succinogenes* has a potent ability to solubilize crystalline chains of cellulose (Halliwell and Bryant, 1963; Shinkai and Kobayashi, 2007). *F. succinogenes* produces primarily succinate (a propionate precursor), and lesser amounts of acetate, *R. flavefaciens* produces primarily acetate and lesser amounts of succinate converts to propionate by *Selenomonas ruminantium* (Weimer at al., 1999).

The predominant ruminal cellulolytic species digest cellulose at rate approximately from 5 to 10% h^{-1}, however, the extent to which native cellulose is utilizes by ruminal microorganisms is limiting by the cellulose association with lignin (Weimer, 1996; Chesson, 1993). Cellulose degradability of forages varies from 25 to 90% (Pigden and Heaney, 1969).

8.2. Hemicelluloses

Hemicellulose concentration varies from 6 to 22% (on dry matter basis) in leaves of grasses and herbs (Schädel at al., 2010). Hemicelluloses are composed of complex heteropolymers that vary considerably in primary composition, substitution and degree of branching, and can be grouped into four classes: xylans, xyloglucans, mannans and mixedlinkage β-glucans (Ebringerova et al., 2005).

Bacteroides (Fibrobacter) succinogenes, Ruminococcus albus, and *Ruminococcus flavefaciens* and same strains of *Butyrivibrio fibrisolvens* and *Bacteroides ruminicola* are considered to be the organisms responsible for most of the degradation of hemicelluloses (Hespell, 1988). Rumen degradation of hemicelluloses varies from 16 to 90%, depending on their composition (Pigden and Heaney, 1969; Coen and Dehority, 1970).

8.3. Lignin

Lignin, a complex phenolic polymer, is indigestible by rumen microbes, but their concentration limits digestibility of structural carbohydrates (Van Soest, 1994). The main reason for reduction of accessibility for the hydrolases secreted by ruminal microbes is the presence of strong covalent bonds between lignin and the cell wall polysaccharides (Chesson, 1993).

9. Interactions between energy and protein metabolism in rumen

Manipulation of rumen fermentation through proper diet formulation changes microbial population in a way that improved efficiency of microbial protein synthesis. A major factor in maximizing microbial protein synthesis is the ruminally available energy and N in the diet. There are many interactions of dietary conditions on bacterial populations and on protein and carbohydrates digestion in rumen. For instance: microbial fermentation releases organic acids that readily dissociate to decrease pH that influence on the microbial ecosystem and determining the selective growth of certain microbial species, and the types and quantities of fermentation products (Russell and Rychlik, 2001).

10. "Synchrony" hypothesis

The purpose of proper nutrition is "nutritional synchrony" refers to provision of dietary protein (N sources, true protein) and energy (ruminally fermented carbohydrates) to the rumen in such a manner that they are available simultaneously in proportions needed by the ruminal microorganisms (Hall and Weimer, 2007). Synchronous nutrient availability should allow more efficient use of nutrients, thus enhancing production of microbial products, increasing nutrient supply to the animal, and potentially improving animal production performance (Sinclair et al., 1993; Hall and Huntington, 2008).

10.1. Production efficiency

A number of studies have been conducted to evaluate the effects of "nutritional synchrony" conception on production efficiency, but the results are not consistently. There are many results which confirm increase in the yield of microbial protein when highly degradable carbohydrates were synchronized with rapidly degraded protein (Kovler et al., 1998; Charbonneau et al., 2006). Result from the *in sacco* study confirms that the better synchronization also affects degradation rate of diet components (Niwinska, 2009; Niwińska and Andrzejewski, 2011). However, nutrient synchrony has generally not resulted in improved animal performance (Yang et al., 2010). The fundamental reason is the following: ruminants animal have the ability to recycle N from blood and saliva to the gastrointestinal tract during periods of dietary protein deficiency or during periods of asynchronous carbohydrate and protein supply. Hall and Huntington (2008) suggested that for the optimal use of "nutritional synchrony" conception we may need to look at the whole animal, not just the rumen, and that a term such as the optimal balance may be more appropriate when considering the complexity of the ruminant animal.

10.2. Product composition

Volatile fatty acids, produced in the rumen, can have a major effect on fat composition of ruminant products. Results of studies with animal models, in tissue culture systems and in clinical research indicate that the functional health-related properties of milk and beef fat appear to be linked to the presence of rumenic acid and vaccenic acid (Parodi, 2005; Lee, 2008; Field et al., 2009). Milk fat provides 30% of fat consumed by humans and is the richest natural dietary source of those valuable fatty acids (Ritzenthaler et al., 2001). Research undertaken over the past decade has indicated, that concentration of those favorable fatty acids in milk fat may be controlled by the starch/fibre ratio in the diet of dairy cows (Niwińska et al., 2011).

10.3. Pollution emissions

An excessive supply of feed nutrients results in an increase in waste excreted to the environment. The pollutants produced by ruminants are nitrogen and methane, their production is dependent on carbohydrate composition of diet. Improved efficiency of microbial protein synthesis is considered as the most important target in reduction emissions of N, while synchronization of carbohydrate and protein supply in the rumen has been suggested as one possible solution to achieve this aim (Kaswari et al., 2007; Reynolds and Kristensen, 2008; Yang et al., 2010). The rumen microbial ecosystem produces methane as a result of anaerobic fermentation. Methane production results in losses of 5 to 12% of gross energy of diet and is estimated to be about 15% of total atmospheric methane emissions (Reid et al., 1980; Moss et al., 2000). The proper selection of carbohydrates in the ration, taking into account the structural and nonfibrous carbohydrates content, can reduce the formation of carbon dioxide, hydrogen and formate the major precursors of methane production in the rumen (Mitsumori and Sun, 2008).

11. Conclusion

The current data indicated, that the world production of dairy products is expected to grow 26% by 2020 and beef production worldwide grew by 30 million tons during 1965-2005 (OECD, 2011; FAO; 2006). The projected increase in cattle production directs our attention to the better utilization of feed resources. Understanding the effects of carbohydrates types and rumen microbial population shifts in response to nutrients contained in different feeds may be valuable to improve production efficiency, to modify the composition of the product and to minimize pollution emissions.

Author details

Barbara Niwińska
Department of Animal Nutrition and Feed Science,
National Research Institute of Animal Production, Kraków, Poland

12. References

An D., Cai S., Dong X. 2006. Actinomyces ruminicola sp. nov., isolated from cattle rumen. Int. J. Sys. Evol. Microbiol. 56:2043–2048.

Anderson K.L. 1995. Biochemical analysis of starch degradation by *Ruminobacter amylophilus* 70. Appl. Environ. Microbiol., 61:1488–1491.

Ariza P., Bach A., Stern M.D., Hall M.B. 2001. Effects of carbohydrates from citrus pulp and hominy feed on microbial fermentation in continuous culture. J. Anim. Sci. 79:2713-2718.

Aspinall G.O. 1970. Polysaccharides. Oxford: Pergamon Press.

Baldwin R.L., Allison M.J. 1983. Rumen metabolism. J. Anim. Sci. 57 (Suppl. 2):461-477.

Bazooyar G., Momany F.A., Bolton K. 2012. Validating empirical force fields for molecular-level simulation of cellulose dissolution. Comp. Theor. Chem. 984:119–127.

Bergman E.N. 1990. Energy contribution of VFA from the gastrointestinal tract in various species. Physiol. Rev. 70:567–590.

Cassida K.A., Turner K.E., Foster J.G., Hesterman O.B. 2007. Comparison of detergent fiber analysis methods for forages high in pectin. Anim. Feed Sci. Tech. 135:283–295.

Charbonneau E., Chouinard P.Y., Allard G., Lapierre H., Pellerin D. 2006. Milk from forage as affected by carbohydrate source and degradability with alfalfa silage based diets. J. Dairy Sci. 89:283–293.

Chesson A. 1993. Mechanistic model of forage cell wall degradation. In: Rorage cell wall structure and digestibility. Ed: H.G.Jung, D.R.Buxton, R.D. Hatfield, J. Ralph. ASA-CSSA-SSSA, Madison, WI, pp. 347-376.

Chesson A., Forsberg C.W. 1997. Polysaccharide degradation by rumen microorganism. In: Hobson PN, Stewart CS (eds). The rumen microbial ecosystem. Blackie, London. 329-381.

Coen JA. Dehority BA. 1970. Degradation and utilisationof hemicelluloses from intact forage by pure culture of rumen bacteria. Appl. Microbiol. 20:362-368.

Courtois J. 2009. Oligosaccharides from land plants and algae: production and applications in therapeutics and biotechnology. Curr. Opinion Microbiol. 12:261–273.

Czerkawski J.W., Breckenridge G. 1969. Fermentation of various soluble carbohydrates by rumen micro organisms with particular reference to methane production. Br. J. Nutr. 23:925-937.

Dado R.G., Allen M.S. 1995. Intake limitations, feeding behavior, and rumen function of cows challenged with rumen fill from dietary fiber or inert bulk. J. Dairy Sci. 78:118–133.

De Boever J.L., Cottyn B.G., Buysse F.X., Wainman F.W., Vanacker J.M. 1986. The use of an enzymatic technique to predict digestibility, metabolisable and net energy of compound feedstuffs for ruminants. Anim. Feed Sci. Technol. 14:203–214.

Eastridge M.L. 2006. Major advances in applied dairy cattle nutrition. J. Dairy Sci. 89:1311-1323.

Ebringerova A., Hromadkova Z., Heinze T. 2005. Hemicellulose. Polysaccharides 1: Structure, Characterization and Use. Springer-Verlag, Berlin, pp 1–67.

FAIP. 2003. The economic value of livestock production in the EU 2003. Farm Animal Industrial Platform. AnNe Publishers, ISBN 90-76642-19-2.

FAO 2006. *Food and Agriculture Organization.* Livestock's Long Shadow. Environmental Issues and Options.

Field C.J., Blewett H.H., Proctor S., Vine D. 2009. Human health benefits of vaccenic acid. Appl. Physiol. Nutr. Metab. 34:979–991.

France J., Dijkstra J. 2005. Volatile fatty acid production. In: J. Dijkstra, J. Forbes, J.M. France (Ed) Quantitative aspects of ruminant digestion and metabolism. 2nd Ed. CAB International, Wallingford, UK, pp.

Gradel C.M., Dehority B.A. 1972. Fermentation of isolated pectin and pectin from intact forages by pure cultures of rumen bacteria. Appl. Microbiol. 23:332-340.

Hackmann T.J., Spain J.N. 2010. *Invited review:* Ruminant ecology and evolution: Perspectives useful to ruminant livestock research and production. J. Dairy Sci. 93:1320–1334.

Hall M.B. 2003. Challenges with nonfiber carbohydrate methods. J. Anim. Sci. 81:3226–3232.

Hall M.B., Herejk C. 2001. Differences in yields of microbial crude protein from in vitro fermentation of carbohydrates. J. Dairy Sci. 84:2486-2493.

Hall M.B., Hoover W.H., Jennings J.P., Webster T.K.M. 1999. A method for partitioning neutral detergent-soluble carbohydrates. J. Sci. Food Agric. 79:2079–2086.

Hall M.B., Huntington G.B. 2008. Nutrient synchrony: Sound in theory, elusive in practice. J. Anim. Sci. 86(E. Suppl.):E287–E292.

Hall M.B., Larson C.C., Wilcox C.J. 2010. Carbohydrate source and protein degradability alter lactation, ruminal, and blood measures. J. Dairy Sci. 93:311–322.

Hall M.B., Pell A.N., Chase L.E. 1998. Characteristics of neutral detergent soluble fiber fermentation by mixed ruminal microbes. Anim. Feed Sci. Technol. 70:23–29.

Hall M.B., Weimer P.J. 2007 Sucrose concentration alters fermentation kinetics, products, and carbon fates during in vitro fermentation with mixed ruminal microbes. J. Animal Sci. 8:1467-1478.

Halliwell G., Bryant M.P. 1963. The cellulolytic activity of pure strains of bacteria from the rumen of cattle. J. Gen. Microbiol. 32:441–448.

Herrera-Saldana R.E., Huber J.T., Poore M.H. 1990. Dry matter, crude protein, and starch degradability of five cereal grains. J. Dairy Sci. 73:2386–2393.

Hespell R.B. 1988. Microbial digestion of hemicelluloses in the rumen. Microbiol. Sci. 5:362-365.

Hobson P.N. 1989. The rumen microbial eco-system. Elsevier Applied Science, London, 1989.

Hungate R.E. 1966. The rumen and its microbes. Academic Press Inc. (London) LTD.

Huntington G.B. 1997. Starch utilization by ruminants: From basics to the bunk. J. Anim. Sci. 75:852–867.

IUB-M. Enzyme nomenclature database. International Union of Biochemistry and Molecular Biology, http://enzyme.expasy.org/

Judkins M.B., Krysl L.J., Barton R.K. 1990. Estimating diet digestibility: a comparison of 11 techniques across six different diets fed to rams. J. Anim. Sci. 68:1405-1415.

Jung H.J.G. 1997. Analysis of forage fiber and cell walls in ruminant nutrition. J. Nutr. 127: 810S–813S.

Kamra D.N. 2005. Rumen microbial ekosystem. Curr. Sci. 89:124-135.

Kaswari T., Lebzien P., Flachowsky G., ter Meulen U. 2007. Studies on the relationship between the synchronization index and the microbial protein synthesis in the rumen of dairy cows. Anim. Feed Sci. Technol. 139:1-22.

Koenig K.M., Beauchemin K.A., Rode L.M. 2003. Effect of grain processing and silage on microbial protein synthesis and nutrient digestibility in beef cattle fed barley-based diets. J. Anim. Sci. 81:1057–1067.

Koike S., Kobayashi Y. 2001. Development and use of competitive PCR assays for the rumen cellulolytic bacteria: *Fibrobacter succinogenes*, *Ruminococcus albus* and *Ruminococcus flavefaciens*. FEMS Microbiol. Lett. 204:361–366.

Kolver E., Muller L.D., Varga G.A., Cassidy T.J. 1998. Synchronization of ruminal degradation of supplemental carbohydrate with pasture nitrogen in lactating dairy cows. J. Dairy Sci. 81:2017–2028.

Lee Y. 2008. Isomer specificity of conjugated linoleic acid (CLA): 9*E*,11*E*-CLA. Nutr. Res. Pract. 2:326–330.

Lynd L.R., Weimer P.J., van Zyl W.H., Pretorius I.S. 2002. Microbial cellulose utilization: fundamentals and biotechnology. Microbiol. Mol. Biol. Rev. 66:506–577.

Lynd L.R., Wyman C.E., Gerngross T.U. 1999. Biocommodity engineering. Biotechnol. Prog. 15:777–793.

Marchessault R.H., Sundararajan P.R. 1993. Cellulose, p. 11–95 *In* G. O. Aspinall (ed.), The polysaccharides, vol. 2. Academic Press, Inc., NewYork, N.Y.

Marounek M., Duškovă D. 1999. Metabolism of pectin in rumen bacteria *Butyrivibrio fibrisolvens* and *Prevotella ruminicola*. Lett. Appl. Microbiol. 29:429–433.

Martin S.A., Russel J.B. 1987. Transport and phosphorylation of disaccharides by the ruminal bacterium *Streptococcus bovis*. Appl. Environ. Microbiol. 53:2388–2393.

Mehrez A.Z., Ørskov E.R. 1977. A study of artificial fibre bag technique for determining the digestibility of feeds in the rumen. J. Agric. Sci. 88:645–650.

Menke K.H., Raab L., Salewski A., Steingass H., Fritz D., Schneider W. 1979. The estimation of the digestibility and metabolisable energy content of ruminant feedingstuffs from the gas production when they are incubated with rumen liquor. J. Agric. Sci. 93:217–222.

Mertens D.R. 1992. Nonstructural and structural carbohydrates. In: Van Horn H.H., Wilcox, C.J. (Eds.), Am. Dairy Sci. Assoc., Champaign, IL, USA. p. 219.

Mitsumori M., Sun W. 2008. Control of Rumen Microbial Fermentation for mitigating Methane Emissions from the Rumen. Asian-Aust. J. Anim. Sci. 21:144-154.

Moss A.R., Jouany J.P.,, Newbold J. 2000. Methane production by ruminants: its contribution to global warming. Ann. Zootech. 49:231–253.

Nie Z., Tremblay G. F., Bélanger G., Berthiaume R., Castonguay Y., Bertrand A., Michaud R., Allard G., Han J. 2009. Near-infrared reflectance spectroscopy prediction of neutral detergent-soluble carbohydrates in timothy and alfalfa. J. Dairy Sci. 92:1702–1711.

Niwińska B. 2009: Effect of carbohydrates in grass silage- based diets on *in sacco* ruminal degradability of barley grain (*Hordeum vulgare* L. cv. Lomerit) ground to different particle sizes. Czech J. Anim. Sci. 54: 260–269.

Niwińska B., Andrzejewski M. 2011. Effect of carbohydrates in grass silage-based diets on *in sacco* ruminal degradability and nutritional value of lupin seeds (*Lupinus angustifolius* L *.cv.* Sonet) ground to different particle sizes. Czech J. Anim. Sci. 56: 231–241.

Niwińska B., Bilik K., Andrzejewski M. 2011.Factors influencing rumenic acid and vaccenic acid content in cow's milk fat. Ann. Anim. Sci. 11:3–16.

Nocek J.E., Russell J.B. 1988. Protein and energy as an integrated system: Relationship of ruminal protein and carbohydrate availability to microbial synthesis and milk production. J. Dairy Sci. 71:2070–2107.

Nowak R.M. 1999. Walker's Mammals of the World. Vol 2. 6th ed. John Hopkins Univ. Press, Baltimore, MD.

OECD-FAO 2011. Organisation for Economic Co-operation and Development-*Food and Agriculture Organization.* Agricultural Outlook 2011-2020; *http://dx.doi.org/10.1787/888932428329.*

Ørskov E.R. 1986. Starch digestion and utilization in ruminants. J. Anim. Sci. 63:1624-1633.

Parodi P.W. 2005. Dairy product consumption and the risk of breast cancer. J. Am. Coll. Nutr. 24:S556-S568.

Pigden W.J., Heaney D.P. 1969. Lignocellulose in ruminat nutrition. Adv. Chem. 95:245-261.

Piva G., Masoero F. 1996. Carbohydrates with different ruminal fermentability: starch and fibrous fractions. Zoot. Nutr. Anim. 22:215-229.

Reid J.T., White O.D., Anrique R., Forstin A. 1980. Nutritional energetic of livestock: Some present boundaries of knowledge and future research needs. J. Animal Sci. 51:1393-1415.

Reynolds C.K., Dürst B., Lupoli B., Humphries D.J., Beever D.E. 2004. Visceral tissue mass and rumen volume in dairy cows during the transition from late gestation to early lactation. J. Dairy Sci. 87:961–971.

Reynolds C.K., Kristensen N.B. 2008. Nitrogen recycling through the gut and the nitrogen economy of ruminants: An asynchronous symbiosis. J. Animal Sci. 86(Suppl.):E293–E305.

Ritzenthaler K.L., McGuire M,K., Falen R., Shultz T.D., Dasgupta N., McGuire M.A. 2001. Estimation of conjugated linoleic acid intake by written dietary assessment methodologies underestimates actual intake evaluated by food duplicate methodology. J. Nutr. 131:1548–1554.

Russell J.B., Rychlik J.L. 2001. Factors that alter rumen microbial ecology. Science 292:1119-1222.

Schädel C., Richter A., Blöchl A., Hoch G. 2010. Hemicellulose concentration and composition in plant cell walls under extreme carbon source–sink imbalances. Physiol. Plantarum 139:241–255.

Shinkai T., Kobayashi Y. 2007. Localization of ruminal cellulolytic bacteria on plant fibrous materials as determined by fluorescence in situ hybridization and Real-Time PCR. Appl. Environ. Microbiol. 73:1646 –1652.

Sinclair L.A., Garnsworthy P.C., Newbold J.R., Buttery P.J. 1993. Effect of synchronizing the rate of dietary energy and nitrogen release on rumen fermentation and microbial protein synthesis in sheep. J. Agric. Sci. 120:251–263.

Smith D. 1973. The nonstructural carbohydrates. Chemistry and Biochemistry of Herbage. Ed: G.W. Butler and R. W. Bailey, Academic Press Inc. (London) LTD.

Stan-Glasek K., Kasperowicz A., Guczyńska W., Piknováb M., Pristaš P., Nigutová K., Javorský P., Michałowski T. 2010. Phosphorolytic Cleavage of Sucrose by sucrose-grown ruminal bacterium *Pseudobutyrivibrio ruminis* strain k3. Folia Microbiol. 55:383–385.

Stewart C.S., Flint H.J., Bryant M.P. 1997. The rumen bacteria. The Rumen Microbial Ecosystem. P. N. Hobson and C. S. Stewart, eds. Chapman & Hall, London, UK.

Sutton J.D., Bines J.A., Morant S.V., Napper D.J., Givens D.I. 1987. A comparison of starchy and fibrous concentrates for milk production, energy utilization and hay intake by Friesian cows. J. Agric. Sci. (Camb.) 109:375–386.

Tamminga S., Van Vuuren A.M., Van Der Koelen C.J., Ketelaar R.S., Van Der Togt P.L. 1990. Ruminal behavior of structural carbohydrates, non-structural carbohydrates and crude protein from concentrate ingredients in dairy cows. Neth. J. Agric. Sci. 38:513–526.

Tilley J.M.A., Terry R.A. 1963. A two-stage technique for the digestion of forage crops. J. Br. Grassl. Soc. 18:104–111.

Van Soest P.J. 1963. Use of detergents in the analysis of fibrous feeds. I. Preparation of fiber residues of low nitrogen content. J. Assoc. Off. Anal. Chem. 46:825–829.

Van Soest P.J. 1983. Nutritional Ecology of the Ruminant. 2nd edn.Orvallis, OR: O & B Books.

Van Soest P.J. 1994. Nutritional ecology of the ruminant. 2nd. ed. Cornell University Press.

Van Soest P.J., Robertson J.B., Lewis B.A. 1991. Methods for dietary fiber, neutral detergent fiber, and nonstarch polysaccharides in relation to animal nutrition. J. Dairy Sci. 74:3583-3597.

Weimer P.J. 1996. Why don't ruminal bacteria digest cellulose faster? J. Dairy Sci. 79:1496-1502.

Weimer P.J., Waghorn G.C., Odt C.L., Mertens D.R. 1999. Effect of diet on populations of three species of ruminal cellulolytic bacteria in lactating dairy cows. J. Dairy Sci. 82:122–134.

Wejsberg M.R., Hvelplund T., Bibby B.M. 1998. Hydrolysis and fermentation rate of glucose, sucrose and lactose in the rumen. Acta Agric. Scand. A. 48:12-18.

Williams A.G. Orpin C.G. 1987. Polysaccharide degrading enzymes formed by three species of anaerobic fungi grown on a range of carbohydrate substrates. Can. J. Bot. 33:418–426.

Wojciechowicz M., Heinrichova K., Ziolecki A. 1982. An exopectate lyase of *Butyrivibrio fibrisolvens* from the bovine rumen. J. Gen. Microbiol. 128:2661–2665.

Yang J.Y., Seo J., Kim H.J., Seo S., Ha J.K. 2010. Nutrient Synchrony: Is it a suitable strategy to improve nitrogen utilization and animal performance? Asian-Aust. J. Anim. Sci. 23:972-979.

Permissions

The contributors of this book come from diverse backgrounds, making this book a truly international effort. This book will bring forth new frontiers with its revolutionizing research information and detailed analysis of the nascent developments around the world.

We would like to thank Chuan-Fa Chang, for lending his expertise to make the book truly unique. He has played a crucial role in the development of this book. Without his invaluable contribution this book wouldn't have been possible. He has made vital efforts to compile up to date information on the varied aspects of this subject to make this book a valuable addition to the collection of many professionals and students.

This book was conceptualized with the vision of imparting up-to-date information and advanced data in this field. To ensure the same, a matchless editorial board was set up. Every individual on the board went through rigorous rounds of assessment to prove their worth. After which they invested a large part of their time researching and compiling the most relevant data for our readers. Conferences and sessions were held from time to time between the editorial board and the contributing authors to present the data in the most comprehensible form. The editorial team has worked tirelessly to provide valuable and valid information to help people across the globe.

Every chapter published in this book has been scrutinized by our experts. Their significance has been extensively debated. The topics covered herein carry significant findings which will fuel the growth of the discipline. They may even be implemented as practical applications or may be referred to as a beginning point for another development. Chapters in this book were first published by InTech; hereby published with permission under the Creative Commons Attribution License or equivalent.

The editorial board has been involved in producing this book since its inception. They have spent rigorous hours researching and exploring the diverse topics which have resulted in the successful publishing of this book. They have passed on their knowledge of decades through this book. To expedite this challenging task, the publisher supported the team at every step. A small team of assistant editors was also appointed to further simplify the editing procedure and attain best results for the readers.

Our editorial team has been hand-picked from every corner of the world. Their multi-ethnicity adds dynamic inputs to the discussions which result in innovative outcomes. These outcomes are then further discussed with the researchers and contributors who give their valuable feedback and opinion regarding the same. The feedback is then collaborated with the researches and they are edited in a comprehensive manner to aid the understanding of the subject.

Apart from the editorial board, the designing team has also invested a significant amount of their time in understanding the subject and creating the most relevant covers. They scrutinized every image to scout for the most suitable representation of the subject and create an appropriate cover for the book.

The publishing team has been involved in this book since its early stages. They were actively engaged in every process, be it collecting the data, connecting with the contributors or procuring relevant information. The team has been an ardent support to the editorial, designing and production team. Their endless efforts to recruit the best for this project, has resulted in the accomplishment of this book. They are a veteran in the field of academics and their pool of knowledge is as vast as their experience in printing. Their expertise and guidance has proved useful at every step. Their uncompromising quality standards have made this book an exceptional effort. Their encouragement from time to time has been an inspiration for everyone.

The publisher and the editorial board hope that this book will prove to be a valuable piece of knowledge for researchers, students, practitioners and scholars across the globe.

List of Contributors

Chuan-Fa Chang
Department of Medical Laboratory Science and Biotechnology, College of Medicine, National Cheng Kung University, Taiwan

Brighid Pappin
Institute for Glycomics, Griffith University, Gold Coast, Australia

Milton J. Kiefel
Institute for Glycomics, Griffith University, Gold Coast, Australia

Todd A. Houston
Institute for Glycomics, Griffith University, Gold Coast, Australia, School of Biomolecular and Physical Sciences, Griffith University, Nathan, Australia

Aurore Richel and Michel Paquot
University of Liege, Belgium

Josef Jampílek and Jiří Dohnal
Faculty of Pharmacy, University of Veterinary and Pharmaceutical Sciences Brno, Research Institute for Pharmacy and Biochemistry (VUFB, s.r.o.), Czech Republic

Petra Kapková and Stephanie Bank
University of Würzburg/Department of Pharmacy and Food Chemistry, Germany

Natércia F. Brás, Pedro A. Fernandes, Maria J. Ramos and Nuno M.F.S.A. Cerqueira
REQUIMTE, Departamento de Química e Bioquímica da Faculdade de Ciências, Universidade do Porto, Porto, Portugal

S. Jasemine Yang and Delbert R. Dorscheid and Samuel J. Wadsworth
James Hogg Research Centre, Institute for Heart + Lung Health, St. Paul's Hospital, University of British Columbia, Vancouver, Canada

M. B. Gorishniy and S. P. Gudz
National University of Lviv Ivan Franko, Lviv, Ukraine

Marko Anderluh
Department of Medicinal Chemistry, University of Ljubljana, Faculty of Pharmacy, Ljubljana, Slovenia

Dominique Schols and Marijke Alen
Department of Microbiology and Immunology, Rega Institute for Medical Research, University of Leuven, Leuven, Belgium

Barbara Niwińska
Department of Animal Nutrition and Feed Science, National Research Institute of Animal Production, Kraków, Poland